THE PHYSICS OF
CONFORMAL
RADIOTHERAPY
Advances in
Technology

Medical Science Series

THE PHYSICS OF CONFORMAL RADIOTHERAPY

Advances in Technology

Steve Webb

Professor of Radiological Physics,
Joint Department of Physics,
Institute of Cancer Research and
Royal Marsden NHS Trust, Sutton, Surrey, UK

Institute of Physics Publishing
Bristol and Philadelphia

British Library Cataloguing-in-Publication Data

A catalogue record for this book is available from the British Library.

ISBN 0 7503 0396 4 (hbk)
ISBN 0 7503 0397 2 (pbk)

Library of Congress Cataloging-in-Publication Data are available

The author has attempted to trace the copyright holder of all the figures and tables reproduced in this publication and apologizes to copyright holders if permission to publish in this form has not been obtained.

Cover illustration reproduced by permission of Williams and Wilkins from the journal *Neurosurgery*, 1988, volume 22 (3) 254–464 by K R Winston and W Lutz.

Series Editors:
   R F Mould, Croydon, UK
   C G Orton, Karamanos Cancer Institute, Detroit, USA
   J A E Spaan, University of Amsterdam, The Netherlands
   J G Webster, University of Wisconsin-Madison, USA

Published by Institute of Physics Publishing, wholly owned by The Institute of Physics, London

Institute of Physics Publishing, Dirac House, Temple Back, Bristol BS1 6BE, UK

US Editorial Office: Institute of Physics Publishing, The Public Ledger Building, Suite 1035, 150 South Independence Mall West, Philadelphia, PA 19106, USA

Typeset in TEX using the IOP Bookmaker Macros

The Medical Science Series is the official book series of the International Federation for Medical and Biological Engineering (IFMBE) and the International Organization for Medical Physics (IOMP).

IFMBE

The IFMBE was established in 1959 to provide medical and biological engineering with an international presence. The Federation has a long history of encouraging and promoting international cooperation and collaboration in the use of technology for improving the health and life quality of man.

The IFMBE is an organization that is mostly an affiliation of national societies. Transnational organizations can also obtain membership. At present there are 42 national members, and one transnational member with a total membership in excess of 15 000. An observer category is provided to give personal status to groups or organizations considering formal affiliation.

Objectives

- To reflect the interests and initiatives of the affiliated organizations.
- To generate and disseminate information of interest to the medical and biological engineering community and international organizations.
- To provide an international forum for the exchange of ideas and concepts.
- To encourage and foster research and application of medical and biological engineering knowledge and techniques in support of life quality and cost-effective health care.
- To stimulate international cooperation and collaboration on medical and biological engineering matters.
- To encourage educational programmes which develop scientific and technical expertise in medical and biological engineering.

Activities

The IFMBE has published the journal *Medical and Biological Engineering and Computing* for over 34 years. A new journal *Cellular Engineering* was established in 1996 in order to stimulate this emerging field in biomedical engineering. In *IFMBE News* members are kept informed of the developments in the Federation. *Clinical Engineering Update* is a publication of our division of Clinical Engineering. The Federation also has a division for Technology Assessment in Health Care.

Every three years, the IFMBE holds a World Congress on Medical Physics and Biomedical Engineering, organized in cooperation with the IOMP and the IUPESM. In addition, annual, milestone, regional conferences are organized in different regions of the world, such as the Asia Pacific, Baltic, Mediterranean, African and South American regions.

The administrative council of the IFMBE meets once or twice a year and is the steering body for the IFMBE. The council is subject to the rulings of the General Assembly which meets every three years.

For further information on the activities of the IFMBE, please contact Jos A E Spaan, Professor of Medical Physics, Academic Medical Centre, University of Amsterdam, PO Box 22660, Meibergdreef 9, 1105 AZ, Amsterdam, The Netherlands. Tel: 31 (0) 20 566 5200. Fax: 31 (0) 20 6917233. Email: IFMBE@amc.uva.nl. WWW: http://vub.vub.ac.be/~ifmbe.

IOMP

The IOMP was founded in 1963. The membership includes 64 national societies, two international organizations and 12 000 individuals. Membership of IOMP consists of individual members of the Adhering National Organizations. Two other forms of membership are available, namely Affiliated Regional Organization and Corporate Members. The IOMP is administered by a Council, which consists of delegates from each of the Adhering National Organization; regular meetings of Council are held every three years at the International Conference on Medical Physics (ICMP). The Officers of the Council are the President, the Vice-President and the Secretary-General. IOMP committees include: developing countries, education and training; nominating; and publications.

Objectives

• To organize international cooperation in medical physics in all its aspects, especially in developing countries.
• To encourage and advise on the formation of national organizations of medical physics in those countries which lack such organizations.

Activities

Official publications of the IOMP are *Physiological Measurement*, *Physics in Medicine and Biology* and the Medical Science Series, all published by Institute of Physics Publishing. The IOMP publishes a bulletin *Medical Physics World* twice a year.

Two Council meetings and one General Assembly are held every three years at the ICMP. The most recent ICMPs were held in Kyoto, Japan (1991) and Rio de Janeiro, Brazil (1994). Future conferences are scheduled for Nice, France (1997) and Chicago, USA (2000). These conferences are normally held in collaboration with the IFMBE to form the World Congress on Medical Physics and Biomedical Engineering. The IOMP also sponsors occasional international conferences, workshops and courses.

For further information contact: Hans Svensson, PhD, DSc, Professor, Radiation Physics Department, University Hospital, 90185 Umeå, Sweden. Tel: (46) 90 785 3891. Fax: (46) 90 785 1588. Email: Hans.Svensson@radfys.umu.se.

CONTENTS

PREFACE

The aim of curative radiation therapy is to deliver as high and as homogeneous a dose as possible to diseased tissue without causing unwanted and unnecessary side effects for the patient. This aim recognises that it is not enough to destroy tumour cells and prolong the life of a patient, but that that life must be of high quality. This goal has been accorded the title of conformal radiation therapy to describe the aim of conforming or shaping the high-dose volume to the volume occupied by diseased tissue. A related goal is sometimes referred to as conformal avoidance whereby the dose distributions are shaped to maximise the avoidance of high dose to normal tissues even if this compromises the dose delivered to the volume occupied by the tumour. In a subject somewhat beset by terminology, some of which is helpful and some of which is less revealing, conformal radiation therapy is sometimes referred to as three-dimensional radiation therapy (3D RT) since there is no doubt that, given the varied and arbitrary shape of tumours in 3D space, 3D RT is a prerequisite for three-dimensional *conformal* radiation therapy (3D CFRT). However, the term three-dimensional radiation therapy can be viewed as encompassing a wider range of therapeutic techniques not all of which are conformal. Annoying as some acronyms may be, the two above are regularly seen written and even heard in everyday conversation in the oncology department.

As the next millennium approaches, 3D CFRT is becoming increasingly more practicable and we know that the practice of radiation therapy in the next millennium has the possibility to be vastly different from that witnessed to date. Most of the technical elements of the 'radiation therapy chain' are available and understood. Anatomical and functional images in 3D of both diseased and normal anatomy can be obtained by x-ray computed tomography (x-ray CT), magnetic resonance imaging (MRI), single-photon-emission computed tomography (SPECT) and positron-emission computed tomography (PET). Methods exist to automatically define structures in three dimensions. The same process can be performed by humans and multimodality imaging applied to radiotherapy is increasing as 3D images become more commonly available. Three-dimensional planning techniques can design beams shaped to the geometrical projection of the target.

These fields can be created rapidly and automatically via the multileaf collimator (MLC) on linear accelerators delivering photons. The interaction of photons with matter is well understood so that 3D dose distributions can be computed. Exotic 3D dose distributions can be shaped 'more conformally' (if qualifying this adverb is permitted) by the superposition of intensity-modulated beams (IMBs). To do this, conventional 'forward planning' must be replaced by 'inverse planning' in which the beams are designed from a specification of the required dose distribution, created by superposition, rather than by forward trial and error. Ways to deliver such IMBs are becoming available either using multiple-static fields, dynamic-leaf movement, scanning beams or special-purpose collimators. Increasingly sophisticated methods to immobilise the patient are being developed. From the 3D dose distributions it is possible to predict the outcome in biological terms of tumour control probability (TCP) and normal-tissue-complication probability (NTCP). Imaging can be used further to confirm the success or otherwise of the radiation therapy. Portal imaging and megavoltage computed tomography can be used as the basis of verifying the delivered dose.

All this has become possible through the efforts of physical scientists and engineers who have put in place the tools to achieve 3D CFRT. The centenary (1995) of the discovery of x-rays has passed and one may question why it has taken so long to reach this point in the development of 3D CFRT. The answer lies in the complex synergy of progress in computer technology, in understanding the principles of 3D medical imaging and in having the detector technology to exploit tomography in practice, in developments in electrical engineering and in control systems. Developments in data transfer and manipulation, in informatics, in image processing and display all contribute to the reason why the present time is the threshold of potential change in clinical practice and outcome of treatment for the patient.

Yet these oversimple remarks paint too naive a picture and require qualification. And these qualifications confront us immediately with a less rosy view of the future and challenge our ingenuity in arenas beyond those of scientific and medical practice. We must observe that whilst the elements of the chain are largely understood, it is a fragile chain. All the elements are very complex. There are many niggling difficulties. Let us cite just one example. Whilst methods to fuse and manipulate multimodality images are well studied and may be used even routinely in some centres, these methods are frustrated by incompatibilities in data formats; it is usually necessary to write some in-house software to register images; the images may be generated in different hospital departments that do not communicate as well as they should; the protocol for ensuring patients are imaged in several ways may not work well in practice; the change of a machine can temporarily set back a technique it was hoped to encourage; a commercial planning system may not have been designed to accept certain modalities; and the list could go on; i.e. even setting aside scientific questions concerning the effect of introducing a new piece of technology, there will be what one might loosely call organisational difficulties, not insuperable but certainly not helpful. This is just one example and one could similarly analyse all the elements of the radiation therapy chain and improvements thereto and, with one's pessimistic hat on, identify potential stumbling blocks.

There is also the inescapable observation that in its developmental phase 3D CFRT is inevitably more expensive than conventional practices and has reached the stage of clinical viability just as there is possibly the greatest pressure ever seen on the healthcare markets in both the USA and the UK to deliver value for money and maybe even to cut costs. So the question arises: will the promise of 3D CFRT be fully exploited? And once again a familiar 'Catch 22' situation arises. 3D CFRT cannot be fully evaluated until many collaborating centres have introduced the techniques into their clinical practice, yet the funding needed to establish these techniques may not be forthcoming until they are proved cost effective. Issues of health economics, statistical evaluation of clinical trials and close scrutiny of the purchaser–provider constraints are clearly as important a part of the agenda of improving cancer treatment as the development of new physical treatment methods. Perhaps, however, it is worth remembering that 3D CFRT has not been developed entirely in an atmosphere of unsupported hope. There is *evidence* that improved conformation leads to greater tumour control and decreased side effects. It should be borne in mind that a failed conventional treatment is a complete waste of resources even if it is less expensive. Also there is plenty of precedent to show that what is expensive and time consuming today becomes tomorrow's routine tool.

These issues are beyond the scope of this book and the expertise of this author but they are highlighted here to place in a wider context the role of the technological developments that are the remit of this work. It is my firm belief that physicists should dream and that they should largely spend their time and use their talents to generate the best solutions to the technical problems confronting them. It helps initially to temporarily ignore the catalogue of potential stumbling blocks, the technical gremlins, the economic realities, the effect of technological change on human practices. The limitations presented by the physics and engineering of each problem should be a prime concern. The next concern is to study the potential extra problems introduced by treating a living patient rather than an inanimate phantom, for example accounting for patient motion, for the fact that all patients are not alike in their geometry or radiosensitivity, for the changes which may take place during a course of therapy, for the practical feasibility of a method compared with its theoretical possibility. However, the physicists must not stop there but, with what mental energies remain, should become involved in those wider questions contextualising their work, mindful that they can provide an informed view but may not be in a position to solve these problems as effectively as other professionals whose full-time remit this is.

In a field as rapidly developing as the physics of 3D RT a textbook is inevitably only a milestone beyond which developments will continue. When *The Physics of Three-Dimensional Radiation Therapy* (the companion Volume of this work) was completed in June 1992 no previous work had brought together all the elements of the chain of physical techniques contributing to the practice of 3D CFRT and I wrote in its preface 'It might be argued that it is too soon to prepare a work reviewing this field. All these aspects of conformal therapy are in a fluid state, continually being developed. In this Volume I have tried to extract the key elements which will probably stand the test of time.' In retrospect I do *not* believe it was too early to make a first stab at a review. The companion Volume documents many principles and techniques which are as valid today as they were in 1992. I have been told that it has been useful for research workers to have had available that detailed review, but perhaps I had not foreseen the enormous growth of interest and hence subsequent activity in this area since that time and it is certainly time to put down a second milestone. In this Volume some of the major themes introduced in the companion Volume are considerably expanded. There is a more general treatment of the importance of medical imaging and the interaction with 3D treatment-planning systems. Optimisation of computer plans has become almost a subspecialty and receives extensive treatment. A major growth area is in the possibility to deliver IMBs and, whilst it must be admitted that this is still in its infancy, there is a mass of literature that has been published since 1992 and which is drawn together and critically reviewed here. The treatment of the multileaf collimator in the companion Volume was largely confined to its technical development. Here its clinical innovation and performance characterisation are reviewed. The physics of planning and delivering radiotherapy is so central to the development of 3D CFRT that Chapters 1–3 necessarily form over half of the material in this Volume. Portal imaging and megavoltage computed tomography have extended their roles beyond simply verification of patient position into the much more complex realm of verification of delivered dose. The elements of transit dosimetry are thus introduced. There have been some important developments in biological modelling which have merited an overview. The subject of proton therapy has marched on apace and a brief update is given on new facilities and techniques. However, other volumes are available already reviewing the complex technology of proton therapy in detail. Finally some appendices review some of the thematic issues such as the decision trees in 3D planning, algorithms for 3D treatment planning, quality assurance, and, in a lighter vein, there is a historical comment on the origins of IMB therapy.

This Volume supplements *The Physics of three-Dimensional Radiation Therapy*. It complements the material there and has been prepared with a small overlap which will hopefully make independent reading. However, the study of both Volumes should provide a fuller understanding. I think the two milestones should stand the test of time and serve their purpose but I know the future holds yet more developments in store which others as well as I will document.

Steve Webb
August 1996

ACKNOWLEDGEMENTS

The Institute of Cancer Research (ICR) and Royal Marsden NHS Trust (RMNHST) collaborate on a large research programme working towards improving radiation therapy leading to achieving conformal radiotherapy. Clinical research is conducted under Alan Horwich in the Academic Department of Radiotherapy and, alongside, the Joint Department of Physics under Bob Ott is developing new techniques to improve the physical basis of conformal radiotherapy. I am grateful for the opportunity to be a part of this programme in this enthusiastic environment.

I am particularly grateful to my colleagues at the ICR/RMNHST who are experts in specific aspects of radiation therapy physics and who have shared their knowledge and experience with me. During the evolution of this book these have included Roy Bentley, Margaret Bidmead, Phil Evans, John Fenwick, Elizabeth Fernandez, David Finnigan, Glenn Flux, Vibeke Hansen, Sarah Heisig, Richard Knight, Penny Latimer, Mike Lee, Philip Mayles, Ian Moore, Amin Mosleh, Alan Nahum, Mark Oldham, Mike Rosenbloom, Carl Rowbottom, Glyn Shentall, Bill Swindell, Margaret Torr and Jim Warrington.

I am grateful to Sue Sugden in the Institute of Cancer Research Sutton Branch Library for help with chasing references and to Ray Stuckey and his photographic staff for help with half-tone illustrations.

I should like to acknowledge colleagues throughout the world with whom it has been my pleasure to meet from time to time and whose work collectively creates the field which is reviewed here. I am particularly grateful to colleagues at the Deutches Krebsforschungszentrum in Heidelberg, led by Wolfgang Schlegel, with whom we have enjoyed a specially close working collaboration.

I should like to thank all the publishers who have allowed figures to be reproduced. The authors are acknowledged in the figure captions. All publishers were contacted with request for permission to reproduce copyright material.

The reference lists are up to date as of July 1996.

The work of the ICR and RMNHST in conformal radiotherapy is supported by the Cancer Research Campaign and the Medical Research

Council. I am also grateful to Mark Carol and his Team at the NOMOS Corporation (Sewickley, Pennsylvania) with whom we closely collaborate on the development of IMB therapy. I am grateful for the support of the Philips IMB Research Development Team (Crawley, UK) led by Kevin Brown.

I thank Kathryn Cantley (Commissioning Editor) at Institute of Physics Publishing for her enthusiasm for this project. I am also grateful to Jacky Mucklow for her very professional work in the production of this book.

The material reviewed here represents the understanding and personal views of the author, offered in good faith. Formulae or statements should not be used in any way concerned with the treatment of patients without checking on the part of the user.

This book is dedicated to Linda and the fight against chronic disease.

It is also in memory of my father, Arthur Stanley Webb, who encouraged all my academic labours and who died in church on 29 November 1996 a couple of weeks before this book was accepted for publication.

CHAPTER 1

DEVELOPMENTS AND CONTROVERSIES IN THREE-DIMENSIONAL TREATMENT PLANNING

1.1. THE RATIONALE FOR DEVELOPING CONFORMAL THERAPY

It has been estimated by the World Health Organisation (WHO) that one third of all cases of cancer could be cured if detected early enough and if given adequate therapy but that 30% of all those dying of cancer do so with locally uncontrollable tumour growth (Ching-Li and Volodin 1993). The WHO also estimates that one third of all cancers, including cancers of the lung, mouth, liver and skin, are preventable, but, until social habits change to yield this outcome, the need for improvements in radiotherapy remains. The history of radiation therapy is one of continuous development of new skills and new approaches (Fraass 1995, Laughlin 1995, Walstam 1995). It has often been said that most of what we regard as 'new' in the field of conformal therapy and three-dimensional (3D) radiation treatment design and execution is only 'new' in the sense of 'recently achieved'. Many of the desirable concepts were understood years ago but it is only with the recent developments in physics, engineering and computing that techniques have become practicable.

It is postulated that a significantly increased tumour control probability will arise if the dose to the tumour is escalated by 20% (Thames *et al* 1992) and indeed dose escalation in treating the prostate is now being attempted (Warmelink *et al* 1994) as is dose escalation for high-grade astrocytomas (Sandler *et al* 1994) and gliomas (Grosu *et al* 1996). The major technical advance in recent years has been the development of conformal radiotherapy in which the high-dose volume is tailored to match the planning target volume (PTV), sparing organs-at-risk (OAR) wherever possible (Webb 1993c). Lichter (1994b) states in the '3D hypothesis' that achieving this and

Figure 1.1. *Two of the most important developments in improving the physical basis of radiotherapy. A PTV is shown adjacent to an OAR (the spinal cord). Radiation is* geometrically shaped *by a MLC (shown schematically) so that primary rays do not fall on the OAR. The intensity of the radiation could also be varied if desired across the field. Ways to do this are discussed in Chapter 2.*

invoking dose escalation would lead to increased cure. Such conformation, resulting from optimising the physical basis of radiotherapy, is a prerequisite to dose escalation since 'conventional' radiotherapy has already evolved to operate at the limits of normal-tissue tolerance and it would be unacceptable to increase dose to normal tissues and specifically organs-at-risk (Lichter 1994b). Achievement of a higher tumour control probability with reduced treatment morbidity is expected to follow the development of conformal radiotherapy (Suit and du Bois 1991). Lichter (1994b) wrote 'I can now state with absolute confidence that the 3D hypothesis is true and that conformal radiation therapy allows dose escalation to occur'.

Great technological developments have been required to work towards this goal which has by no means yet been reached (Lichter 1996). One major advance has been in faster and more practical methods to tailor the geometrical shape of the beam to the projected area of the planning target volume in the direction of viewing of the beam. A second very important advance is in attempting to tailor the spatial distribution of the intensity of beams in either one or two dimensions (figure 1.1) (Webb 1993b). These developments are at the leading edge of the field and make heavy demands on treatment planning and methods of treatment delivery. In this chapter we shall largely concentrate on planning issues, specifically reviewing recent advances in optimising treatment planning and associated controversies. Chapter 2 will focus on the improvements which have become possible for delivering conformal radiotherapy with photons.

The history of radiotherapy is one of continual development (Halnan

1995a,b). Treatment planning has come a very long way since its early inception (Tsien 1955, van De Geijn 1965). The earliest systems for treatment planning were very simple systems for dose calculation and display in a plane, calculations performed on the first generation of digital computers. Today, 3D treatment planning, whilst still a generic title open to some interpretation (Fraass and McShan 1995), is a practical reality in many centres.

Three-dimensional treatment planning is moving from the experimental and research arena into routine clinical practice (Chaney and Pizer 1992, Kessler *et al* 1992, Lichter *et al* 1992, 1993, Sailer *et al* 1992, Schlegel 1993). The first evidence is appearing for reduced morbidity from employing conformal therapy. For example Schultheiss *et al* (1993) have shown conformal therapy of the prostate has led to reduced acute grade-2 toxicities and Sandler *et al* (1993) have shown reduced rectal complications in a large series of prostate patients (539 cases) treated conformally. However, Tait *et al* (1993), Mayles *et al* (1993), Horwich *et al* (1994) and Nahum (1994, 1995) have not found correlations between acute toxicity (recorded via a patient-created quality-of-life questionnaire) and improved conformation although a reduction of late morbidity is expected. Details have been reported by Tait *et al* (1997). Wachter *et al* (1996) *have* reported a correlation between late radiation damage to the rectum and irradiated rectal volume, when no correlation occurred for early damage (also recorded via a patient-created quality-of-life questionnaire). There is still scope for improving the clinical justification of improving the technical basis of radiotherapy. The expectation is that a reduction in late toxicity will correlate with improved dose conformation. As this process of clinical evaluation accelerates many controversies arise.

Not least, as with the introduction of any proposal for improving the outcome of any medical therapy, there is a 'chicken-and-egg' dilemma. Namely, many clinicians are unwilling to change a therapy until the benefits of doing so have been proved. The data to complete this proof are, however, necessarily sparse until such changes have been made, at least on a trial basis, and the results observed over a considerable period of time.

Two solutions have emerged. The more conservative solution is to conduct comparative planning studies (paper exercises) with predictions of clinical outcome based on biological models. A number of such studies are now beginning to be published. The critics may point to the sparcity of models and clinical data on which they may be based as well as to the potential for not including the simulation of all the practical aspects of therapy, such as the possibility of patient movement and internal organ motion.

The second way forward is for a small number of centres to pioneer potentially improved therapy, carefully monitoring the situation and becoming gradually bolder as confidence grows (Fraass 1994, Lichter 1994a,b). This is not easy when there is a surrounding climate of the requirement to reduce or at least contain costs (Emami 1994, Kutcher and Leibel 1994). In fact, to make a political statement, it is hard to reconcile the opposing demands of improving patient treatment within a framework of decreasing resources and the short-term vision of some governmental cultures.

New therapies are often more time consuming and expensive in the first instance, although in time may become cheaper. Their development is only possible with the full support of a good technical research team. The justifiable emphasis on increased vigilance over safety issues also conflicts with cost containment.

Fortunately one can set against this the argument that expensive new treatment which improves the outcome of therapy is in the long run more cost effective than cheaper therapy with a poorer outcome. It is a matter of deciding over what length of time one conducts the appropriate accounting tasks. There is growing evidence in the USA that the increased public awareness of the potential improvements from conformal therapy is leading to a greater demand, and if we accept that market forces play a role, this works in favour of the effort to improve the physical basis of radiotherapy.

The rest of this book is therefore predicated on the belief that improving the physical basis of radiotherapy is worthwhile and will in the long term lead to more cost-effective radiotherapy, simpler working practices and improved clinical outcome for the patient (Kutcher and Mohan 1995). The physicist must accept that some steps forward have to be an act of faith and that there will be a temporary climate of reasonable doubt. As the clinical data show benefit, confidence in new technological therapies will grow. If we study the history of radiotherapy (Mould 1993, Robison 1994) we might conclude that developments have always been made against this background but that the present time promises a quantum leap in treatment outcome.

1.2. METHODS FOR AND CONTROVERSIES IN DETERMINING THE CONTOURS OF TARGET VOLUMES, ORGANS-AT-RISK AND THEIR BEAM'S-EYE-VIEW PROJECTIONS

One of the most important parts of the chain of tasks involved in planning and delivering radiotherapy is the determination of the PTV and the nearby OAR to be spared. Modern practice determines these volumes from 3D image datasets, most usually computed tomography (CT), sometimes magnetic resonance imaging (MRI) and with the possibility of single-photon emission computed tomography (SPECT) (Marks *et al* 1994) and positron emission computed tomography (PET) (Wong *et al* 1994). The volumes are constructed from a series of contours representing the outline of the structures in tomographic planes which intersect the volumes. Special-purpose computer packages have been developed to perform this task. One such is TOMAS from the German Cancer Research Centre (DKFZ)† which is used in conjunction with the VOXELPLAN treatment-planning package (Pross 1993, 1995, Pross *et al* 1994). At the University of North Carolina, automated image feature detection is under development (Chaney and Pizer 1992). All 3D treatment planning programs must have some such facility with a variety of outlining techniques. The determination of volumes for treatment planning can be one of the most time-consuming parts of the process and a time bottleneck. To obtain a good definition of the target volume CT slices should be closely spaced. Slices should also be obtained over a longitudinal extent greater than the extent of disease to allow for the possibility of using non-coplanar fields. However, this leads to two obvious problems: (i) a large scanning time and (ii) a large amount of work for the clinician to outline the contours. Algorithms for automatic segmentation are a subject of much recent development. For example, in the large pan-European AIM Project A2003 known as COVIRA (Elliott *et al* 1992, Kuhn 1995) several algorithms for automatic segmentation, specifically 'volume growing' algorithms have been shown to speed up this process significantly (Neal *et al* 1994a,b).

Creating the PTV and OAR contours is still one of the most time-consuming parts of 3D treatment planning, hence all this activity. Some form of autocontouring is now considered essential even if it is crudely to copy a contour from slice to slice and edit the result. In the USA a major study involving nine institutions is about to get underway to assess, among other things, the reproducibility of determining planning target volumes (Purdy 1994a,b).

Creating a contour relies on the fidelity of the underlying image data. The simplest method of autocontouring is to direct the computer to select all those voxels within some specified range of CT numbers and identify regions with CT ranges. These methods are error prone as they do not take into account the relationships between pixel (or voxel) values. More sophisticated algorithms 'grow' regions from a starting pixel (or voxel) taking into account the connectivity between elements. The use of texture can also be taken into autocontouring. 'Snakes' are a new concept finding application in autocontouring. Volume rendering has also advanced and the use of composite volumes rendered in 3D can assist the determination of the beam's-eye-view of a treatment portal (see also section 1.2.5.).

It has been stated often that the most significant recent advance in the improvement of radiotherapy has been the ability to be more sure of the extent of tumour in three dimensions and the ability to take precautionary measures to accommodate residual uncertainty. In this respect 3D multimodality medical imaging 'drives' the physical basis of conformal

† TOMAS=Tool for Manual Segmentation; DKFZ=Deutches Krebsforschungszentrum, Heidelberg.

radiotherapy. An understanding of the formation and use of medical images is necessary to make advances in 3D radiotherapy planning (Webb 1988), since improvements in imaging have led to improvements in planning therapy (Sauer and Linton 1995). Some aspects of the generation and use of 3D medical imaging datasets are discussed in Chapter 7. It has also been argued that the definition of the planning target volume should also take into account both the previous history of surgery (Lukas 1996) and of chemotherapy (Dunst 1996) both of which determine the residual volume of tumour cells.

1.2.1.   The interaction with the human observer

If only one imaging modality has been used (this would generally be CT) the following problems arise in determining the contours.

(i) Inter- and intra-observer variability (Goitein 1995). The same observer may create a slightly different outline at sequential attempts. Different observers may well make different decisions. This could arise even if automated contouring algorithms are used, since these generally depend on observer choices (such as selecting starting seed values for automatic region-growing) and automatic algorithms should always include the option for human editing. Since 'the correct contour' is almost never known (the exception might be outlines of bony structures), the question arises of determining which outline is to be adopted from a number of candidates. Austin Seymour (1994) has made a study of inter-physician variability in specifying tumour volumes from 3D image data. Bucciolini *et al* (1996) showed changes in the dose–volume histogram (DVH) for the PTV when this was separately and independently determined by five different radiotherapists.

(ii) The magnitude of any displacement error in contour determination must be distinguished from its importance. For example making an *underestimate* of the size of the spinal cord in cross section by say 2 mm could lead to severe complications whereas making an *overestimate* would be relatively harmless.

The influence of, and the burden to, the human observer has been a driving force behind the development of automatic algorithms for image segmentation, such as region growing. However, whilst automation removes the burden and the clinician variability, it is only as successful as its training allows. Even when automation is used, a human should always *review* the segmented volumes before proceeding to the next stage of planning.

An interesting development from the German Cancer Research Centre is the use of 'fuzzy logic' to determine the target volume. An observer is asked to specify a minimal target volume in which the tumour certainly lies, a maximal target volume outside which no tumour cells are expected and a computer algorithm evaluates the optimum target volume to use in terms of the probability function for the presence of clonogenic cells together with a model for tumour control probability (TCP) and normal-tissue-complication probability (NTCP) (Levegrün *et al* 1995, 1996), the aim being to maximise TCP for minimal NTCP. Studies showed that the ratio of the maximal PTV to minimal PTV could be as large as 3.4:1 showing huge inter-observer variability (Van Kampen *et al* 1996).

1.2.2. The influence of organ movement

As well as considering the physical limitations on accuracy, such as spatial resolution, the accuracy of the outlining algorithm or manual contouring, account must also be taken of biological movement when determining the target volume. This may be gross movement of the patient with respect to the fields or internal movement of organs with respect to each other. The former may be involuntary or a setup error (see section 1.2.3). For example, even in the head where, when planning radiotherapy, it is generally assumed that structures are well-constrained spatially (unlike say in the abdomen), it should be remembered that the brain elastically pulses with heartbeat and breathing. This pulsing has been demonstrated by making movies of gated CT images from the Mayo Clinic Dynamic Spatial Reconstructor and also from gated MRI images (Nitz *et al* 1992, Poncelet *et al* 1992). The brain also 'floats' in its cushioning fluid. These considerations may present ultimate limits to the accuracy of stereotactic brain radiotherapy (figure 1.2) particularly with protons. Breuer and Wynchank (1996) made magnetic resonance (MR) images of normal patients with patients prone and supine and observed movement of the brain with respect to external skull-based landmarks of between 1.5 and 4 mm.

Whilst methods of immobilisation of the head might be considered fairly well established (Verhey *et al* 1982, Thornton *et al* 1991), the situation for other body sites is more uncertain. Several studies of organ movement in the pelvis have shown that even with external immobilisation devices, the prostate can move (Forman *et al* 1993). For supine patients Beard *et al* (1993) showed that the prostate moves in the anterior–posterior (AP) direction by up to 1 cm. Melian *et al* (1993) showed for prone treatment positions that the AP movement could be up to 3 cm with up to 1.5 cm lateral movement. These data were obtained by studying the projections of the PTV outlines determined from CT data. The movement correlated with rectal and bladder filling over a seven-week course of radiotherapy. They referenced the daily position of the prostate determined by CT to a pre-treatment CT scan. There is a clear message that the determination of the PTV cannot ignore these movements if the therapy is to be tailored conformally. The movements are *not* determinable from studying bony landmarks. Ten Haken *et al* (1994) showed, by comparing 3D images obtained from spiral CT with

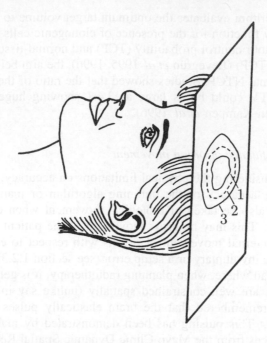

Figure 1.2. *The determination of the volume to be treated is one of the most important considerations in radiotherapy. The data are obtained from multiple cross-sectional images. In this illustration one of such a series of slice images is shown. The inner contour (1) represents the GTV. The middle contour (2) represents the CTV which includes the GTV plus a margin for the spread of microscopic disease. The outer contour (3) is the PTV which encompasses the CTV with a margin to account for potential movement error and setup error. It is important to realise that the margins do not have to be isotropic. Indeed for some tumours (e.g. of the prostate) it is possible to make use of specific knowledge of the anisotropy of potential movement. Also one should note that there may be inter- and intra-observer variability in determining the GTV and CTV and these volumes may also be redefined if multimodality information is available.*

breathholding and conventional CT without, that the errors associated with breathing can be as large as those associated with density-correction dose algorithms. Turner *et al* (1994) showed that during a course of treatment for bladder cancer, the bladder PTV required a 2 cm margin to account for potential movements.

There would appear to be at least four solutions.

(i) Monitor the prostate motion. Balter *et al* (1993) surgically implanted three or four radio-opaque markers around the prostate, imaged them before each treatment and adjusted the patient position accordingly. They determined there was a 'natural axis' for prostate motion. This group incidentally found quite small movement of the prostate ($\leqslant 3$ mm relative to the skeleton). Balter *et al* (1994a) used surgically implanted platinum microcoils in liver and lung to assess tumour movement due to breathing.

(ii) Add margins: Pickett *et al* (1993) worked out the 'ideal safety margins' which should be added, finding that the margin was different for different parts of the prostate surface. These margins were evaluated from a study of a population and then the solution was to simply apply them to each new patient. In the study of Roach *et al* (1994) the beam's-eye-views (BEVs) of a gross tumour volume (GTV) were computed for six fields in a coplanar treatment technique. The fields were then expanded for a range of margins from 7.5 to 22.5 mm in steps of 2.5 mm and for each specific margin a 3D treatment plan was prepared. Then, using the best available estimates for: (a) extracapsular spread to create the clinical target volume (CTV), (b) organ movement and (c) setup uncertainties, the margins required for each border of the prostate were determined in order that the GTV and/or CTV always resided within a specific (e.g. 90%) isodose surface. It was found that these margins were anisotropic. The 3D dose distribution using the overall anisotropic margins was superior to that using a regular fixed 2 cm margin in terms of improved rectal sparing. Jones *et al* (1995) also added margins based on a study of the movement of the prostate and setup uncertainties. Rosenwald *et al* (1995) presented an automatic method of adding margins in 3D for conformal therapy.

(iii) Regulate rectal and bladder filling. Leibel *et al* (1994) also recognised the problem of coping with movement of the prostate and seminal vesicles between the time of CT scanning (or simulation) and each treatment fraction. They identified the biggest cause for concern was systematic variation due to different states of rectal and bladder filling and therefore adopted the method of instructing the patient to empty the bladder a fixed time before each imaging or treatment session. Then, treating the patient prone, they predicted that random variations would be of much less importance. Immobilisation of the prostate target by controlling the position of the rectum has also been proposed. However, the suggestion that a rectal stent could be used to restrict the prostate motion is criticised by Kutcher *et al* (1995) on the reasonable grounds that it would be hard to tolerate over 45 fractions.

(iv) Correlate with breathing cycle. Lawrence (1994) has identified the problem of determining volumes from a single planning CT set which does not average over any breathing cycle. The associated uncertainty affects the computation of normal-tissue complications. Balter *et al* (1994b) replanned patients with volumes extracted from CT data: (a) with normal breathing and (b) with breath holding at normal inhale and normal exhale. They noted significant differences in the location of internal organs, some (e.g. liver and kidney) moving by as much as 1 cm. They are actively considering gated therapy to cope with these problems.

It can be seen that these four strategies fall into two classes. Strategy (ii) accepts inevitable movement and plans to the envelope of that movement.

This is the conventional approach to forming the PTV. Strategies (i), (iii) and (iv) attempt to monitor or control movement and to tailor daily radiotherapy to the current location of the CTV.

An MRI study by Korin *et al* (1992) showed that the liver makes a superior–inferior excursion of some 1.7 cm during quiet breathing and some 3.9 cm during deep breathing; motion in the AP or left–right directions was much smaller. This suggests the need for anisotropic margins. Casamassima (1994) has reported that the kidney and liver move with a mean value of 3 cm during respiration and showed CT images in which a pulmonary nodule moved by approximately 1–2 cm during respiration.

With regard to OAR, Gademann (1996) argues that these should be defined without any margin to account for movement but this is debateable as OAR could move into the high-dose volume due to internal organ movement. Zimmermann *et al* (1996) has reported increased NTCPs for rectum and bladder when account was taken of the movement of these organs with respect to the treatment volume.

Tissue movement presents a limitation on the achievement of conformal radiotherapy. In general the technology of new treatment deliveries is running ahead of the problem of solving the patient and target immobilisation. The problems are not all solved. Sailer and Tepper (1995) write so truly 'For all of us, it is much easier to deliver precise and accurate radiation treatment to a phantom than to a patient'. It would be wonderful to crystal-ball gaze to a day when a non-invasive 3D imaging modality could daily track the location of the CTV or therapy could be gated to organ movement. In this context an interesting development from Schweikard *et al* (1996) actually allows the patient to move (within limits) and the movement is observed by stereoscopic x-ray cameras which feed the data in real-time to a robot arm holding a small linear accelerator (linac). The arm position compensates for the patient movement. This is suggested as a serious alternative to frame-based stereotactic fixation for the treatment of brain tumours.

1.2.3. Setup errors

The precision of conformal therapy is compromised by two separate factors which may result in the PTV not being in the desired 3D treatment volume. The first is organ motion discussed above. The second is the possibility of setup error. This is the geometric misregistration between the position in which the patient is setup and that which is the desirable position, i.e. that in which the patient is planned and for which the location of the high-dose treatment volume is known with respect to the known coordinate system and landmarks. In turn there are two separate contributions to setup error: systematic error and random error. Consider figure 1.3, in which each dot represents the location of the patient relative to the planned position at any

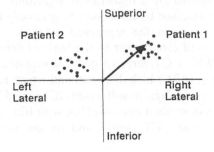

Figure 1.3. *A schematic diagram illustrating the difference between systematic and random setup errors. Each dot represents the position of a patient (simplified to within a plane) at a specific treatment fraction. The arrow represents the origin to the mean position of the dots and represents the systematic setup error for patient 1 about which the random errors cluster. (From Kutcher et al 1995.)*

given treatment fraction. The dots cluster about a position which is not at the origin, indicating that there is a systematic deviation as well as a random deviation. For any one single patient all that can be determined is the single systematic deviation and both the mean and standard deviation of the random deviation. Such measurements are obtained using electronic portal imaging devices. If a *group* of patients is analysed then the systematic variation may be different for each patient and it would be possible to calculate the mean and standard deviation of the systematic deviation. Kutcher *et al* (1995) have reviewed the literature on setup errors for pelvic treatments. Unfortunately these data show rather inconsistent values between the different studies.

Kutcher *et al* (1995) present the interesting technique of modelling the setup errors and calculating the effect on the DVH. First the mean DVH due to random setup errors is calculated. Then the systematic setup error is included by sampling the distribution of systematic errors for a patient population. Each sample is represented by a vector and then another dose distribution is computed for the radiation fields shifted by the negative of the vector. For each sample a new DVH is computed, creating an ensemble of DVHs. By comparing these with the nominal DVH with no systematic setup error an estimate of the confidence in the DVH can be obtained.

1.2.4. The use of multiple-modality imaging

If more than one imaging modality has been used, the datasets must be registered. Landmark-based correlation or surface matching are popular methods (Fazio *et al* 1994). Others include chamfer matching (Kooy *et al* 1994). Before any outlining is attempted, MRI images must be de-distorted (Finnigan *et al* 1995, Moerland *et al* 1995, Prott *et al* 1995) and 3D datsets must be locked together by a registration algorithm. The accuracy of the registration can be assessed by inspecting how closely known corresponding bony landmarks appear in the registered datasets. These tasks precede contouring which itself cannot ever be the basis of determining the accuracy of registration. Whilst CT directly gives a measure of electron density, needed for planning, MRI which does not, and which has comparable spatial resolution, can better distinguish soft tissues (Chen and Pelizzari 1995). MRI can also give slices other than transaxial but with isotropic spatial resolution. Until spiral CT scanning, CT sagittal and coronal slices had non-isotropic spatial resolution.

As well as the points discussed in section 1.2.1, new difficulties arise with contouring registered datasets.

Since each modality may display a different physical property—for example function rather than anatomy—a tumour may appear very different in two different modalities. Hence the *PTV* will be chosen differently. Several studies have shown that PTVs in the brain determined from CT data are changed when MRI data for the same patient are available (e.g. Rampling *et al* 1993, Rosenman *et al* 1994, Pötter *et al* 1996). The same problem should not arise with OAR. Wong *et al* (1994) showed that x-ray CT and PET with fluoro-deoxyglucose (FDG) can give complementary information with the extent of tumour being different for each 3D modality. Gademann *et al* (1993) have studied the impact of MRI on the determination of the target volume in tumours of the brain and base of skull. From a study of 60 patients it was shown that CT was only able to detect 70% of the findings confirmed by MRI. In 45 of the 60 tumours the gross tumour extended in more MRI slices than CT slices even though the volume was not always larger in MRI. This implied that in CT the target was too short in the craniocaudal direction but too large in lateral extent. At DKFZ, MRI and CT are therefore used complementarily. Levy *et al* (1994) also demonstrated that target volumes for stereotactic proton irradiation of arteriovenous malformations were changed when multimodality image data were available. Schmidt *et al* (1996) argued that an additional advantage of MRI is that sectional images in any direction can be obtained with the same resolution.

Kamprad *et al* (1996) used correlated SPECT and PET functional images registered to anatomical MRI images to show the importance of taking account of functional information. High-grade gliomas and low-grade astrocytomas were imaged by SPECT with the radiopharmaceutical I^{123}-alpha-methyl-tyrosine (IMT) and the IMT concentration in the tumour area was measured and compared with the corresponding concentration in the opposite hemisphere. The mean uptake ratio was 1.83 and 1.63 for each tumour type respectively, indicating that the PTV should be increased from that which corresponded to just the anatomical information. Feldman *et al* (1996) also showed increased uptake of IMT with SPECT in glioblastoma and astrocytoma tumour regions and also increased areas of F^{18}-FDG uptake imaged with PET. They concluded that both SPECT and PET showed functional volumes larger than anatomical volumes with the former clearly superior. Fusion images were prepared in which part of the image was contributed by separate imaging modalities. The functional images also displayed the response to treatment.

1.2.5. Beam apertures

Assuming now that many of the problems above can be overcome, the next task which is faced in treatment planning is the construction of appropriate beam apertures. Modern 3D treatment-planning systems have the facility to display a BEV of structures from different orientations of the beam with respect to the patient (Bendl 1995, Bendl *et al* 1994). Generally this view consists of sets of shaded 3D surfaces which in some cases can be 'switched on and off' so that overlapping surfaces can be viewed separately (see Chapter 7). Until recently the construction of beam apertures has relied on manual contouring of the projection of the PTV. However, as multileaf collimators become more available and the possibility arises to use a large number of beam orientations, manual outlining becomes less and less feasible. Also it is potentially error prone and leaves to human judgement decisions about what to do when the projection of a PTV overlaps that of one or more OAR.

Brewster *et al* (1993) have developed a technique for automatic generation of beam apertures. The method comprises four stages: (i) the surface representation of PTV and OAR structures is formed in tiled form (no need for shaded surfaces) from the structure outlines on a set of CT slices; (ii) the vertices of the triangles are projected through coordinate transformation from patient-scanning space to the 2D space of the aperture and a digital picture is formed in which a pixel is shown bright if it contains one or more vertices; (iii) the outline of the projected structure is formed by edge detection (this stage includes the possibility to add (or subtract) a constant margin; such an outline is constructed for as many structures as required); (iv) the aperture contour is determined from the extended target outlines. In particular at this stage there is the possibility to exclude from the aperture any part of the projected PTV which overlaps with part of the projection of one or more OAR. The code is sufficiently flexible to handle even the situation where the projection of an OAR is entirely within the projection of the PTV.

Brewster *et al* (1993) applied this technique to determining the apertures for six-field irradiation of the prostate when excluding the rectum from the BEV was considered important. They also showed examples of a complete splitting into two of the aperture irradiating the brain PTV when the brainstem and spinal cord overlapped the view of the brain.

Once the PTV, OAR and the BEV for all beams have been determined the next challenge to planning is to determine the radiation beamweights.

The development of techniques to select beams and optimise radiation beamweights has received a great deal of attention and is a fast growing field of research. The next sections review the options.

1.3. TREATMENT PLANNING AND A TUTORIAL ON TREATMENT-PLAN OPTIMISATION BY SIMULATED ANNEALING

1.3.1. *What is treatment-plan optimisation?*

The aim of radiotherapy of local disease is to achieve a high dose to the planning target volume, simultaneously keeping the dose to OAR as low as possible. When it is intended to use a small number of rectangular fields with or without wedges and blocking, the treatment-planning technique is to try a number of different beam weightings, compute the dose distribution, evaluate the plan and then repeat the task until satisfied that the plan meets prescribed criteria. This is known as forward treatment planning. It is the 'traditional method' and is still in widespread use. It might be called 'human optimisation' but in fact the resulting plan has no more status than 'acceptable'.

As we move towards an era of increased automation and precision in 3D radiotherapy, a number of desirable treatment options will increasingly be called upon, including:

(i) the use of a larger number of fields;

(ii) the use of non-coplanar fields;

(iii) the use of fields shaped to the BEV of the PTV by a multileaf collimator (MLC);

(iv) the introduction of intensity-modulated beams (IMBs) (figure 1.4) via:

- multiple exposures with different static MLC settings;
- dynamic sweeping of MLC leaves;
- apparatus for tomotherapy;
- electronically-steered time-modulated pencil beams;
- the moving-attenuating-bar technique.

As these technologies become increasingly more common (the methods for *delivering* IMBs are discussed in detail in Chapter 2), it becomes quite impossible to create treatment plans by forward treatment planning because:

(i) there are just too many possibilities to explore and not enough human time to do this task;

(ii) there is little chance of arriving at the optimum treatment plan by trial-and-error;

(iii) if an acceptable plan could be found, there is no guarantee it is the best, nor any criteria to specify its precision in relation to an optimum plan.

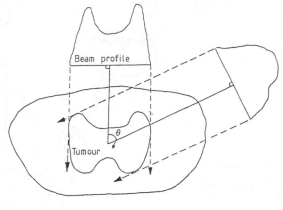

Conformal

Figure 1.4. *The concept of intensity-modulated beams. Two (of a large set of) such beams with 1D intensity-modulation are shown irradiating a 2D slice. The beams combine to create a high-dose treatment volume spanning the PTV which has a concave outline in which might lie OAR. Such uniform high-dose treatment volumes cannot be achieved using beams without intensity-modulation. Planning such treatments relies on tools such as simulated annealing for solving the inverse problem.*

For this reason the treatment-planning process has to be radically changed to solve the problem 'given a prescription of desired outcomes, compute the best beam arrangement'. Stated this way the problem is solved by *inverse* treatment planning (Brahme 1994a,c, Mackie *et al* 1994b). The task has to be solved by a computer with human guidance rather than by human experience alone (figure 1.5).

Brahme (1994b) has emphasised, however, that the concept of an 'optimum plan' is possibly a misnomer since it only has meaning when the optimisation criteria are specified. For example, from a purely practical viewpoint, an optimum treatment could be that with very few fields which can be executed quickly. An optimum plan for a given machine may not necessarily be the optimum if one could replace that machine with a different one, perhaps with different energy. However, when physicists speak of optimisation what they generally mean is a method to achieve the best outcome for the patient in terms of predicted tumour control and few complications and often these are reflected in the goal to obtain the best 3D dose distribution. For the purposes of this chapter, this is what we shall mean by optimisation. Optimisation is today a 'fashionable topic' but those interested in the history of radiotherapy might look at Newton's (1970) paper 'What next in radiation treatment optimisation?' to appreciate that this problem has been under consideration for a long time. Like many developments in medical physics, the availability of rapid computational facilities has resulted in renewed efforts to solve old problems.

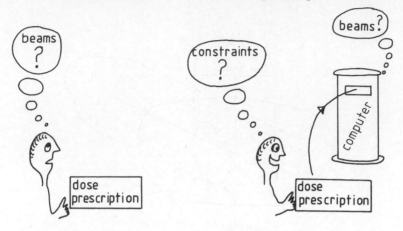

Figure 1.5. *The different approaches of 'forward' and 'inverse' treatment planning. On the left a human planner is presented with an ideal dose prescription and uses experience to work out the best distribution of beamweights. On the right the approach is reversed. The planner is presented with a prescription and constraints on the problem and uses a computer to predict the optimum treatment plan which solves the inverse problem.*

1.3.2. Classes of inverse-treatment-planning techniques

Inverse treatment planning was first discussed by Brahme in the early 1980s. Since then, several groups of workers have developed inverse-treatment-planning tools. Inverse-treatment-planning techniques have been developed largely in response to solving the problem of generating optimised IMBs. Whilst methods are not exclusive to this problem, most of the discussion of optimisation in sections 1.3–1.12 is posed in this context. Inverse-planning techniques fall into two broad classes.

(i) Analytic techniques. These involve deconvolving a dose-kernel from a desired dose distribution to obtain the distribution of desired photon fluence in the patient and forward projecting this fluence with attenuation factors to create profiles of beam intensity. Because of the formal similarity with x-ray slice imaging, these methods have sometimes been called inverse computed tomography (Brahme 1995d). The first tools were for 2D planning, later extended to 3D.

(ii) Iterative techniques. Both linear-programming algorithms (e.g. the Simplex) and the Cimmino algorithm have been exploited together with simulated annealing.

Some optimisation tools combine a mixture of analytic and iterative-refinement stages. In particular negative beamweights, created by analytic inversions, have to be removed in some way. Analytic inversions are considered later in this chapter.

Figure 1.6. *Drawing of a face composed only of the superposition of straight lines each from a different orientation and with a different blackness. (From Birkhoff 1940.)*

Several authors (e.g. Cormack 1995) have recently pointed to an interesting paper by Birkhoff (1940) in which he showed that an arbitrary 2D drawing could be described by the superposition of a series of straight lines from different directions each with different uniform darkness. The total darkening at any point thus becomes the sum of the darkenings of each line intersecting that point (figure 1.6). For Birkhoff's creation of arbitrary pictures some of the lines need to have 'negative blackness' (i.e. act like an eraser) in order to obtain a solution (see also Appendix D). The problem illustrated has some similarity to the CT reconstruction problem and also the inverse-computed-tomography problem of radiotherapy planning. The impossibility of delivering negative beamweights frustrates the analogy. Whereas in the absence of noise and with sufficiently many projections the CT reconstruction solution always exists, the same cannot be said for the solution to the inverse-therapy-planning problem (Brahme 1995e). It is for this reason that there is a large number of possible solutions developed by different authors.

There has been some elegant mathematical work on the so-called inverse problem, discussed at length in Chapter 2 of the companion Volume (Cormack 1987, 1990, Cormack and Cormack 1987, Cormack and Quinto 1989, 1990) considering the problem as a Radon problem. The paper by Cormack and Quinto (1990) is a useful summary of these analyses. These papers give a great insight into the problem but practical solutions have subsequently required computational rather than formulistic solution and it is on these that we concentrate in this chapter.

1.3.3.   Simulated annealing; general concepts and properties

We choose to describe this technique in some detail both (i) because it has been widely used and (ii) because it embodies and illustrates many of the features common to the whole class of inverse-planning techniques (Webb 1995b,c, 1996).   Simulated annealing is an iterative optimisation technique.   It is a method of finding the global minimum of some function when the state-space of that function may possess multiple local minima. Because of this property it is regarded as superior to other iterative methods which may become trapped in a local minimum.   Simulated annealing has been used in a variety of different fields with many applications (see e.g. Metropolis *et al* 1953, Kirkpatrick *et al* 1983, Geman and Geman 1984, Jeffrey and Rosner 1986, Press *et al* 1986, Willie 1986, Radcliffe and Wilson 1990, Press and Teukolsky 1991).   It was first introduced into the field of medical physics (coded aperture reconstruction of images of radioactive distributions) by Professor Barrett and colleagues in Tucson, Arizona, who applied the method to minimising a quadratic cost function (Barrett *et al* 1983, Paxman *et al* 1985, Smith *et al* 1985).   The present author continued this application into SPECT (Webb 1989a) and then into radiotherapy treatment planning (see section 1.3.7) (Webb 1989b, 1990a,b, 1991a,b, 1992a,b,c, 1993a, 1994a,b).   Mageras and Mohan (1992) have further extended the method and application with other cost functions and techniques to accelerate convergence. A tutorial review has been written by Rosen (1994).   Rosen *et al* (1995) have compared four different simulated annealing schemes for the same clinical cases and found that the schemes differed in their speed of convergence and success.

1.3.3.1.   A simple analogy of a walker descending from a hilltop to the lowest point in a valley.   To understand simulated annealing and how it may be useful in inverse radiotherapy planning let us first consider an analogy. Imagine a landscape which consists of a series of undulations, some gentle, others more rugged. It comprises hills and valleys, gulleys and arrêts. There is an especially large mountain, the largest in the region. There is a walkers' refuge on top of this largest mountain. There is an especially deep valley with a well at its lowest point. From the refuge it is not possible to see, nor to know the whereabouts of, that deepest valley with its well.

Now imagine a walker in the refuge is instructed to walk to the well. He cannot have any idea how to complete this task but he has one obvious clue. He had better start off walking down the mountain since he knows *a priori* that the refuge is higher than the well. So it seems a good idea to start off walking downhill. Now let us turn this into the language of physics. We know height represents potential energy (let us call this V) and $V_{well} < V_{refuge}$. The instruction our walker has in mind is, in the language of physics, 'lower your potential energy, i.e. lower V, in fact *minimise* $(|V - V_{well}|)$'.

Our walker continually changes direction *locally* at each step turning in the direction in which the step is a move downhill. It helps to imagine the walker is in a complete fog or very short-sighted. He certainly cannot see over nearby hills if he can see them at all. All he knows is that the best course of action at any one moment is to move downwards. He cannot know whether the spot to which he is aimed is the lowest in all the terrain or just the lowest in his immediate vicinity. In fact, contrary to achieving his goal of reaching the well, the simple instruction to only move lowering V will lead to our walker finding himself in a local valley. As physicists we would say the walker reaches a *local* minimum potential V. If our walker refuses to ever walk uphill he will stay put in that local minimum.

So how can he ever get to the well at the foot of the deepest valley? This well represents the *global* minimum potential $V_{min} = V_{well}$. There is a way. He must grow long legs and start off with big steps exploring the whole landscape. From time to time he must climb uphill to get out of a local depression in the landscape, but clearly he must not go uphill as often as downhill or the net effect will be to make no progress towards the well. He must go uphill less often than downhill and as his journey continues the probability of going downhill must start to rise and the probability of going uphill must fall. It helps if his legs gradually shrink so the steps become ever smaller. Adopting this scheme he will eventually reach the well albeit taking a longer time than if the direct path to the well were known *a priori*. There are other schemes. He could alternatively resolutely refuse to go uphill but could sample his step size from a distribution so wide that all possibilities to leap over local hills are available. This would be like tunnelling through the hills.

What is described is exactly what is achieved by the algorithm known as simulated annealing. Our walker's itinerary is an iterative process. Most often, moves are downhill but occasionally uphill moves are made to avoid trapping in local minima. Initially large steps are made with many uphill steps as well. As time progresses steps shorten and become predominantly downhill. This algorithm is designed so that when such changes are made iteratively to some system, the changes will converge to a global minimum in some function which depends upon them. The iterations will not become trapped in one of a number of local minima of such a function.

The name 'simulated annealing' arises from the process by which metals are annealed. If the temperature of a melt falls too fast then amorphous states can arise, whereas annealed metals form from slow cooling. The computational method is now used in many fields.

1.3.3.2. The optimisation problem and simulated annealing in general language. Let us try to describe the most general problem in language which will allow us to see the relevance to the problem of radiotherapy optimisation. From this it will be easy to turn the description into the mathematical terms needed to code the algorithm.

The optimisation problem is to arrive at a set of values for some variables (we might call these the 'driving variables') which determine some other dependent parameters (we might call these the 'driven variables') and which correspond to the global minimum in some third quantity (which we call a 'cost function') dependent in turn on these other dependent parameters (the driven variables). Firstly we identify that, at each iterative step, taking some action (e.g. changing one or more physical driving variables) has a known or predictable effect on some other dependent driven quantity. In the above example the (driving) action taken at each step is that the walker changes his position coordinates and the deterministic dependent (driven) outcome is a change in the potential energy of the walker. The third quantity, the cost function, is the difference between the current value of this potential energy and the global minimum potential energy. It is the minimisation of this cost which is the goal of the optimisation. Simulated annealing addresses this optimisation problem by making successive changes to the driving variables and (via the driven variables) calculates the change in the cost function. Driving variable changes are accepted if they lead to decreased cost and are only accepted with a small probability if they lead to increased cost (to perform the 'hill climbing'). The probability is controlled by the size of the change in the cost function with respect to a reference; this feature is best explained mathematically.

1.3.3.3. General mathematical description of the optimisation problem and simulated annealing. Now let us cast the whole general algorithm mathematically. Suppose at any iteration n the set of I driving variables are

$$x_1^D(n), x_2^D(n), x_3^D(n), \ldots, x_I^D(n). \tag{1.1}$$

The dependent set of J driven variables are

$$x_1^d(n)[x_1^D(n), x_2^D(n), x_3^D(n), \ldots, x_I^D(n)],$$

$$x_2^d(n)[x_1^D(n), x_2^D(n), x_3^D(n), \ldots, x_I^D(n)],$$

$$x_3^d(n)[x_1^D(n), x_2^D(n), x_3^D(n), \ldots, x_I^D(n)],$$

$$\vdots$$

$$x_J^d(n)[x_1^D(n), x_2^D(n), x_3^D(n), \ldots, x_I^D(n)]. \tag{1.2}$$

The cost function $C(n)$ is a function $f[\ ]$ of these J driven variables and some set of K specified parameters

$$p_1(n), p_2(n), p_3(n), \ldots, p_K(n) \tag{1.3}$$

i.e.

$$C(n) = f[x_1^d(n), x_2^d(n), x_3^d(n), \ldots, x_J^d(n), p_1(n), p_2(n), p_3(n), \ldots, p_K(n)]. \tag{1.4}$$

The *change* $\Delta C(n)$ in the cost function between iterations n and $(n-1)$ is then

$$\Delta C(n) = C(n) - C(n-1). \tag{1.5}$$

If $\Delta C(n) < 0$ the change in the driving variables is always accepted. If $\Delta C(n) > 0$ the change is accepted with a probability

$$\exp\left(-\frac{\Delta C(n)}{k_b T(n)}\right) \tag{1.6}$$

where the denominator $k_b T(n)$ has the same dimensions as $\Delta C(n)$ and is itself a function of n. The variable $T(n)$ may be thought of as a temperature if k_b is the Boltzmann constant and $\Delta C(n)$ has units of energy.

So called 'classical' simulated annealing iteration proceeds by starting with a large temperature (high probability of accepting 'uphill' changes leading to $\Delta C(n) > 0$) and decreases the temperature slower than $(1/\log(n))$ so gradually reducing this probability to zero. Driving variables are selected so that the changes derive from a Gaussian distribution. Alternatively in so called 'fast simulated annealing' the temperature is lowered faster and the driving variables are selected from a wide Cauchy distribution with the width of the distribution collapsing as iteration proceeds.

1.3.3.4. The hill descender's problem in mathematical terms. So now let us re-examine the problem of our hillwalker in these mathematical terms. There is only one driving variable ($I = 1$), namely the geographical position vector r. i.e.

$$x_1^D(n) = r(n). \tag{1.7}$$

This determines the single driven variable ($J = 1$), potential energy

$$x_1^d(n)[x_1^D(n) = r(n)] = V(n)[x_1^D(n) = r(n)]. \tag{1.8}$$

The cost C is a function of this one driven variable and one ($K = 1$) specified parameter V_{well} i.e. $C(n) = |V(n) - V_{well}|$. The goal of optimisation is to vary r until a minimum in C is obtained using the rules above.

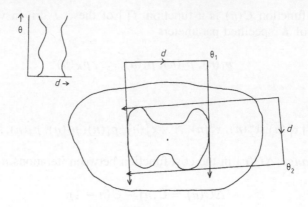

Beam sinogram

Figure 1.7. *The concept of the beamweight sinogram. The butterfly-shaped region is the PTV and a number of beams with intensity modulation are arranged to span the PTV. The angle θ labels each beam and the parameter d labels each element of each beam (sometimes called a bixel). If the beamweights are stacked, as shown in the upper left, then the resulting image is a beamweight sinogram by analogy with medical imaging.*

1.3.3.5.   IMB optimisation by simulated annealing.   To give the discussion substance in the radiotherapy context, let us imagine a number of beam orientations have been selected and the problem is to determine the optimum set of beamweights for beams with intensity modulation. These are the 'driving variables', i.e. it is required to compute the beamweight sinogram (figure 1.7). The analogy with medical imaging is shown in figure 1.8. The 'driven variables' are the dependent dose-point values.

A large number of combinations of beamweights are explored in some iterative manner (see section 1.3.5). Define a quadratic cost function C_n at the nth iteration (i.e. nth choice of such beamweights) to be:

$$C_n = \left[\left(\frac{1}{M} \right) \sum_r I(r)(D_p(r) - D_n(r))^2 \right]^{0.5} \qquad (1.9)$$

where $D_p(r)$ means the prescribed dose value at some point r in the patient and $D_n(r)$ denotes the computed dose at the same point at the nth iteration and the sum is taken over a large number M of specified dose points. $I(r)$ is a term which allows the user to weight the importance of contributions to the cost from different parts of the body. For example it may be more important to have the dose conform in the PTV than in the OAR or conversely protection of the OAR may be the desired goal at the expense of some dose inhomogeneity in the PTV (so-called 'conformal avoidance'). The choice of the parameters $I(r)$ greatly influences the outcome of the optimisation (Webb 1994b, 1995c,d). The aim of the optimisation is to compute the

Analogy

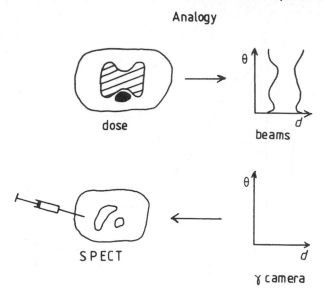

ALL 2D images

Figure 1.8. *The analogy between inverse treatment planning, when the 1D beams are intensity modulated, and emission tomography imaging. In the former, the dose prescription is known on a 2D matrix (top left) and it is required to compute the beamweight sinogram (top right). In the latter, a gamma camera makes a measurement of the projection sinogram (bottom right) of the distribution of activity and this is reconstructed (bottom left) from these data. Simulated annealing has been used to solve* both *problems, the latter problem providing the stimulus and motivation for introducing simulated annealing into radiotherapy treatment planning. The reversal of the direction of the arrows signifies the (approximately) inverse relationship between the two problems.*

lowest value of C_n since this corresponds to the calculated dose distribution best matching the prescription in a least-squares sense (figures 1.9 and 1.10).

This suffices for the purpose of discussion although it is pointed out that this is only one of many possible functions that can be minimised and that the quadratic cost function does not actually have any local minima when optimising beamweights. Bortfeld and Schlegel (1993) have shown that the same quadratic dose cost function does have local minima when optimising beam directions. We return later (see section 1.3.4) to discuss other cost functions including those, embodying biological endpoints, which almost certainly do have local minima.

For some general cost function it is not possible to know ahead of time the form of the multidimensional surface in state space of the function for

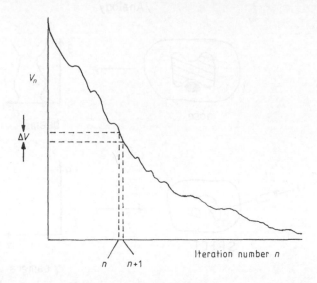

Figure 1.9. *A general cost function V_n in one dimension which may have local minima as shown as well as a global minimum.*

all possible beam arrangements. As different beam arrangements are tested, the cost function will change (figure 1.11). The problem is to compute a very large number of elemental beamweights which combine to give a dose distribution evaluated on all the dose pixels contributing to the cost function.

Sometimes the cost function will be larger than at the previous iteration and sometimes smaller. At first sight it would seem obvious to accept only each new beamweight set that *lowers* the cost function and those that achieve this reduction are indeed substituted for the previous best estimate. However, in simulated annealing a mechanism for accepting changes which lead to an *increase* in the cost function is also built in. This mechanism allows the system to climb out of local minima in the cost function, should there be any, and eventually reach the global minimum. If the cost function changes by ΔC_n in progressing to the nth from the $(n-1)$th iteration, and ΔC_n is positive, then this change is accepted with a small probability

$$\exp(-\Delta C_n / k_b T) \tag{1.10}$$

where k_b is the Boltzmann constant and T is the temperature (figure 1.12).

In practice $k_b T$ is simply a quantity ascribed the same dimensions as ΔC_n. Initially in the iterative process the temperature is large allowing many 'wrong-way' changes of the cost function (and thus a wide exploration of state space). In so-called classical simulated annealing provided the temperature is reduced slower than, or as, $(1/\ln(n))$ this guarantees progression towards the global minimum (Press *et al* 1986).

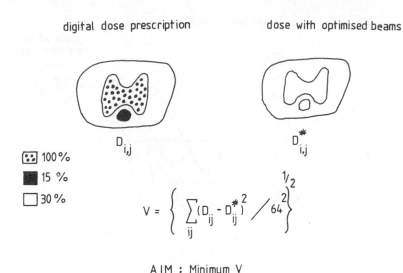

digital dose prescription   dose with optimised beams

$D_{i,j}$ $D^*_{i,j}$

[··] 100%

■ 15 %

□ 30 %

$$V = \left\{ \sum_{ij} (D_{ij} - D^*_{ij})^2 \Big/ 64^2 \right\}^{1/2}$$

AIM : Minimum V

Figure 1.10. *The left part shows a possible dose prescription (D is $D_p(r)$ in the text). The aim is to have a high dose in the PTV (shown polka-dotted) and a low dose in the OAR (shown black). The right part shows the 'running estimate' of the dose on a 2D grid specified by i, j (D^* is $D_n(r)$ in the text) and this example illustrates the computation of a quadratic dose cost function on a 64^2 grid. The aim of simulated annealing is to minimise the cost function V.*

The method gets its name from the process by which metals are annealed. If the temperature falls too fast then amorphous states can arise, whereas annealed metals form from slow cooling. Consider also a second analogy with a skier descending a slope (figure 1.13). Imagine the slope is the graph of the cost function and for the purposes of illustration imagine the cost function is one-dimensional. The skier starts at the top (large value of the cost function) and wishes to reach the hotel at the bottom (global minimum of cost function).

In general the skier must aim to go downhill (of course). But if our skier is overprincipled and refuses ever to suffer a potential energy rise (i.e. go uphill), then that skier will be trapped behind any snowbump blocking their path (see figure 1.13) and will not reach the bottom. Just as a skier requires some momentum to rise over the snowbump, so the optimisation requires the mathematical step described in equation (1.10) for overcoming local minima.

1.3.4. *The power of simulated annealing; cost functions*

The great power of simulated annealing also lies in its flexibility. The cost function can be as simple or as complicated as one likes. The

Simulated   annealing

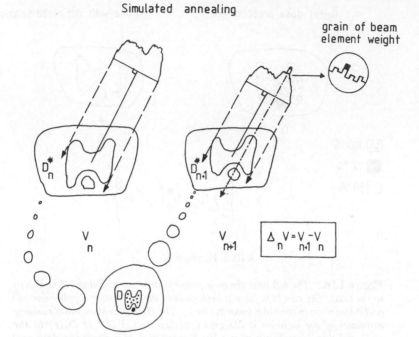

Figure 1.11. *The figure shows how the addition of a grain changes the cost function. On the left at the nth iteration the current estimate of the dose distribution is D_n^* and the cost function is V_n, computed by comparing the running estimate of the dose distribution with the prescription D (shown as 'bubbles'). Then (right) one particular element of beamweight has a grain added (shown black) and the new dose distribution becomes D_{n+1}^* and the new cost function is V_{n+1}. The change of the cost function ΔV_n controls what happens next.*

quadratic cost function in dose (equation (1.9)) has been a popular choice since it has an intuitive meaning, namely it reaches a minimum when a planned dose distribution matches a dose prescription. Webb (1991a, 1994b) has commented on the need to tune the algorithm via a suitable choice of importance parameters. As the name 'cost function' suggests, what is achieved depends on what the user is prepared to pay for. A purely quadratic cost function for beamweight optimisation has no local minima and the matrix linking 3D dose to a set of beamweights could, in principle, be analytically inverted (Niemierko 1996a). However, such inversions would be lengthy and ill-conditioned. The enormous matrices would be complicated to construct and would require huge computer storage. The analytic inversion may well be impossible, would almost certainly lead to the inclusion of negative beamweights, and simulated annealing still provides a convenient way to solve the problem and build in the

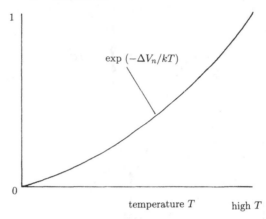

Acceptability of positive-potential changes

probability of acceptance $= \exp(-\Delta V_n / kT)$

$\Delta V_n =$ potential change due to grain placement
$k =$ Boltzmann's constant
$T =$ temperature

Figure 1.12. *The graph controlling the acceptability, or otherwise, of 'wrong-way' (uphill) changes in the cost function. When ΔV_n is positive, equation (1.10) gives the probability with which such changes should be accepted. At high temperatures the probability will be high and vice-versa.*

constraint to positivity and flexible tuning. In the case of beamweight optimisation with the quadratic dose cost function, simulated annealing at zero temperature reduces to constrained iterative least-squares optimisation. If beam orientation is optimised with the quadratic dose cost function, non-zero-temperature simulated annealing must be implemented.

An alternative cost function, still based on dose, is to use a quadratic term in the PTV and a linear term in the OAR, representing integral dose to the OAR and the rest of the body B. Oldham and Webb (1993, 1994, 1995a,b,c) have used

$$C_n = a_1 \frac{C_{PTV}}{C_{PTV_{st}}} + a_2 \frac{C_{OAR}}{C_{OAR_{st}}} + a_3 \frac{C_B}{C_{B_{st}}} \qquad (1.11)$$

where C_{PTV} has the same form as in equation (1.9) and

$$C_{OAR} = \sum_r D_n(r) \qquad \text{with } r \text{ in OAR} \qquad (1.12)$$

and

$$C_B = \sum_r D_n(r) \qquad \text{with } r \text{ in B} \qquad (1.13)$$

and the subscript *st* indicates the starting value when all beamweights are uniform. a_1, a_2 and a_3 are constants which control the optimisation, allowing

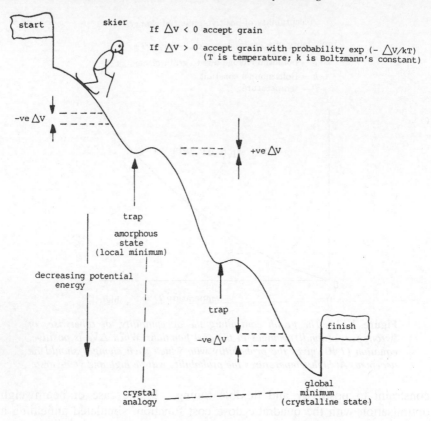

If $\triangle V < 0$ accept grain

If $\triangle V > 0$ accept grain with probability exp $(- \triangle V/kT)$
    (T is temperature; k is Boltzmann's constant)

Figure 1.13. *The analogy between minimising a cost function by simulated annealing and the technique of a skier descending a slope (see text for discussion).*

different initial importance to be ascribed to each region of interest in the fit.

Since dose is only a surrogate for biological outcome, some workers have argued that, instead of trying to match a dose prescription, it is better to optimise biological outcome. The argument applies equally to analytic and iterative techniques. Using models, supported by observational data, TCP and NTCP can both be computed knowing the 3D dose distribution in the PTV and OAR. Formally, the TCP at the nth iteration is

$$\text{TCP}_n = f_1[D_n(r)] \qquad \text{with } r \text{ in PTV} \qquad (1.14)$$

where f_1 represents the function linking inhomogeneous dose and TCP (see Chapter 5). Similarly the NTCP at the nth iteration is

$$\text{NTCP}_n = f_2[D_n(r)] \qquad \text{with } r \text{ in OAR} \qquad (1.15)$$

where f_2 represents the function linking inhomogeneous dose and NTCP (see Chapter 5).

It is then possible to specify different cost functions for optimisation involving biological outcome, e.g.

(i) maximise TCP subject to some maximum-allowed NTCP;

(ii) minimise NTCP subject to some minimum-allowed TCP;

(iii) maximise the probability of uncomplicated tumour control TCP× $(1 - \text{NTCP})$.

Note, however, that a physical parameter cannot both be a constraint and a goal for optimisation simultaneously. Not all these objectives are endorsed. For example option (iii) might correspond to a too high NTCP. Clinicians usually set an upper bound to the NTCP as a constraint and (i) is often the preferred clinical rationale in determining a treatment.

It would even be possible to construct cost functions involving both dose and biological outcome. Cost functions based on TCP and NTCP almost certainly have local minima (Deasy 1996).

The argument in favour of optimising biological outcome is that this is of course the aim of radiotherapy; dose being a traditional surrogate. The argument against is that the biological models are relatively new and somewhat empirical and the data supporting them are based on limited observations. To an extent there exists a diversity, workers in the UK preferring to optimise dose and compute TCP and NTCP *a posteriori*, whilst some in the USA take a bolder approach and bypass dose. There is plenty of scope for debate on the issue which has received some attention already (Webb 1992a, Mohan *et al* 1994, Niemierko 1996b). There is a real need for detailed studies comparing predicted NTCP with the observed actual incidence of complication. Such studies are rare, but recently Graham *et al* (1994), for example, have shown a very good correlation for pneumonitis by following up 60 patients who had been planned with a 3D planning system.

1.3.5. *Classical and fast simulated annealing*

Imagine the system of beamweights is changed at each iteration by the addition of a 'grain' g of beamweight to one particular beam element. In classical simulated annealing the system is perturbed only slightly at each iteration, sometimes with a Gaussian generating function for g

$$\exp(-g^2/T_n) \tag{1.16}$$

where T_n is the temperature at iteration n. The author's early work had a generating function which was simplified to a delta-function choice of either a positive or negative grain g, initially constant and then reduced in size towards the closing stages of iteration. The cooling proceeds either as, or slower than

$$\frac{T_0}{\ln(1+n)}.$$ (1.17)

In fast simulated annealing (Szu 1987, Szu and Hartley 1987) the grains are generated by a Cauchy distribution

$$\frac{T_n}{(T_n^2 + g^2)}.$$ (1.18)

The cooling proceeds as

$$\frac{T_0}{(1+n)}.$$ (1.19)

The faster cooling (and hence quicker computational times) is allowed because the form of the Cauchy distribution generates occasional large grains which allow the system to tunnel out of a local minimum. Depending on the cost function there may be no need for classical hill climbing.

1.3.6. *Practicalities of implementing simulated annealing for inverse treatment planning*

Firstly, it should be made clear that simulated annealing substitutes for only one part of the planning process, the calculation of beamweights by otherwise traditional forward planning. So it requires to be fed exactly the same data as the forward-planning problem including:

(i) specification of the PTV and OAR contours, derived from high-quality 3D imaging data (e.g. x-ray CT, MRI, SPECT and PET);

(ii) specification of the dose to each dose-space voxel per unit beamweight (the 'beam model');

(iii) relevant biological models and data if a biological cost function is used;

(iv) prescription/constraints on the treatment plan and/or on the beamweights themselves.

Similarly, the results of the calculation (3D dose maps) must be evaluated by the same tools as would be used to evaluate the outcome of forward treatment planning, i.e. DVHs, display of isodoses, 3D shaded-surface display of dose, dose ribbons, display of dose superposed on anatomical sections, *a posteriori* TCP and NTCP calculations (if simulated annealing was dose-based) etc.

Hence simulated annealing should properly form part of an integrated 3D treatment-planning system. When first introduced into the armamentarium of planning tools (Webb 1989b) this was not possible. A stand-alone implementation demanded that all these tasks were worked on independently for the application. How this was done is not central to understanding simulated annealing but is described at length in a suite of papers (Webb 1989b, 1990a,b, 1991a,b, 1992a,b,c, 1993a).

Ideally a 3D planning system should be capable of both forward and inverse treatment planning, the results of which can be compared using identical tools for evaluation. Ideally such a planning system should be capable of planning using geometrically shaped fields *without* intensity modulation by both forward- and inverse-planning techniques so that the same planning problems can be tackled a number of ways to deduce when inverse planning is likely to be advantageous. This ideal system should also allow different cost functions to be switched in at will by software switches. Ideally proton planning should also be available as well as planning with a number of photon energies so that comparative plans can be made.

This ideal is a dream. It does not exist in any 3D planning system. Most of the scientific development of inverse-planning algorithms has taken place in a 'stand-alone' context. Manufacturers, not surprisingly, also tend to focus on a limited subset of the above requirements, given limited resources and generally they have responded to a specific need. So, for example, most 3D planning codes do not include planning intensity-modulated radiotherapy at all since, for 100 years, open and wedged fields have been the norm and even multileaf collimation is fairly new. There have been no comparative studies of IMB therapy versus open-or wedged-field therapy using the same planning cases in the same computational environment although there have been comparisons for specific cases using more than one environment. If IMB therapy is to be taken more seriously at the practical level these obstacles must be removed.

The following features for implementing simulated annealing must be included.

(i) Both positive and negative grains must be sampled so that there is a mechanism to 'undo' structures which may be created early in the optimisation but which may need to be removed once a wide search of state-space has been made.

(ii) In classical simulated annealing the grain size should be reduced at the later stages of iteration to fine-tune the solution. In fast simulated annealing the Cauchy distribution should become increasingly narrow;

(iii) Beamweights must be constrained positive. Any candidate change which passes the acceptance criterion for annealing is nevertheless rejected if it would lead to a beamweight going negative. This is an important step. Analytic inversion techniques on the other hand generally generate negative beamweights and then artificially these have to be set to zero *a posteriori*, somewhat massaging the status of the result. A constrained iterative solution never gets into this difficulty.

(iv) Attention must be paid to the computational aspects of the calculation of the cost function since this is at the heart of each iteration. This part must be computer-optimised. For example, there are tricks whereby the change in a quadratic cost function can be computed without evaluating the full function for each of two successive iterations (Webb 1989a,b). Also when large 3D arrays are being manipulated care must be exercised in handling the order of sequencing through the dimensions of the array.

(v) Other constraints can be applied to beamweights, e.g. a requirement for smoothness in intensity-modulated beam profiles. The application of constraints generally leads to it being impossible to reach a zero of the cost function (which would require disallowed negative beamweights). The optimisation will achieve some finite global minimum which in turn will depend on the constraints.

Although simulated annealing appeals for its wide flexibility, this very same ability to be tailored in many ways has a drawback. In the author's experience it is necessary to experiment somewhat in choice of grain sizes, number of iterations and the tuning of the importance parameters in the cost function, which in turn depends on the desired clinical outcome. If the cost function is expected to have local minima, and uphill moves are to be included, then attention must be paid to the cooling scheme and the initial temperature (which determines both the initial and the total number of 'wrong-way' changes accepted). The early papers by Barrett *et al* (1983) did not address these points. The choice depends to some extent on the nature of the cost function in parameter space (Barrett 1994). Yet this function is not well known *a priori*, hence the need to experiment (Silverman and Adler 1992).

Barrett (1994) has commented that problems such as the one used here for illustration may have cost functions with a very broad region with shallow curvature in the vicinity of the global minimum and that consequently different schemes for optimisation can arrive at different parts of this 'floor', the global minimum being hard to reach. Put another way, there can be a large number of possible beam arrangements which correspond to much the same final dose distribution, the inversion from 3D dose to beamweights being ill-conditioned. This has already been advanced as a potential disadvantage of analytic matrix inversion. It opens up further flexibility for searching for solutions with further constraints such as minimising the number of IMBs.

1.3.7. *Implementations at the Institute of Cancer Research (ICR) and Royal Marsden NHS Trust (RMNHST)*

The work at ICR/RMNHST has proceeded in several stages.

(i) The first application was to compute the one-dimensional (1D) intensity-modulated beam profiles which would create a 2D dose distribution

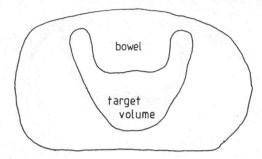

Figure 1.14. *A 2D planning problem where the aim is to obtain a high-dose treatment volume within the irregular PTV shown whilst simultaneously sparing the dose to the bowel (OAR). This is a clinical planning problem first presented by Chin.*

for a PTV with a concave outline in which might lie OAR. A series of model problems and clinical cases from the literature were solved, first with a simple beam model (Webb 1989b) and then with a model including scattered radiation (Webb 1991b). This work occupied the period between the middle of 1988 and the middle of 1990 and at that time there was a lot of interest in the development of radiotherapy with intensity-modulated beams even though it was almost impossible to deliver radiation with this feature at that time (figures 1.14, 1.15 and 1.16).

(ii) The second application was to compute the beamweights for fields defined by a multileaf collimator (MLC) (Webb 1991a). The main advance was to recognise that each field could be divided into two, one part seeing only the PTV and the other part seeing both the PTV and the OAR. The application was for those problems where it was not possible to create beams whose primary radiation only intersected the PTV. For example, when irradiating the prostate, the rectum is in the field of view and indeed often overlaps the PTV. Simulated annealing was used to find the optimum set of pairs of beamweights. A number of 'difficult' geometric model problems were studied where the PTV wrapped around the OAR (figure 1.17).

(iii) The third application was to compute the 2D intensity-modulation across the MLC-shaped apertures for optimal 3D conformal dose distributions (Webb 1992a). Again a number of 'difficult' geometric model problems were studied and this time the results were evaluated *a posteriori* by TCP and NTCP models. Applications (ii) and (iii) were developed in 1990 and 1991 (figure 1.18) when commercial MLCs were just becoming available in the UK.

The second application (with the original classical simulated annealing replaced by fast simulated annealing and the cost function of equation (1.11)) has been implemented in a clinical context (Oldham and Webb 1994, 1995a,b,c, Oldham *et al* 1995a,b,c,d,e) (early 1992 onwards) as part

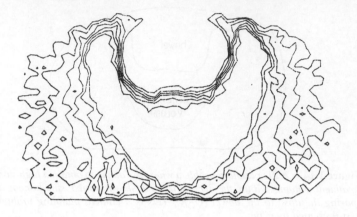

Figure 1.15. *The resulting optimised 2D dose map for the problem in figure 1.14 created by simulated annealing. The innermost dose contour is the 90% contour and the others (moving outwards) are 80%, 70%, 60%, 50% and 40% respectively. The dose map is shown as a greyscale image together with the beam sinogram in Webb (1991b).*

of the European Union-funded AIM project A2003 called COVIRA in collaboration with the German Cancer Research Centre (DKFZ) at Heidelberg and the IBMUK Scientific Research Centre at Hursley. The stand-alone system accepted CT data from the image handling package TOMAS (see section 1.2) and the method has been implemented within the 3D treatment-planning package called VIRTUOS which has grown out of VOXELPLAN (Bendl 1995, Bendl *et al* 1994). Oldham *et al* (1995a,b,c) made a comparison between the performance of the optimisation algorithm for MLC-shaped fields and the performance of a human planner when optimising clinical cases of prostate cancer. Oldham and Webb (1995b,c) demonstrated some limitations in the optimisation of such treatments arising when the PTV included part of the rectum OAR. Oldham *et al* (1995c,e) showed that this optimisation technique could generate the best wedge angles. Although the improvements were small they were achieved for almost no cost (optimisation as implemented in VOXELPLAN is fast) and they did not call for more complex treatments.

The first application has been further studied to decide which method of delivering intensity-modulated beams may be most practical. Some preliminary results have been presented (Webb 1994a). A fuller study (Webb 1994b) showed that, with careful tuning, IMBs could be determined which would be deliverable by either multiple-static MLC-shaped fields or by using a special temporally modulating slit collimator (so called tomotherapy). The delivery of IMBs by these methods is discussed in Chapter 2.

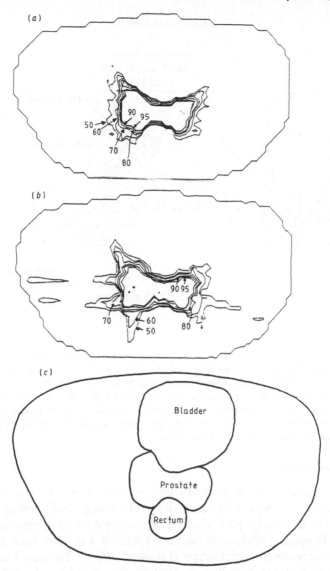

Figure 1.16. *The isodose plot for the plan created according to two separate conditions: (a) 120 beams; (b) 9 beams. The cost function has importance parameters $I_{PTV}(x, y) = 200$ and $I_{OAR}(x, y) = 10$. The isodose distribution is tightly conformal in both cases. In case (a), corresponding to delivery by tomotherapy (see Chapter 2), the isodose lines are a little tighter than in case (b) which is nevertheless still an acceptable distribution. The isodoses shown are 95%, 90%, 80%, 70%, 60% and 50%. The regions of interest (ROIs) are shown in (c). The bean-shaped ROI is the PTV (prostate); the anterior ROI is the bladder and the posterior ROI is the rectum, a planning problem posed by Bortfeld et al (1994) and also solved by his planning technique.*

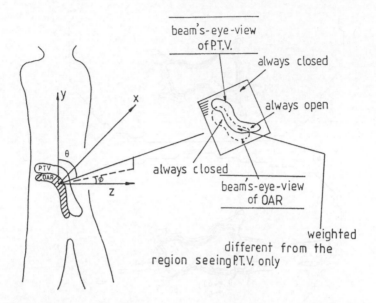

Figure 1.17.   *This figure shows how to optimise the separation between the DVH for PTV and OAR when multileaf-collimated beam's-eye-view ports are used. The multileaf collimator is shown to the right with the projected opening area corresponding to the views of the PTV and the OAR. These two curves overlap. The question arises of how to handle the overlap region. A different beamweight is applied to the aperture viewing PTV only and the overlap aperture viewing PTV and OAR.*

1.3.8.   Conclusion

Simulated annealing is a powerful and flexible tool for treatment plan optimisation. It was first used for optimising radiotherapy at the ICR/RMNHST, closely paralleled by developments at Memorial Sloan Kettering Hospital (MSK—see section 1.4). It has also been used for optimising stereotactic radiotherapy (Lu *et al* 1994). Simulated annealing has desirable features including controlling the positivity of beamweights and allowing a wide choice of cost function and flexibility to weight the importance of constraints in different regions in the patient. This same flexibility, however, demands considerable experimentation to determine optimum operating conditions. The first applications used classical simulated annealing which is conceptually relatively simple but time consuming. Workers are now moving towards fast simulated annealing but still encountering the need to 'tune' the algorithm (Webb 1994b). The method has been recently compared with the performance of iterative 'genetic algorithms' by Langer *et al* (1994).

Modulated fields

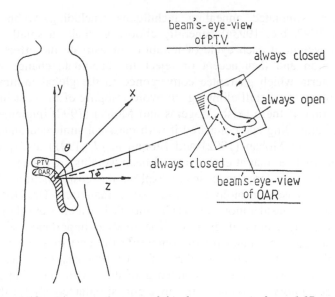

Figure 1.18.   *The main features of this figure are as in figure 1.17 except that the projected area of the PTV is now shown as a continuous solid curve. Some radiation passing through this also intersects the OAR. The beam intensity is modulated continuously across this aperture. Different intensity modulations arise at each orientation of the treatment gantry.*

1.4. PROGRESS IN 3D TREATMENT-PLAN OPTIMISATION BY SIMULATED ANNEALING AT MEMORIAL SLOAN KETTERING HOSPITAL (MSK), NEW YORK

Mohan (1990) presented initial results of the work on simulated annealing at MSK, New York at the 10th International Conference on the Use of Computers in Radiotherapy. Mageras and Mohan (1992) and Mohan *et al* (1992) have investigated techniques for accelerating the convergence of the simulated annealing optimisation technique for determining the beamweights for a large number of non-coplanar-shaped fields. Mageras and Mohan (1992) investigated a model problem with 54 beams, arranged as three non-coplanar rings of 18 fields equispaced in 20° azimuth, one ring being transaxial and the other two at ±30° in latitude. The problem was to optimise a prostate treatment. Mohan *et al* (1992) chose fields, not necessarily coplanar, so that no parallel-opposed fields occurred. These studies distinguished, for each orientation, two separate components of the field, namely that 'seeing' only PTV and that 'seeing' PTV and OAR. The radiation beamweights for the former part-fields were traded off against those for the latter creating inhomogeneous dose distributions in the prostate. In this respect the formalism was identical to that of Webb (1991a, 1992b, 1993a).

Classical simulated annealing techniques (including Webb 1989a,b, 1991a,b, 1992a,b,c, 1993a) generally choose to make a small change to one beamweight at each cycle of iteration, investigate the effect on some cost function and then accept or reject the candidate change according to the criteria which guarantee convergence to the global minimum cost. This involves some 'hill-climbing' to avoid trapping of the cost function in local minima, if these exist. Mageras and Mohan (1992) implemented fast simulated annealing and compared it with classical simulated annealing.

Mageras and Mohan (1992) and Mohan *et al* (1992) also proposed a radical alternative, that at each cycle all beamweights are changed. Initially the changes were large to coarsely explore the cost function. The cost function was based on biological response to radiation. The cost function was based on considerations of NTCP and TCP which were derived from dose data using empirical models. The relative importance of TCP and NTCP was also accounted for in computing the cost or score function to be optimised. They proposed three distinct schemes for generating the candidate changes (classical simulated annealing with Gaussian and fixed element generators—varying all beamweights simultaneously for the former and one beam only for the latter per iteration—and fast simulated annealing with a Cauchy generator—varying all beamweights simultaneously per iteration) and each scheme requires specifying a function which updates the temperature which controls the number of allowed 'hill-climbs' and the absolute size of the changes at each cycle.

Their results showed that with the two schemes in which all beamweights are simultaneously varied at each cycle, there is no need for 'hill-climbing' (i.e. zero-temperature runs produce much the same result). This is because the wide search of the configuration space corresponds to 'tunnelling' through from local minima as well as descent towards the global minimum.

Although, for a model problem, the final cost function, TCP and NTCP end up much the same with these methods as with 'one-change-per-cycle' simulated annealing, the beamweights can be quite different, reflecting the wide exploration of configuration space and indeed this allows them to greatly reduce (by eliminating beams with small weights) the number of fields needed (from 54 to 13 in the model problem) without radically altering the result.

In some problems, such as these, achieving the global minimum may not be necessary since a local minimum has almost the same cost. Simulated annealing is, however, still necessary to determine whether a local minimum is close to a global minimum. Even when the solutions are close, one may be preferred over the other because it is less complex to deliver. This is of course a failure of setting up the problem to include treatment complexity in the first place.

An additional advantage of fast simulated annealing with multiple beamweight changes per iteration is the increased speed of convergence by a factor of ten or so. Although some effort is expended to determine the optimum cooling scheme, they showed that the temperature may be set to zero with much the same result, tunnelling and descent 'doing all the work'.

Mohan *et al* (1992) found that 'uneducated application of constraints' could thwart the optimisation altogether. Mohan *et al* (1992) state (although this is not formally proved) that optimisation problems in radiotherapy involving TCP and NTCP must have multiple minima and so simple descent methods cannot be relied on. It certainly seems intuitively obvious that since TCP and NTCP are non-linear functions of dose, and since OAR have volume effects, cost functions based upon them will have local minima (Deasy 1996). There is scope for demonstrating the existence of local minima in specific circumstances and with specific cost functions.

The work at MSK and at RMNHST is so closely parallel that some detailed points are extracted for comparison. Both approaches have the following features in common.

(i) Both emphasised that there is a need for clear graphics to show the enormous wealth of information obtained by 3D treatment-planning systems.

(ii) Both calculated cost on a uniform dose grid.

(iii) Both pointed out that when using biological response data it is important not to trust absolute values of TCP and NTCP but that the use of relative values is acceptable.

(iv) Both ensured that there are no parallel-opposed beams. The beam locations were prespecified and a separate check must be done to see that the beam positions are practical without collisions.

(v) Both used the biological data collated by Emami *et al* (1991) and Burman *et al* (1991).

(vi) Webb (1992a) used the NTCP model of Niemierko and Goitein (1991) and the effective-volume-at-maximum-dose method of reducing the DVH. Mohan *et al* (1992) used the Lyman equations and the effective-dose-to-whole-volume method of reducing the DVH. These are equivalent provided the NTCP is small (see Chapter 5).

(vii) Both methods incorporated the concept of 'part-fields' seeing either PTV alone or PTV and OAR and created 3D dose distributions which were highly conformal.

Some differences between the work reported from the two centres include the following.

(i) The work at ICR/RMNHST preceded that at MSK and used classical simulated annealing in this developmental phase. The work at MSK built on this application and developed fast simulated annealing (FSA). Later work at ICR/RMNHST independently used FSA.

(ii) Mohan *et al* (1992) did not compute optimisations with intensity-modulated beams; Webb (1992a,b, 1993a) provided this feature.

(iii) Mohan *et al* (1992) pointed out the difficulty of comparing different optimisation schemes in the literature when some are based on biological cost and others are not. Mohan *et al* (1992) included biological response *a priori* in the cost function whereas Webb (1991a,b, 1992a,b,c, 1993a) constructed a cost function based solely on dose and calculated biological response *a posteriori*. More recently the ICR/RMNHST work has included biological optimisation (Webb and Oldham 1993). The use of physical and biological objective functions has recently been reviewed by Brahme (1995a). The topic is one of considerable debate (Mohan 1994, Mohan *et al* 1994).

1.5.   OPTIMISATION AT DELFT UNIVERSITY OF TECHNOLOGY

Smit (1993) has also made use of both classical and fast simulated annealing combined with downhill simplex optimisation to investigate optimisation in multidimensional space. In this work each beam is represented by eight parameters, two field angles, two weights for an open field and a 60°-wedged field respectively at these angles, the orientation angle of the wedge and the same latter three parameters for a 'mode 2' field which 'views' only the PTV and excludes the view of the OAR. The concept of 'part-fields' is discussed above in section 1.3.7.

The optimisation was based on maximising the probability of uncomplicated tumour control, the so-called P_+ (see also sections 1.3.4 and 1.10). First it was shown that this parameter P_+ could have multiple local maxima in the $8N$-dimensional state-space of the variables for N beams. This was demonstrated by keeping several of the parameters fixed and varying some of the others. Thus it was deemed necessary to use simulated annealing which would eventually reach a global maximum value of P_+. Smit (1993) studied two problems. The first was a model geometry also studied by Webb (1991a) and the second was the goal of optimising the treatment of a lung tumour with lung and spinal cord as OAR. This second problem was defined by CT data. It was shown that fast simulated annealing provided quicker and more advantageous solutions.

Another very interesting conclusion was that when several optimisations were performed for the same geometry but optimising by different methods, the same treatment angles appeared but very different beamweights, with much the same P_+ values. It must be emphasised that these optimisations optimised on the space of all the $8N$ parameters simultaneously. From the results it was deduced that there was no coupling between parameters and it might be possible to first optimise on beam directions and then subsequently optimise beamweights. This of course would reduce computation time enormously. It is also the basis of the kind of optimisation discussed in section 1.3 where fixed directions were assumed. It also points to the fact that at least with this objective function the outcome may be fairly insensitive to beamweight but more sensitive to beam direction. The subject of optimising beam directions is discussed later in this chapter.

De Vroed (1994) continued this work in Delft but with modifications. The dose model was extended from the simple exponential of Smit (1993) to include tissue inhomogeneity corrections, computed on a radiological pathlength model from CT data, as well as a simple model of beam penumbra. Instead of maximising P_+, the TCP was maximised subject to some upper limit on the NTCPs of the involved OAR. Three clinical cases were studied for tumours in the lung, prostate and oesophagus respectively. Optimised problems were constrained so that parameters remained within feasible ranges and were then solved by simulated annealing in two ways: (i) by making no *a priori* assumptions and solving globally for all the free variables; (ii) by defining a tolerated beam weight (TBW) for a given beam direction as the maximum weight a single open beam can have whilst the NTCPs do not exceed their maximum values. Then, making maps of TBW as a function of beam orientation, the 'best orientations' were selected by inspection *and then kept fixed* whilst a so-called 'partitioned problem', i.e. the optimisation of the remaining variables, was solved.

It was found that for each of the three clinical problems the solution of the global problem was able, more or less, to find the preferred beam directions from the TBW map, but with slightly reduced TCPs. This suggested once again that the global problem could be reasonably partitioned.

1.6. PROJECTION AND BACKPROJECTION ISSUES IN ANALYTIC INVERSE TREATMENT PLANNING FOR OPTIMISING IMBS

Simulated annealing is an optimisation technique which is equally applicable to the computation of the beamweights of intensity-modulated or open (uniform or wedged) beams. Optimisation techniques are unequivocally required for computing IMBs where only inverse planning is applicable. Forward planning would be quite impossible. For this reason developments in optimising 3D treatment planning have gone hand in hand with the concepts of IMB therapy. Hence the next seven sections of this chapter concentrate exclusively on optimising therapy with IMBs. Towards the end of this chapter we shall return to the debate on when IMBs are needed and place these state-of-the-art innovations in the context of other improvements in radiotherapy treatment planning.

The first methods ever presented to calculate the intensity-modulated beam profiles for a desired dose distribution were analytic. Brahme (1988) presented the technique for 2D dose distributions and Boyer *et al* (1991) generalised the method to 3D. The method is predicated on the assumption that the dose distribution can be expressed as the convolution of a point-dose kernel and a density function of point irradiations ('kernel density'). The technique comprised two steps.

(i) Deconvolve the point-dose kernel from the desired dose distribution ('the prescription') to obtain the density function of point irradiations. This deconvolution could be performed by an analytic technique as discussed by Lind and Källman (1990), Källman *et al* (1992) and Boyer *et al* (1991).

(ii) Forward project the kernel density function, from dose-space to beam-space, taking account of (possibly spatially variable) photon attenuation and in the appropriate geometry (which may be parallel-beam, fan-beam or cone-beam) to obtain the map of spatially-varying beam intensity. For the 2D problem these are 1D profiles of photon intensity (sometimes called profiles of incident energy-fluence); for the 3D problem they are 2D maps.

The method has inbuilt positivity constraints. If iterative deconvolution is performed, the kernel density is constrained positive everywhere during iteration. If the kernel density is instead formed by direct deconvolution then it may contain negative values and then, when the projections of the density function are formed in step (ii), the beam profiles must be constrained positive everywhere. Equations for all these steps are given in the companion Volume, Chapter 2.

One may now observe that there are two ways of forming the resultant dose distribution from the derived quantities. Either:

(i) convolve the kernel density with the point-dose kernel itself, or

(ii) backproject the spatially-varying incident beam profiles from beam-space into dose-space, taking into account photon attenuation and summing the dose delivered by each profile to form the complete dose distribution.

The results of these two operations will not be identical (see figure 1.19). Firstly the result of operation (i) will not exactly match the starting desired dose distribution because of the positivity constraints on the deconvolution to form the kernel density and possibly also the finite number of beams in use. Secondly the result of operation (ii) will not exactly match the starting desired dose distribution and also not exactly match the result of operation (i) because of the fundamental principle that projection and backprojection are not simply inverse operators. When the kernel density is *projected* into beam-space the density for each point along a projection ray is weighted by a positive exponential term modelling the (possibly spatially varying) photon attenuation from that point to the point where the ray exits the patient. The exponentially weighted contributions from each ray point in dose-space are then summed to form the projection value. Conversely *backprojection* takes each single beam-element value and adds it back to points down the corresponding beamline with a negative exponential term. It may be appreciated that these two operations are not direct inverses of

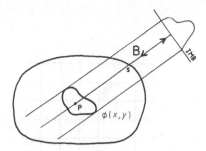

Figure 1.19. *This illustrates that projection and backprojection are not inverse operators of each other. The first stage in analytically constructing an IMB is to deconvolve the point-dose kernel from the 2D map of dose to obtain a 2D distribution of fluence density $\phi(x, y)$. The IMB is constructed by forward projecting this distribution to beam-element-space, weighting the contribution from each point by its attenuation. For example, the contribution from point P is weighted by the attenuation between point P and the patient surface S so that points nearer the surface have less attenuation. This is done for each ray within each IMB and for a series of IMBs at different gantry angles. The delivered dose can then be calculated by backprojecting these IMBs, appropriately attenuated, and summing the result. This will not give the desired dose distribution in one step because of the non inverse nature of the operators and it is necessary to perform some iteration as discussed in sections 1.8 and 1.9.*

each other. However, the authors of these techniques and more recently Liu *et al* (1993) have shown that the resulting dose distributions are quite accurate in the high-dose regions even if these stand on a pedestal of lower dose.

Liu *et al* (1993) have addressed a detail of the projection and backprojection which concerns the discretisation of the problem. The general equations relating dose, kernel density and elemental beam intensity are written down as continuous functions. Following their analysis let us consider the 2D dose distribution in relation to 1D line integrals. Since the dose matrix is discretely specified on a matrix of pixels, the line integral projections involve tracing each ray through the 2D matrix representing dose-space. The simplest and sometimes-used default is to simply group those dose-space pixels corresponding to each projection element and include, without interpolation in the forward projection, all the kernel density values at pixels through which the line passes. Then the same is done in reverse during backprojection, no account being taken of exactly where through the pixel each ray passes.

Liu *et al* (1993) present two possible improvements:

(i) the contribution in both forward and back projection between a dose-space pixel and a ray connecting to a beam-element ('bixel') is weighted by the length of intersection of the ray and the pixel, called 'longitudinal-

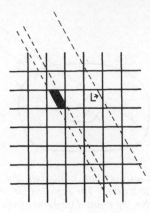

Figure 1.20. *Illustrates the difference between longitudinal-distance weighting and area-overlap weighting. The solid grid represents some dose-space pixels whilst the dotted lines represent rays coming from beam-elements (bixels). The ray to the right is the central ray from a particular bixel and the length L is the length of intersection of this ray with a particular pixel. Conversely, to the left, is shown the intersection of all the rays from another bixel and the area of overlap with a dose-space pixel is shown shaded.*

distance weighting', or

(ii) the contribution in both forward and back projection between a dose-space pixel and a ray connecting to a beam-element is weighted by the area of overlap of the ray of finite width and the pixel, called 'area-overlap weighting'.

These concepts (figure 1.20) are very familiar in the analogous theory of image reconstruction from projections where the corresponding ray sums are known as line integrals or alternatively strip integrals (Herman 1980). Liu *et al* (1993) investigated several model problems in both ways and came to the conclusion that:

(i) the resulting dose distributions by area-overlap weighting are smoother and more accurate;

(ii) the area-overlap weighting method for the dose calculation gives a more accurate result for a 64^2 dose matrix than a longitudinal-distance weighting method even with 512^2 dose matrix;

(iii) the overall calculation time including both forward projection to calculate the beam profiles and backprojection to calculate the resultant dose distribution *with comparable precision* is shorter for the area-weighting method.

This discussion takes no account of any variation in fluence across each bixel. In practice such variations do exist for certain methods of delivering IMBs due to the penumbra introduced when adjacent bixels are closed by metal vanes. This problem requires detailed analysis (Webb 1996).

1.7.   DOSE CALCULATION FOR INVERSE PLANNING OF IMBS

The dose-calculation algorithm is invoked many times in inverse treatment planning especially when iterative techniques are used. Hence it becomes a key determinant of the overall time to complete a treatment optimisation. For this reason many authors have worked with a simple beam model (e.g. Bortfeld and Boyer 1995, Bortfeld *et al* 1990, 1992, 1994, Webb 1989b) and only included scatter as a subsequent refinement.

Holmes (1993) and Holmes and Mackie (1993a) have elegantly shown how 3D dose distributions can be computed, including the full effects of scatter, by a method of filtered backprojection. The well known expression for computing the dose in 3D, from each beam, from a convolution of terma (total energy released per unit mass) $T_E(x, y, z)$ and the energy deposition kernel $h(E, x, y, z)$ is

$$D_E(x, y, z) = \int_{x'} \int_{y'} \int_{z'} T_E(x', y', z') h(E, x - x', y - y', z - z') \, dx' \, dy' \, dz'$$
(1.20)

where the terma is specified including the inverse-square law, photon exponential attenuation and mass-attenuation coefficient $\mu/\rho(E, r)$ (see the companion Volume, equations (2.49)–(2.51)) as

$$T_E(r) = \frac{\mu}{\rho}(E, r) E \Gamma_E(r)$$
(1.21)

where $\Gamma_E(r)$ is the photon fluence at location r given by

$$\Gamma_E(r) = \Gamma_E(r_0) \left(\frac{r_0}{r} \right)^2 \exp \left(-\int_{r_0}^{r} \frac{\mu}{\rho}(E, l) \rho(l) \, dl \right)$$
(1.22)

where r_0 is a reference distance from the source (usually the skin entrance) and l represents the vector of the photon track to the point r. $\Gamma_E(r_0)$ is the IMB function for this beam orientation.

For N dose points (x, y, z) and N interaction points (x', y', z'), of the order N^6 calculations are required in equation (1.20), too many to be in the inner loop of an optimisation algorithm for inverse treatment planning whose aim is to compute the set of IMB functions $\Gamma_E(r_0)$ from a specification of $D_E(x, y, z)$. The ideal method of IMB calculation would be to fully invert equation (1.20) and only then be concerned with taking into account the practicalities of fitting the set of $\Gamma_E(r_0)$ functions to the specification of the collimation of a particular method of delivering IMBs. Holmes and Mackie (1993a) showed (the many steps are not repeated here) that it is possible to rewrite the calculation as

$$D_E(x, y, z) = f(z) \int_{x'} \int_{y'} \frac{\mu}{\rho} E \Gamma(x', y', 0) A^*(E, x - x', y - y') \, dx' \, dy'$$
(1.23)

where

(i)

$$f(z) = \exp(-\mu z) \left(\frac{z_0}{z_0 + z} \right)^2 \tag{1.24}$$

takes care of all the depth (z) dependences, including photon attenuation and inverse-square law and

(ii) $A^*(E, x-x', y-y')$ is a new kernel, which includes (a) an exponential factor $\exp(+\mu(z - z'))$ extracted from the photon attenuation exponential and multiplied by $h(E, x-x', y-y', z-z')$ to form $h^*(E, x-x', y-y', z-z')$ and (b) integration of the modified kernel $h^*(E, x - x', y - y', z - z')$ over z'.

Equation (1.23) has the form of a 2D convolution followed by an attenuated backprojection and therefore exhibits a functional form similar to the algorithm for computing SPECT images by 2D convolution followed by attenuation-corrected backprojection. Holmes and Mackie (1993a) used the algorithms encoded in the much-used Donner Library to implement equation (1.23) for a number of situations. First they showed it can generate the depth–dose curves for single beams (and they compared the accuracy with the prediction of the full 3D convolution equation (1.20)). Then they used it to predict the 3D dose distributions when a large number of 1D intensity-modulated beams were combined to give a PTV with concave outline, again showing good agreement with the results from the longer method of computation. The computational speed-up (factor of at least ten) was impressive.

All inverse-planning algorithms must address the issue of the dose model and because the algorithms themselves are generally complex, oversimple models have often been used. As time goes by and computer hardware becomes faster we may expect a time to come when patients can be planned using direct Monte Carlo techniques, thus eliminating the requirement to construct beam models, use convolution kernels etc. Of course this in itself does not solve the inverse problem; it only gives a method to more accurately calculate a 3D dose distribution once the 'forward' parameters have been decided, such as beam orientations, weights, modulation etc.

Among codes used for forward Monte Carlo calculations are EGS4, ETRAN, ITS PEREGRINE and MCNP (Mackie 1995, Manfredotti *et al* 1995). Electron dose calculations are more accurate at present than photon dose calculations. Photon dose calculations have been made including modelling collimator scatter and the energy spectrum of linac-generated photons. An EGS4-based Monte Carlo code is being developed by a collaboration (known as the OMEGA project) between the University of Wisconsin and the National Research Council of Canada (Mackie *et al* 1994c, Mackie 1995).

A recent development has been the project known as RAPT. RAPT is a medical application of parallel high-performance computing technology being developed with support from the European Union.   3D dose calculations are made by the EGS4 Monte Carlo code using as input, specified CT data and details of the treatment beams. The calculations are performed on a cluster of DEC Alpha workstations. The interaction with internal tissue inhomogeneities is modelled. Dose calculations are being performed at the Parallel Applications Centre in Southampton, England for model geometries and verified experimentally at the Ospedali Galliera in Genoa, Italy (Surridge and Scielzo 1995).

1.8.   DETERMINISTIC ITERATIVE OPTIMISATION OF IMBS: HOLMES' SOLUTION

Holmes *et al* (1993a, 1994a,b) present the details of their inverse-planning algorithm for the tomotherapy method of delivering IMBs described in Chapter 2. This is based on an iterative inversion of the equation linking the 3D dose distribution D to the set of beamweights w via a matrix M representing the fractional dose deposited per unit beamweight

$$D = Mw. \tag{1.25}$$

The iterative solution is

$$w_{i+1} = w_i - M^{-1}(D_i - D_p) \tag{1.26}$$

where i labels the iteration number, p refers to the dose prescription and M^{-1} is an inverse dose-calculation operator which may not be exactly the inverse of the dose-deposition matrix M (see below). Physically this equation says that the ith estimate of the set of beamweights is updated to the $(i + 1)$th estimate by the addition of weights which correspond to operating on the dose residual $(D_i - D_p)$ with M^{-1}. This equation is the iterative solution which minimises a quadratic cost function between the dose prescription D_p and the final achievable dose distribution D_i. This method is just one of a class of quasi-Newton methods which all have a functional form similar to equation (1.26).

We may note that *if* negative beamweights were allowed and *if* the inverse dose operator were able to be analytically specified, this equation would actually yield the desired optimum result in one step. However, it is generally true that the weight updates for the unconstrained problem would lead to negative beamweights. For this reason the solution must be constrained and involve multiple iteration steps.

Holmes *et al* (1993a) create the inverse dose-calculation operator M^{-1} noting a formal similarity between this problem and SPECT reconstruction (see also section 1.7).   Bortfeld and Boyer (1995) have also developed this analogy. Consider the 3D problem as a series of 2D problems. Equation (1.23) then has only one integral on its right-hand side (r.h.s.). The incremental changes in beamweight were computed by projecting the dose residual into beam-space (using the inverse of function $f(z)$ in equation (1.24)), followed by filtering (convolving) the projection with a filter $I(f)$ made up of (in frequency (f) space) the inverse of the 1D dose kernel $A^*(f)$ (section (1.7)) (to take out the blurring due to secondary particle transport), a $|f|$ frequency ramp (to take out the geometrical blurring due to backprojection), a Butterworth low-pass filter $\sqrt{(1 + [f/f_0]^{2m})}$ to control the oscillations in the beamweight profiles ($2m$ controls the filter shape and f_0 is the filter cutoff frequency), an exponential term to correct for primary photon attenuation and the inverse of the mass-attenuation coefficient. They call this the 'fast inversion filter' $I(f)$, i.e.

$$I(f) = \left[\left(\frac{\mu}{\rho} \right)^{-1} \times \exp(+\mu t^\dagger) \right] \times \left[\frac{|f|}{A^*(f)\sqrt{(1 + [f/f_0]^{2m})}} \right]. \quad (1.27)$$

$t^\dagger$ is the mean depth to the centre of the tumour. At each stage of implementing equation (1.26) for updating the weights, positivity was constrained.

Holmes *et al* (1993a) use this method as the basis of computing the intensity profiles needed for the proposed machine for tomotherapy (see Chapter 2). Interestingly it takes some 20 h on a DECSTATION 5000/200 to compute 32 beam profiles with five iterations. It is not true that these semi-analytic methods are necessarily faster than iterative simulated annealing.

In summary the main features of the algorithm were:

(i) to obtain a first estimate of w_i by projecting the dose prescription D_p;
(ii) to invoke positivity on the set w_i;
(iii) to compute the ith estimate of dose D_i by a forward dose calculation (see section 1.7 equation (1.23));
(iv) to compute residual dose $(D_i - D_p)$;
(v) to operate on the residual dose with M^{-1} to compute the change in beamweights;
(vi) to loop back to stage (ii) and repeat loops until the desired result is achieved subject to some external limits on dose in the PTV and in OAR.

Holmes and Mackie (1993b) studied the importance of avoiding amplification of high frequencies which produce highly non-uniform beam profiles with low clinical utility. They found these arose for two computational reasons: (i) poor representation of the area of overlap of a pixel and a finite-width projection ray (see also section 1.6); (ii) overcoarse pixellation of the dose matrix. They improved (i) by writing a projection routine with supersampling to improve on that in the Donner Library and

(ii) by using a 64^2 dose-space resolution early in the iteration and changing to 256^2 for later iterations ('upsampling'). Similar problems arise in the image reconstruction analogy but there due to physical photon noise rather than computational features. It is well known that filtering projections amplifies noise; this is the basis of noise amplification in CT.

Holmes *et al* (1993b) verified the prediction of the inverse-planning algorithm experimentally for two simple rotationally symmetric dose distributions. They chose these because, in the absence of a computer-controlled MLC, they could construct a metal compensator and use it for all 72 profiles used to simulate continuous rotation. The problems posed were to achieve: (i) a high-dose annulus with a hole in it; (ii) a high-dose circle with a second circular region of 20% higher dose. Both phantoms were also circular. They used film dosimetry and found the experimental and theoretical dose distributions agreed to within 5% maximum error. A useful conclusion was that although local errors in fabricating beam modulation occurred (due to the fabricating from a set of finite-sized rods) nevertheless the large number of beam directions 'washed out the error'.

Optimisation solutions divide into two classes, those which are deterministic and those which are stochastic. The former are more efficient and apply when the problem has only a single minimum in the cost function. The latter, generally slow, apply when there are local minima to be avoided *en route* to the global minimum. This class includes the well known method of simulated annealing.

Holmes and Mackie (1992) studied 2D optimisation of targets with concave outlines irradiated by a large set of 1D modulated beam profiles, the same problem as studied earlier by many authors such as Brahme, Lind, Källman, Bortfeld and Webb. Holmes and Mackie (1992) postulated that provided the number of independent beam elements (the product of the number of beams and the number of elements per beam) was greater than or equal to the number of pixels occupied by the target, the problem would have no local minima and be addressable by deterministic search methods. Whilst it is highly desirable from a sampling viewpoint that this criterion be achieved, the existence of local minima is more related to the form of the cost function as discussed in section 1.3.

Holmes and Mackie (1992) showed that the resulting optimised beam profiles are identical irrespective of the starting beam profiles when the sampling criterion is met. The work at RMNHST (Webb 1989b) first investigated the problem reported by Holmes and Mackie (1992) and showed the improved conformation as the number of beams was increased in agreement with the later observations of Holmes and Mackie (1992). Further details on all these aspects of inverse treatment planning and appropriate dose models may be found in Holmes (1993). Holmes (1993) and Holmes and Mackie (1994) provide a formal comparison of the inverse-treatment-planning algorithms developed by themselves, by Bortfeld's group and by Brahme's group (see also section 1.9).

1.9. DETERMINISTIC ITERATIVE OPTIMISATION OF IMBS: BORTFELD'S SOLUTION

Bortfeld and Boyer (1995) have elegantly shown the analogy between SPECT imaging and reconstruction and the solution of the inverse problem in radiotherapy. The attenuated Radon transform can be inverted by a process of filtering the projections of dose and backprojecting these filtered projections (Bortfeld 1995). The sum of the backprojections gives the desired dose. However, the filtered projections will contain unphysical negative beamweights as discussed in detail in the companion Volume, chapter 2. Bortfeld and Boyer (1995) propose these can be removed by either zeroing or by the addition of a constant fluence (either constant per projection or globally constant for all projections). The former will avoid over-irradiating OAR but will reduce dose homogeneity to the PTV whilst the latter preserves dose homogeneity to the PTV but may overdose OAR. Bortfeld and Boyer (1995) provide an elegant solution for simple model dose distributions such as those comprising circular regions or those which can be decomposed into a series of triangles. This gives a deep insight into the possibilities and limitations of the method of creating dose by filtered backprojection. However, to solve for more general prescriptions it is necessary to iteratively refine the projections created by this first step.

Bortfeld *et al* (1992, 1994) have further developed their iterative algorithm for computing intensity-modulated 2D beams for 3D conformal radiotherapy when the 3D target volume has a surface with a concavity. Their earlier work (Bortfeld *et al* 1990) had presented the corresponding solution for a 2D plan (with 1D intensity-modulated profiles) (see the companion Volume, pp 94–5).

The 3D solution creates a first estimate of the 2D fluence profile $\Phi_\theta^{(0)}(x_s, y_s)$ in the direction θ (x_s and y_s indicate coordinates in the plane normal to the central axis of the beam through the isocentre; x_s is in the transverse plane). This is done by projecting the 3D dose prescription $D_p(r)$ along fan lines but weighting it by a factor which applies a higher weight to rays that pass through regions of the target with a larger average distance from the source and vice versa (similar but not identical to equation (2.46) in the companion Volume). The projections were filtered by a 1D function $h(x_s)$ and finally a positivity constraint C^+ was applied. Formally

$$\Phi_\theta^{(0)}(x_s, y_s) = C^+ \left[\left(\frac{P_\theta[D_p(r)]}{P_\theta^\Delta[D_p(r)]} \right) (P_\theta[D_p(r)] \otimes h(x_s)) \right]. \quad (1.28)$$

Here there are two projection operators. The first, P_θ, simply projects all the dose values at r along each ray-line. The second P_θ^Δ projects the same values but each weighted by a function Δ which is the product of the tissue-maximum ratio and the inverse-square-law (ISL) factor. It may be seen that equation (1.28) contains a ratio which applies a higher weight to rays that pass through regions of the target with a larger average distance from the source.

In turn this led to a first estimate of the 3D dose distribution by backprojecting the fluence weighted by photon attenuation and ISL and summing over all beams (cf equation (2.47) in the companion Volume). Formally from Bortfeld *et al* (1994)

$$D^{(k)}(r) = \sum_\theta \Phi_\theta^{(k)}(x_s, y_s)\Delta. \tag{1.29}$$

The parameter k labels the iteration number; for the first time $k = 0$ of course.

This first estimate of the 2D fluence profile was then iteratively refined using the projection of the *difference* between the 3D dose prescription $D_p(r)$ and the first estimate of the delivered 3D dose $D^{(k)}(r)(k = 0)$ weighted by a function f_s which controlled the relative importance of constraints in the PTV and the OAR. The resulting fluence profile became the next estimate and the process cycled to convergence. Formally

$$\Phi_\theta^{(k+1)}(x_s, y_s) = C^+[\Phi_\theta^{(k)}(x_s, y_s) - P_\theta^\Delta(D^{(k)}(r) - D_p(r))f_s]. \tag{1.30}$$

The parameter f_s was unity in the PTV and a value between 1 and 10 in the OAR if, and only if, the dose was above a constraint, and zero elsewhere.

Bortfeld *et al* (1994) formulated the mathematics as above in more intuitive procedural terms than the formulation by Bortfeld *et al* (1990) although the methods are the same (see also discussion in the companion Volume, pp 94–5). The method also bears a formal resemblance to that of Holmes *et al* (1993a), being one of a class of quasi-Newton methods. Indeed Holmes and Mackie (1994) present a formal comparison of methods. Bortfeld *et al* (1990) wrote their algorithm (equation (1.30)) in the form

$$w_{i+1} = w_i - (1/N)S^{-1}M(D_i - D_p) \tag{1.31}$$

where i labels the iteration number, p refers to the dose prescription, w refers to the vector of beamweights, M is defined in equation (1.25) and S is an approximation to the Hessian MM^T whose diagonal elements are the total squared contribution of dose per unit beamweight in the target and sensitive structures. N is the number of beams. By comparing equations (1.26) and (1.31) the formal similarity between the independently formulated methods of Bortfeld and Holmes is apparent.

Finally, as also pointed out by Holmes and Mackie (1994) and Holmes (1993), the method of Brahme (1988) and Lind and Källman (1990) (see the companion Volume, pp 88–94) can also be viewed in the same formalism. This method deconvolves the point irradiation kernel from the required dose distribution to create the distribution of 'density of point irradiations'. Lind and Källman (1990) formulate this as an iterative deconvolution (the companion Volume, equations (2.42)–(2.45)) which has a similar shape to equations (1.26) and (1.31). The actual beamweight elements are then formed by a forward projection into beam-space of the density of point irradiations (the companion Volume, equation (2.40)).

Mohan *et al* (1994) and Jackson *et al* (1994) have investigated the performance of the Bortfeld inversion algorithm for two classes of disease; prostate and lung cancer. Two patients in each class were planned and the results compared, using DVH tools, isodose display and computation of TCP and NTCP, with the best which a human planner could achieve without invoking intensity modulation. It was found that the use of intensity-modulated beams could improve on the human planner for the prostate, especially when the volume of normal rectum included in the PTV was small. Because rectum is the dose-limiting organ and has a small volume effect ($n = 0.12$), Mohan *et al* (1994) conclude that the Bortfeld method which optimises the *dose* distribution also optimises TCP at constant NTCP. The use of intensity-modulated beams also renders the optimisation less sensitive to choice of beam angles and eliminates the need for other beam modifiers such as wedges. Chen *et al* (1994, 1995) showed that in the Bortfeld algorithm the inclusion of scatter in both the design of IMBs and in the final dose calculation was important. Investigation of the TCP and NTCP indicated that dose escalation with nine IMBs to 86 Gy might be feasible (Reinstein *et al* 1994).

On the other hand the human planner created a superior plan without the need for intensity-modulated beams to that created by the Bortfeld method for cases of lung cancer when the plans were evaluated by biological outcome. This was because the normal lung, which is the dose-limiting OAR, has a large volume effect and this knowledge can be used by the human planner. However, the inverse-treatment-planning method only tries to optimise the dose distribution and Mohan *et al* (1994) argue that in doing so it takes no account of the dose-volume effect of NTCP. Indeed analytic inversion methods based on dose can never do this and if the aim is altered to be that of optimising biological outcome the only solution is to use Monte Carlo techniques such as simulated annealing (Wang *et al* 1994). This conclusion is a consequence of the deficiency in the objective function.

A further observation was that the analytic-iterative inversion worked better for small PTVs. As shown above (equation (1.30)), the Bortfeld method removes negative beamweights at each stage of iteration. Mohan *et al* (1994) tabulate the number and importance of such negative weights at the last iteration. This is useful information since it characterises the mismatch between the dose prescription and what can be achieved by the inversion method. It was found therefore that the lung cases led to a large percentage of negative beamweights compared with the prostate cases.

The code developed by Bortfeld *et al* (1994) makes use of an elemental beam function which is a 'primary-only' model, being based on the inverse-square-law and tissue-maximum ratio (embodied in the function Δ). The reason for this is that the inverse-planning algorithm requires repetitive projection between beam-space and dose-space which is computer costly (equations (1.29) and (1.30)). Chen *et al* (1994, 1995) have examined this technique. They have taken a version of the Bortfeld code and adapted it so it will use a photon pencil-beam convolution algorithm in the stages leading to the production of the intensity-modulation values. Thus scatter is taken into account not only in this stage but also in the final 3D dose calculation from the finally arrived at set of IMBs. They named the code in its original form OPT3D and with this modification OPT3DSCAT.

Both codes were used to plan for the irradiation of a cubic target volume in a demonstration cubic volume of tissue and for four patients with prostate cancer where the dose-limiting organ was the rectum and appropriate dose limits were employed. The resulting 3D dose distributions were quantified via DVHs and measures of TCP and NTCP. The cubic volume was irradiated with four fields perpendicular to the cube sides. So, unsurprisingly, the OPT3D code developed almost flat beam profiles. However, the OPT3DSCAT code developed profiles with horns at the field edges to compensate for scatter loss from the irradiated volume (figure 1.21(*a*)). The corresponding dose distribution had a flatter shape with the OPT3DSCAT code (figure 1.21(*b*)). From the patient cases it was deduced that the plan developed by the OPT3DSCAT code delivered a more uniform target dose and so higher TCP and at the same time the dose to the rectum was reduced with reduced NTCP. The authors thus concluded that IMB generation codes using inverse planning should make use of a beam model which properly accounts for scattered radiation.

Ésik *et al* (1995) implemented the Bortfeld method for planning a single case in the cervical and upper mediastinal regions and came to the conclusion that with appropriate dose limits and setting of penalty functions it was possible to create a plan with adequately homogeneous dose coverage of the PTV whilst simultaneously sparing OAR (spinal cord and lungs). This study did not give a benchmark of what could be achieved by fixed fields and only considered one case but nevertheless demonstrated the power of inverse planning.

In sections (1.6)–(1.9) repeated comment and discussion on the inaccuracies introduced into inverse planning by use of a too simple beam model have been given. The issue is complex. Simple models have been used in the developmental phase of this subject because IMB computation

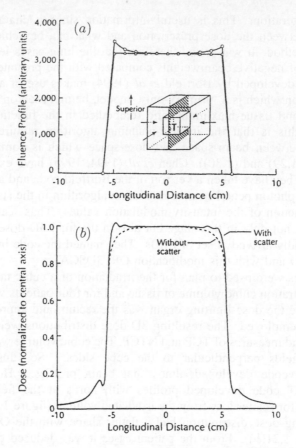

Figure 1.21. *(a) The insert shows a cubic volume T to be irradiated within a cubic phantom. Four beams were specified with central axes normal to the cube sides and the intensity-modulated fluence profiles were computed by the code developed by Bortfeld et al (1994). The lower curve shows the IMB developed when scatter was not included in the beam model and the upper curve shows the IMB developed when scatter was included. It can be seen that the latter case developed a less flat profile with horns at the field edge. (b) The corresponding distribution of dose in a profile through a sagittal plane passing through the isocentre. The dose is more uniform when scatter has been included in the beam model. (From Chen et al 1995.)*

is already more complex than forward planning (where the same issues of beam model arise) due to the multistage nature of inverse planning. Ideally the beam function should represent the true mode of delivery but as shown by Webb (1996) this is impossible since the form of beam collimation cannot be known until the IMBs are computed. Hence, in view of this Catch-22 situation, there arises another iterative stage. IMBs must be computed with as representative a beam model as possible and then the dose distribution must be recalculated with the exact collimation used to form the IMB in practice. These issues are revisited in Chapter 2.

1.10. OPTIMISATION USING IMBS, A BIOLOGICAL COST FUNCTION AND ITERATIVE TECHNIQUES

Söderström and Brahme (1993), like Mohan *et al* (1992), optimised dose distributions so that the resulting *biological outcome* was optimised. The 'cost function' optimised was the probability of complication-free tumour control P_+ where

$$P_+ = \text{TCP} - \text{NTCP} + \delta(1 - \text{TCP})\text{NTCP}. \qquad (1.32)$$

The parameter δ controls the statistical independence of benefit (TCP) and injury (NTCP). Note if $\delta = 1$ the probabilities are statistically independent and

$$P_+ = \text{TCP}(1 - \text{NTCP}). \qquad (1.33)$$

If, however, at the other extreme $\delta = 0$ then the probabilities are statistically dependent and

$$P_+ = \text{TCP} - \text{NTCP}. \qquad (1.34)$$

In between there are other possibilities for δ. The basis of the method of Söderström and Brahme (1993) was to adjust beamweights, compute dose, predict P_+ and make changes until P_+ was optimised. However, some workers might argue that this is *not* the best biological parameter to optimise. It might be better to optimise TCP subject to some ceiling on NTCP for example (see section 1.3.4).

The method of Söderström and Brahme was developed for the situation of treatment using intensity-modulated beams and, although not limited to 2D, the method was presented in its 2D form. Each beam (meaning portal of radiation from the accelerator at some particular gantry orientation) was regarded as a set of smaller beam-elements, whose 'weight' was to be optimised. The beam-elements were considered sequentially for each beam and then for all beams. One cycle of iteration of beam-element weight was complete once all beams and all elements within each beam had been adjusted once. For each beam-element in turn the radiation weight was adjusted until P_+ was maximised. Then the next beam-element was considered and so on. The process was relaxed by incrementing each beam-element weight by only a fraction r of the increment which for that beam-element would have maximised P_+. This ensured the iterative algorithm was well behaved and converged to a solution. Some 35 or so iterations were carried out.

No statement was made about the presence or absence of local maxima. It was implicitly assumed that the method would converge to the best solution. On completion the result was an optimised dose distribution, a set of intensity-modulated beam profiles, a value of P_+ and statistics about regions of interest (PTV and OAR).

Söderström and Brahme (1993) used the simple formula

$$P(D) = 2^{-\exp[e\gamma(1-D/D_{50})]} \qquad (1.35)$$

for the probabilities TCP and NTCP where D_{50} is the dose for 50% probability, D is dose and γ is the normalised gradient of the dose–response curve. The structural organisation (series or parallel or some combination) was also taken into account. They argued that these parameters may be obtainable by predictive assays on biopsy specimens and if not available, should be found from a library of radiobiological parameters (such as the Kutcher/Burman/Emami data). The results from these authors must therefore be seen against a background of belief in this model, the data supporting it and the chosen cost function.

Söderström and Brahme (1993) applied the method to two clinical cases for a small number of beams and showed that a high degree of dose conformation can be achieved. They also showed that the same philosophy can be used to select the orientations of beams from a range of possibilities, i.e. beam orientations can be adjusted to maximise P_+. However, this is (the authors admit) not yet really a practical proposition because of the huge amount of computer time involved. Each set of beam orientations has itself to be optimised for P_+ and then this process is looped for a range of possible combinations.

Gustafsson *et al* (1994) have introduced a generalised pencil-beam algorithm for inverse planning optimisation in radiotherapy. The dose was expressed in terms of a Fredholm integral relating incident radiation fluence and the pencil-beam kernel appropriate for each radiation component. The new feature of this work was the ability to combine radiation fluences of different modalities; photons, electrons, neutral pions etc. The optimisation was fed with the precalculated radiation pencil-beam kernels and then the algorithm computed the incident radiation fluence profiles so the resulting dose satisfied certain constraints. As with the earlier work of Söderström and Brahme (1992) the resulting dose distributions were converted into TCP, NTCP and P_+, the probability of uncomplicated tumour control, and the optimisation aim was to maximise P_+. The solution, formulated in tensor notation, has a formal similarity with earlier solutions proposed by Lind and coworkers although the first of these earlier formulations used a quadratic norm in dose rather than a biological endpoint as the cost or objective function. The solutions were constrained to positive fluence. This new algorithm is able to optimise beam energies, beam directions and beam modalities as well as beam intensities, a situation which is referred to as 'generalised conformal therapy'. In practice because of the enormous space of the possible solutions, it is usual to restrict some of these and search for optimised solutions from a limited set of possibilities.

For example, Gustafsson *et al* (1994) themselves fixed the number of beams and the modalities and the incident energies and used their algorithm to optimise the intensity distribution of radiation in 2D planning.

They have carried out a very instructive planning study. The clinical case studied was a PTV comprising a cervix tumour and locally involved lymph nodes. This PTV is butterfly-shaped with a concave outline in the folds of which lie organs-at-risk, rectum and bladder and small bowel. The 2D dose-space was divided into 64^2 pixels with each pixel of size 3.9 mm. A statistically dependent model for P_+ was used with $P_+ = \text{TCP} - \text{NTCP}$. Four specific field geometries were planned. Each comprised four fields at 90° to each other: anterior–posterior (AP), posterior–anterior (PA), right and left laterals. The fields themselves had different specifications as follows:

(i) all 'open' i.e. not intensity-modulated, with the intensity of the AP field three times that of the (equal intensity) other three;

(ii) the fields were uniform but their intensities were varied;

(iii) the fields were divided each into three 'bixels' (beam-element pixels) in an intelligent way, i.e. the bixels were not of uniform size but depended on what tissues were in the line of sight of each;

(iv) each of the four beams was fully intensity-modulated with 64 bixels per beam.

The resulting values of P_+ for the four conditions were: (i) 0.343; (ii) 0.387; (iii) 0.894; (iv) 0.898. So we see that there is a big increase in P_+ using intensity-modulated beams. However (at least for this geometric planning problem), there is *not* a big difference in P_+ between the case of using coarse-spatial-scale intensity modulation and using fine-spatial-scale intensity modulation. If this proves to be a general conclusion for other treatment geometries then this has great importance for determining the choice of treatment delivery technique since case (iv) requires the delivery techniques discussed in detail in Chapter 2 whereas case (iii) can be delivered by fairly conventional treatment machines. This study helps to emphasise that optimising conformal radiotherapy is not simply a matter of determining best beamweights. The problem includes determining the optimum number and direction of beams and also considering the practicalities of beam delivery (see also Brahme 1994a,b,c, 1995b). The different solutions of Gustafsson *et al* (1994) are not equally sensitive to movement error. For example, case (iii) above might be rather sensitive to movement errors because of the field-matching requirements through the tumour in the AP and PA directions which if misaligned may cause hot and cold spots. The effect may average out over treatment fractions but this points to the need to have detailed studies of movement errors in IMB therapy.

Söderström and Brahme (1995) have made an interesting study of the number of radiation beams required to optimise conformal radiotherapy.

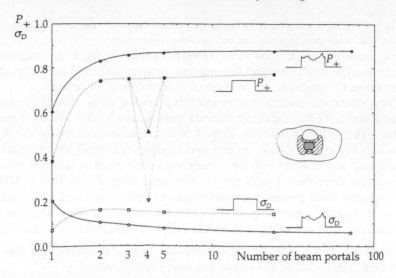

Figure 1.22. *The variation of complication-free control P_+ (solid squares and circles) and the standard deviation of the mean dose to the target volume σ_D (open squares and circles) when increasing the number of beam portals for optimised uniform and non-uniform delivery of dose. Optimisations were performed for a case of cervix cancer with uniform beams (squares) and non-uniform beams (circles). Also shown in the diagram is the P_+ obtained when using a standard four-field box technique (open triangle) and a four-field box technique where the lateral fields have been used as boost fields on the gross tumour volume (solid triangle). (Reprinted by permission of the publisher from Söderström and Brahme 1995; ©1995 Elsevier Science Inc.)*

They considered three clinical cases and optimised the parameter P_+ using the above methods for a range of numbers of beam portals. They separately considered the cases of uniform and non-uniform profiles. They plotted P_+ as well as the homogeneity of dose to the PTV as a function of this number of portals for the three clinical cases and for both types of beam.

They concluded that the optimum value of P_+ is reached for a small number of beams, three in the case of uniform fields and five for non-uniform fields. This behaviour held for all three clinical cases. They also observed that the optimum P_+ was always considerably larger for non-uniform fields than for uniform fields. In fact P_+ would, for uniform fields, *never* reach the values obtained for non-uniform fields, even for an infinity of uniform fields. The plateau of P_+ corresponded with obtaining a dose homogeneity of the order of 5% in the PTV. They thus concluded that, except for very special cases of PTVs with concavities in which OAR may lie juxtaposing the PTV, it was not necessary to arrange a larger number of fields than five (figure 1.22).

They also showed that for a small number of intensity-modulated fields, such as one, two or three the value of P_+ depended critically on the *orientation* of the fields with respect to the patient. However, for five fields or more it was shown that P_+ was virtually independent of orientation and the fields could be equispaced in 2π around the patient. This is naturally a very useful conclusion (Brahme 1995c). Söderström *et al* (1995) give more details and also conclude that these observations may make it unnecessary to use over-sophisticated treatment techniques.

This study and its conclusions have been quite strongly criticised, first in an editorial by Mohan and Ling (1995) and leading subsequently to an exchange of correspondence (Söderström and Brahme 1996, Mohan *et al* 1996). Mohan and Ling (1995) argued that the conclusion that few beams can give as good results as many is counter-intuitive and contrary to conventional practice. They also questioned whether the conclusions would have been the same if the beams had been chosen allowing non-coplanar orientations. Their major criticism was that the optimised objective function P_+ gives equal weight or importance to a gain of $x\%$ in TCP and a decrease of $x\%$ in NTCP. They accepted the concept of using P_+ provided the NTCP is below the tolerance level but argued that the score function should decrease rapidly and non-linearly with increasing NTCP once the dose levels are escalated to the point at which the tolerance NTCP has been exceeded (a feature not embodied in P_+). Arguing that current practice is already operating at the limit of tolerable NTCP, Mohan and Ling rejected the use of P_+ as an optimising function.

Söderström and Brahme (1996) defended their use of P_+ on the grounds that the NTCP was generally less than 0.1 in their studies and that in determining NTCP, parameters should be set which take care of the *severity* of the complication. They argued that different NTCP endpoints should be used for each patient and that clinical judgement should still be used to determine these. They also argued that plans optimised on the basis of P_+ would still be manually reviewed by a clinician before acceptance. They believed that the inclusion of a full 3D choice of beam directions would not have substantially altered their conclusions. Söderström and Brahme (1996) also introduced the argument that in some of their plans the overall TCP was improved even though the dose in parts of the PTV fell since this was compensated by rises in other parts of the PTV.

Mohan *et al* (1996) replied that the observation that a few well-chosen IMBs can be sufficient is partly because of the choice of the P_+ optimisation function and partly because the plans they considered did not involve dose escalation (so NTCP values were naturally low). They argued that the optimum number of beams will depend on the desired target dose level to achieve adequate TCP. They stuck to their guns that the use of P_+ in dose escalation optimisation involves the equal balancing of importance of TCP and NTCP and that this is wrong. Mohan *et al* (1996) believe the penalty function should increase sharply once the normal organ receives a tolerance dose as they themselves used in earlier work (Mohan *et al* 1992). They argued that, had Söderström and Brahme considered dose escalation, they would have seen continued improvement with increasing the number of beams. They rejected the claim that the clinician would always be available to 'manually reject' optimised plans that fell short of expectations, on the grounds that the whole point of computer optimisation was to avoid this human intervention.

Mohan *et al* (1996) also disagreed that cold spots in a PTV can be compensated for by hot spots elsewhere in the PTV. They also had doubts about the confidence in predicting TCP and believe that a technique (such as increasing the number of beams) which improves dose homogeneity to the PTV is desirable since, with approximate homogeneity, TCP is a function only of mean dose.

The issue of 'how many beams?' is clearly far from solved (see also section 1.16).

1.11.   OPTIMISATION OF 2D PLANNING OF IMBS BY MINIMUM NORM, MAXIMUM ENTROPY AND GENETIC METHODS

Yuan *et al* (1994c) have addressed the inverse-planning problem in 2D with a primary-only dose model. They studied two clinical problems, one of a target volume in the brain depicted by MRI data and with a concave outline containing an OAR, and the other of a bladder tumour with the femoral heads and rectum as OAR.

Yuan *et al* (1994c) developed a general formalism relating dose at a point to elemental beamweight in a 2D field; however, their actual implementation was of the reduced formulation. Two new techniques were introduced. In the first the norm of the total beam energy was minimised; i.e.

$$\left[\sum_{i=1}^{i=I} \sum_{m=-M_i}^{M_i'} B_i^2(m)\right]^{0.5} \tag{1.36}$$

was minimised, where $B_i(m)$ is the mth beam element in the ith beam and there are I beams altogether. The inner sum is taken over the totality of beam elements for that beam. The minimisation was subject to the calculated dose distribution matching the prescribed dose distribution, the non-negativity of beam elements and the critical OAR not receiving doses above specified values. They claim that because the matrix relating beamweight to dose is very large, non-square, sparse and with a non-unique solution, of all the possible solutions the one with minimum norm is a natural choice.

In the second formulation Yuan *et al* (1994c) maximised the entropy

$$H = -\sum_{i=1}^{i=I} \sum_{m=-M_i}^{M_i'} B_i(m) \log B_i(m) \tag{1.37}$$

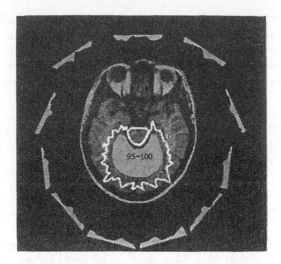

Figure 1.23. *A conformal dose distribution with a concave outline produced by the maximum entropy technique. The target dose only varies by 2.04% across the target slice and this is achieved with just 11 beams with intensity-modulation, the profiles for which are shown around the slice. (From Yuan et al 1994c.)*

subject to the same conditions as above. In both cases, the solution for the beam elements started from the filtered projections of the dose distribution in the direction of the beams. The dose conformation for both clinical problems studied was more accurate using the maximum entropy techniques (figure 1.23). Yuan *et al* (1994c) drew the strong parallel between the problem of inverse treatment planning and the medical imaging problem of determining an image from its projections, concepts also presented by Webb (1989b) and Bortfeld *et al* (1990).

Yuan *et al* (1994a) extended this work to implement the scaled-gradient-projection (Bortfeld *et al* 1994) inverse-planning technique for the familiar 'butterfly-shaped PTV' planning problem introduced by Brahme (1988) and studied further by Webb (1989b). Using just eleven IMBs very good dose homogeneity (better than 2%) was obtained in the PTV.

Yuan *et al* (1994b) have used these inverse-planning techniques to show that with a fixed number of beams, equispaced in 2π arranged around the patient, there is an optimum angular rotation of the whole set (they call this 'offset') relative to the patient, assessed in terms of improved speed of convergence. There has as yet been no assessment of the improved dose conformation consequent on this offset.

This group have also developed genetic algorithms for treatment-plan optimisation (Vance *et al* 1994). The algorithm includes biological cost and dose-volume considerations and aims to maximise a 'fitness function'; so-called 'fit' solutions 'reproduce'. Haas *et al* (1995) have shown that in simple test circumstances a genetic algorithm outperforms an iterative least-squares solution. However, more iterations are required and real treatment-planning problems were not studied. Sandham *et al* (1995) have reviewed the whole work of the Glasgow group. Ezzell (1996) has also developed the use of genetic algorithms for optimising treatment plans.

1.12. ITERATIVE OPTIMISATION OF IMBS AT THE UNIVERSITY OF WISCONSIN; THE POWER OF COST FUNCTIONS AND AN ELECTROSTATIC ANALOGY

In section 1.3.4 it was explained that iterative solutions to the inverse-planning problem can be controlled effectively by the choice of cost function. A quadratic cost function defined only within the PTV and OAR can be minimised to give quite uniform dose to these regions whilst having absolutely no control over what dose is delivered to the rest of the body. Conversely if the cost function is defined everywhere and an attempt made to minimise it, it is self-evident that the dose cannot be so accurately controlled in the PTV and OAR because of the nature of the photon interactions with tissue. In short one 'gets what one pays for'. Webb (1994b) showed specific effects of tuning the optimisation by adjusting the importance of separate components in the cost function.

Deasy *et al* (1994a,d) have come to the same conclusions. The MINOS optimisation code from the Systems Optimisation Laboratory of Stanford University (Murtagh and Saunders 1983) was used to compute the 1D beamweight profiles which combine to give a conformal 2D dose distribution. The problem solved was that introduced by Webb (1989b), a high dose in a 'C-shaped' region with a cold square in the cusp of the 'C'. MINOS was configured to implement a quasi-Newton algorithm. Deasy *et al* (1994a,d) minimised the function

$$\text{cost} = (1/2)\alpha \left[\alpha_T \sum_i (d_i - d_i^p)^2 + \alpha_O \sum_i (d_i - d_i^p)^2 + \alpha_B \sum_i (d_i - d_i^p)^2 \right]$$
$$+ \beta \left[\beta_T \sum_i |d_i - d_i^p| + \beta_O \sum_i |d_i - d_i^p| + \beta_B \sum_i |d_i - d_i^p| \right]$$
$$+ \gamma [\gamma_T \max_i |d_i - d_i^p| + \gamma_O \max_i |d_i - d_i^p| + \gamma_B \max_i |d_i - d_i^p|]$$

$$(1.38)$$

where the subscripts T, O and B refer to target, OAR and the rest of the body respectively, d_i is the computed dose to the ith dose element and d_i^p is the prescription dose to the ith dose element. The 12 constants α, β, γ and $\alpha_{T,O,B}$, $\beta_{T,O,B}$ and $\gamma_{T,O,B}$ exercise the 'control' over the outcome. For

example, Deasy *et al* (1994a,d) found that using the quadratic norm only
(β and γ both zero) and with $\alpha = \alpha_T = \alpha_O = 1$ and $\alpha_B = 0$, dose
was uniform in the PTV to about 5%, in the OAR was less than 10% of
the maximum dose in the PTV, but hot spots some 18% above target dose
occurred in the (uncontrolled) rest of the body. Deasy *et al* (1994a,d) give
several examples of how the distributions change depending on these 12
parameters. For example, using a quadratic cost function and weight to the
body only 0.001 of that in the PTV and OAR, the dose distribution in the rest
of the body is much smoother and is always less than the dose in the PTV.
There are no surprises and the numerical results display the same behaviour
observed by Webb (1989b). The weighted-cost-function technique is a very
powerful way to control the outcome of inverse planning. Webb (1994b)
has recently presented further evidence that well-tuned inversion can give
highly conformal dose distributions.

Deasy *et al* (1994b,c) have developed the interesting analogy between
the solution of the tomotherapy problem and a problem in electrostatics.
If dose $d(x)$ becomes analogous to electrostatic potential and the photon
kernel density $\rho(x)$ becomes analogous to charge density then dose relates
to photon kernel density

$$d(x) = \int \frac{\rho(x')\,\mathrm{d}x'}{|x - x'|} \tag{1.39}$$

in the same way as potential relates to charge density. Deasy *et al*
(1994b,c) solved this convolution equation iteratively. The beamweights
are the projection of kernel density as in Brahme's original exposition. The
iteration steps are needed to cope with maintaining positive beamweights
because otherwise for a general potential, negative charges would be required
which in the tomotherapy analogy could project to give negative unphysical
beamweights. The steps are (dropping the x-argument for simplicity and
using subscript i for iteration number):

(i) introduce a pseudo-dose distribution $d_{pse,i}$
(ii) invert

$$\rho_i = \mathbf{A}^{-1} d_{pse,i} \tag{1.40}$$

where i labels the iteration number and $\mathbf{A}^{-1}$ is the inverse of the matrix
formed by the terms in equation (1.39). At the start the pseudo-dose
distribution is the real dose prescription

$$d_{pse,i} = d_{pre,0} \tag{1.41}$$

(iii) the beamweights w_i are computed by projection

$$w_i = \mathbf{P}(\rho_i)\pi/M \tag{1.42}$$

where **P** is a projection operator and there are M contributing 1D beamprofiles to the 2D dose distribution.

(iv) the beamweights are constrained positive

$$w_i^+ = \max(w_i, 0) \tag{1.43}$$

(v) the current resulting estimate $d_{res,i}$ of the dose distribution arising from these chopped beamweights is computed

$$d_{res,i} = \mathbf{C}w_i^+ \tag{1.44}$$

where matrix **C** is the transpose of matrix **P**.

(vi) iterative relaxation is performed. The pseudo-dose distribution at the next estimate becomes

$$d_{pse,i+1} = d_{pse,i} + \omega(d_{pre,0} - d_{res,i}) \tag{1.45}$$

where ω is a relaxation parameter. The iteration then cycles back to step (ii). Eventually when $d_{pse,i}$ is not changing with further iterations, the resulting dose distribution $d_{res,i}$ equals the prescription $d_{pre,0}$ and corresponds to the final calculation of beamweights w_i^+.

Deasy *et al* (1994b,c) show results for the same 'C' problem using 19 intensity-modulated beams and 20 iterations with $\omega = 1.6$. The dose conformation is very good, although in developing the analogy they have ignored photon scatter and attenuation as well as secondary charged particle motion so the results are somewhat artificial.

We now turn away from considering the fine-scale optimisation of intensity-modulated beams. We shall return to the theme of intensity-modulated radiotherapy in Chapter 2 when we consider techniques for delivery.

1.13. OPTIMISATION OF DYNAMIC THERAPY

Rosen *et al* (1992) studied a clinical case of treating a pancreatic tumour and compared the results of a dynamic rotation treatment with different conditions:

(i) simple unoptimised unwedged uniform-dose-rate rotation,

(ii) optimised (i.e. dose-rate varying with angle) unwedged rotation,

(iii) optimised wedged rotation (i.e. three rotations, one with no wedge, two with a 60° wedge in its two normal orientations).

To simulate continuous rotation, static fields, equispaced in $0-2\pi$ at variable angular separation (from 1° to 20° i.e. 360 to 18 beams) were used.

Optimisation was by linear programming. The conclusions were:

(i) 36 beams adequately represented a continuous rotation,

(ii) both optimised techniques were better than unoptimised (Rosen *et al* (1992) used NTCP measures as well as positive dose-difference metrics),

(iii) there was no clear improvement with the inclusion of wedges in the optimisation process.

They discussed the often-emphasised point that measures of NTCP must only be considered comparatively, not absolutely.

1.14. OPTIMISATION OF BEAM ORIENTATIONS

Bortfeld and Schlegel (1993) have provided a new algorithm for optimising 2D conformal dose distributions. Specifically they addressed the problem of determining the best beam *orientations*. They considered separately how to do this when the beams are (i) uniform and then (ii) modulated in intensity. For the former they used a simple beam model in which the depth–dose profile was purely exponential, the beam geometry was parallel and the lateral dose profile equalled the beam intensity profile. (They were able to account for the inverse-square law by using a modified attenuation coefficient.)

They computed the discrete 2D Fourier transform of the 2D dose distribution from each beam at each gantry orientation using steps of $2°$. It turned out that only 320 frequency components sufficed to describe each dose distribution. In making this transform they assumed that the beam was narrow, i.e. that the patient contour can be considered locally normal to the direction of incidence of the beam. Because the Fourier transform is linear, the transform of the 2D dose distribution from multiple beams is the sum of the separate transforms of the dose distribution from each beam at different orientations.

To determine the optimum orientations Bortfeld and Schlegel (1993) constructed a quadratic cost function, the root mean square (RMS) difference between the achievable 2D dose distribution and the 2D dose prescription. By invoking Parseval's theorem, the discrete form of this cost function can be converted into an objective function in Fourier space. They showed that this cost function must contain local minima (i.e. be 'non-convex') and so they naturally chose to minimise this cost function by the method of simulated annealing.

Bortfeld and Schlegel (1993) extended the method to cope with beams with 1D modulated intensities but then had to invoke the $\mu = 0$ assumption and allow negative beam intensities.

Several model problems were programmed all with circular phantom outlines. The main conclusion was that, for uniform beams, when the number of beams was small (two or three) the algorithm suggested certain preferred orientations in which the beam edges were parallel to local straight edges in the target volume. However, for target volumes with no obvious asymmetry, the algorithm suggested that beams should be uniformly distributed in 2π. For target volumes with concave outline, when modulated intensities are required, and even in the presence of OAR, the algorithm again showed that the uniform distribution in 2π was best when there were seven beams. This is fortunate for studies such as those of Webb (1989b) who used simulated annealing to optimise the beam profiles but completely ignored the optimisation of beam orientations and *a priori* chose a uniform arrangement in 2π. It is worth quoting from the paper 'For modulated beams ... it is not in general useful to avoid beam orientations through organs at risk'. The same is not true of BEV-shaped beams with no intensity modulation. Recent work at the University of Michigan (McShan 1990, Lichter 1990) seeks to select beam orientations on the basis of avoiding irradiating OAR.

The method is restricted to a least-squares objective function based on dose alone. Biological response is not modelled. The method relies on the 'weakly attenuating' property of tissue for high-energy x-rays.

Pickett *et al* (1994) have addressed the problem of determining optimum orientations of coplanar beams for therapy of the prostate in an empirical observational manner. They made a series of six-field plans. Each plan had an AP and PA field as well as four oblique fields whose inclination to the horizontal was varied through 20° to 45° in steps of 5°. Plans were prepared to irradiate the PTV defined from CT data obtained when each of ten patients had an empty rectum and full bladder. Plans were prepared with and without tissue inhomogeneity corrections. DVHs were prepared for rectum, bladder and femoral heads. From this wealth of data Pickett *et al* (1994) deduced that as the angle of inclination to the horizontal increased rectal and bladder doses increased whilst femoral head doses decreased. They concluded that the optimum angle of orientation of the oblique beams was 35°, thus giving acceptable PTV coverage with acceptable doses to the three OAR.

Gokhale *et al* (1994) presented a simple concept for selecting beam orientations. In this work they determined the 'line of least resistance' to radiation from the patient surface to the target and hypothesised that these defined the best orientations. The usefulness of this is not clear given that the approach only deals superficially with the problems posed by OAR in the BEV. However, the approach suggested beams for the problems studied which were similar to those which a human planner would have adopted.

A radically alternative approach was adopted by Sailer *et al* (1993, 1994) who, instead of determining the optimum beam orientations from a large set, *started* with either a tetrad or a hexad of non-coplanar beams in which the exit portal of each beam does not significantly overlap the entrance portal of another beam. By calculating the volume of tissue enclosed by different isodose surfaces they showed that there was significant reduction in normal-tissue irradiation compared with both coplanar-four- and coplanar-six-field treatments. In practice the tetrad and hexad orientations of the beams could

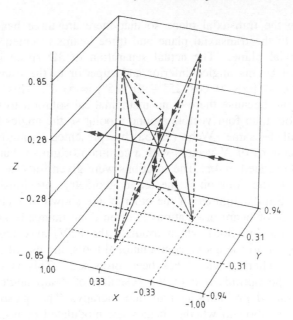

Figure 1.24. *The configuration of N = 4 vectors in space (the so-called tetrad). The vectors point to and from locations on the unit sphere centred on the origin* (0, 0, 0) *where vectors converge. Vectors have been represented by solid lines with arrows. The arrows point both ways along the vectors since each line passing through the origin comprises both entrance and exit directions and the radiation propagates in either of two opposing directions. Dotted construction lines join sets of vector tips by equilateral triangles in the* (y, z) *planes at* $x = \pm\frac{1}{3}$. *Solid unarrowed lines are projections of vectors into these planes.*

be varied as a locked set in relation to the patient. This moves away from what Fraass (1995) has called 'the tyranny of the axial plane'.

The tetrad is the geometry of the methane molecule. Imagine a regular pyramid constructed of equilateral triangle sides. Construct the three dropped perpendiculars on the face of each side intersecting at a point. The four beams enter at the vertices of the pyramid and are directed at this point of intersection in the opposite face. The four beams mutually intersect in 3D space with an included angle of $90° + [\sin^{-1}(1/3)]° = 109.471°$. Viewed from the axis of any one beam, the projections of the other three in the plane perpendicular to the viewing direction appear to be angled at $120°$ to each other (figure 1.24).

The hexad is constructed differently from six beams. Starting with two sets of three beams 120° apart in the transaxial plane, one set is first rotated relative to the other by 30°. Thus when viewed from the inferior to the superior direction (or vice versa) there is always an entrance or exit every 30°. In an alternating way, each beam entrance is then rotated 25° superior or inferior to the transaxial plane so that there are three beams oriented 25° superior to the transaxial plane and three beams oriented 25° inferior to the transaxial plane. The actual separation in 3D space between the closest pair of beams angled inferior and superior to the transaxial plane is then $\Omega = \cos^{-1}[\cos 30° \cos^2 25° - \sin^2 25°] = 57.8°$. This arrangement was worked out because there is no mathematical solution to the problem of placing more than four vectors around a point so the angles between the vectors are all the same. Webb (1995a) has presented an algorithm which gives the directions of N vectors with a common origin so that the vectors are maximally spaced in 4π. The solutions were given for $N = 2, 3, \ldots, 8$. When $N = 4$, the solution is the tetrad geometry, previously discussed (figure 1.24). Similar figures for the other N are shown in Webb (1995a).

The selection of beam orientations is often done manually attempting to find those beam directions which irradiate the PTV, maximally avoiding OAR. In addition to the above computational tools, the use of a 'tumour's-eye-view' or 'spherical view' has been proposed. This is discussed in Chapter 7. The optimisation of the selection of beam angles is one of the great unsolved problems in radiation therapy. The question must be considered in relation to whether beams are modulated or not. Ideally an automatic method should be generated to indicate the optimum choice with respect to minimising some cost function.

Because it can be difficult to visualise the relative orientation of beams selected in this way some 3D treatment-planning systems then provide a 'physician's-eye-view' or 'observer's view' in which the beams are shown in relation to the shaded-surface outline of the patient (e.g. Bendl 1995, Bendl *et al* 1994) (see also Chapter 7).

Bohsung *et al* (1996) have developed a relatively simple but fast algorithm for assisting with the choice of beam directions. The algorithm was implemented within the VOXELPLAN 3D treatment-planning system developed at Heidelberg. It is called GEOOPT and is specifically implemented within the VIRTUOS module. GEOOPT inputs surface information for the PTV and OAR as provided by the TOMAS module for segmentation. The user selects the isocentre and then, for all possible beam directions, determines the surface triangles of each triangulated object which are visible from the source. These triangles are then perspectively projected on to a plane perpendicular to the beam's central ray. Then, the relative overlap area of the beam, irradiating the PTV, and the ith OAR, $A_{rel,i}(\phi_g, \phi_c) = A_{o,i}(\phi_g, \phi_c)/A_i(\phi_g, \phi_c)$ is calculated where $A_{o,i}(\phi_g, \phi_c)$ is the absolute overlap area and $A_i(\phi_g, \phi_c)$ is the projected area of the ith OAR. Thus the variable $A_{rel,i}(\phi_g, \phi_c)$ represents the relative overlap of the PTV and the ith OAR for the beam specified by gantry and couch angles ϕ_g, ϕ_c.

If the projections of the ith OAR and the PTV have no overlap then the relative minimal distance between the projected borders of the PTV

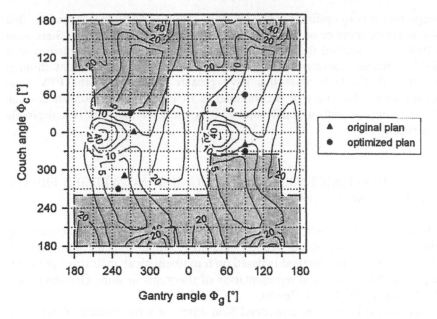

Figure 1.25. *A contour plot of the relative effective area $A_{rel,eff}(\phi_g, \phi_c)$ for a specific clinical case. The contour lines connect points of the same effective area. Beam entries of the original plan created by an experienced treatment planner, rather than the direction-optimising algorithm, are marked by solid triangles. The solid circles label the directions given by an algorithm for selecting optimum directions. Entries inside the grey regions are unreachable because of collisions between gantry and couch. (From Bohsung et al 1996.)*

and the ith OAR, $d_{rel,i}(\phi_g, \phi_c) = d_i(\phi_g, \phi_c)/d_{max,i}$ is evaluated where $d_{max,i} = \max[d_i(\phi_g, \phi_c)]$.

The direction of a beam is considered to be a 'good' direction if it leads to a small value for the relative overlap or a large value for the relative distance. If there is more than one OAR then the figures-of-merit can be combined including weighting for the importance w_i of each OAR, i.e. $A_{rel,eff}(\phi_g, \phi_c) = \sum_i w_i A_{rel,i}(\phi_g, \phi_c)$. This quantity can be plotted with respect to the two angles specifying beam directions (figure 1.25).

Note that this is a simple algorithm which gives a first indication of good beam directions. It does not involve any dose calculation and it takes no account of whether an OAR is distal or proximal to the PTV. It also does not account for the relationship between good directions and the number of beams involved in the plan. One could imagine more sophisticated cost functions being developed for this purpose. The algorithm was considered preferable to, say, the spherical view (see also Chapter 7) because it specifically takes into account the finite size of the beam.

It is reasonably obvious that the smaller the number of beams, the more important it is to optimise their direction. Another fairly trivial result is that the required number of beams depends on the prescription dose. Stein *et al* (1996) have studied the problem of optimisation of beam orientation with IMBs. Rather counterintuitively they actually encouraged beams through OAR with IMBs because this gave greater control over the PTV dose distribution. This may be a consequence of using a large number of such beams because with a small number it is unlikely that this is a desirable concept.

1.15. SHEROUSE'S METHOD FOR COMBINING WEDGED FIELDS BY ANALYSIS OF GRADIENTS

Sherouse (1993, 1994a,c) has provided a very elegant way of determining the best way to combine wedged fields for maximising the uniformity of dose to a PTV. The method is based on a mathematical analysis of gradients and is a rediscovery and representation of theorems in some German papers by Sonntag (1975a,b, 1976a,b).

As is well known, an unwedged field entering a flat surface of tissue has a dose gradient G_a pointing back towards the source along the axis of the beam. If the same field is wedged, the isodose lines form an angle θ_w to the normal to the central axis (call this the transaxial direction). The magnitude of the transaxial gradient of dose is $|G_t(\theta_w)| = |G_a| \tan(\theta_w)$. The resultant dose gradient is the vector sum of G_a and G_t and is of course at right angles to the isodose line as it crosses the central axis (figure 1.26).

Now imagine that several wedged fields are combined; the fields have central axes intersecting at an isocentre. Sherouse (1993) asserts that the necessary and sufficient condition for homogeneous dose over the volume of intersection of two or more beams is that the total vector sum of the dose gradients of the beams be zero everywhere in the volume of intersection. This then suggests a method to determine the combination of beam wedge angles, collimator angles and relative beamweights as follows.

(i) The vector sum $G_{a,total}$ of the gradients of open fields is formed.

(ii) The direction of this vector is reversed to make $G_{t,total}$. This new vector represents the vector sum of all the gradients which must be introduced by the selected combination of wedges and weights so as to cancel out the vector of the gradients of open fields.

(iii) The vector $G_{t,total}$ is resolved along the perpendiculars to the central axes of the unweighted beams to give the exact transaxial gradient required from each constituent beam. This is then combined with the axial gradient G_a of the open field to give the wedge angle for that beam. If the constituent fields are to receive different beamweights these drop out of the mathematical analysis.

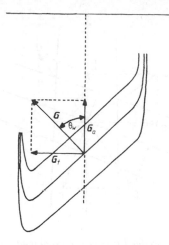

Figure 1.26. *A representation of the dose gradient for a single wedged field. The vector G_a is the axial component of the gradient pointing back to the source. The vector G_t is the transaxial component of the gradient. If the wedge angle is θ_w then $|G_t| = |G_a| \tan(\theta_w)$ by definition. (After Sherouse 1993.)*

Sherouse (1993) applies the method to several simple situations in both 2D and 3D. The method is limited of course by the assumption that the incidence of each beam is perpendicular to the patient surface, and that 'solving for zero gradient' at a point will make the distribution uniform throughout the volume. It was shown that the method is remarkably able to achieve quite high uniformity, only falling down at the edges of the distribution of high dose.

To give some substance to this otherwise rather abstract description we repeat Sherouse's analysis for a 2D wedged pair. Consider (figure 1.27) two beams with equal weight entering at 90° and 90° − θ_h to the transaxial direction. The open fields have simply axial gradients G_a pointing back to the sources. The resultant gradient $G_{a,total}$ has magnitude $2|G_a| \cos(\theta_h/2)$ and angle $90 − (\theta_h/2)$ (vector ad in figure 1.27) so it is required to introduce an equal and opposite gradient $G_{t,total}$ (ap in figure 1.27). This vector decomposes to two transaxial components of magnitude $|G_a| \cot(\theta_h/2)$ (aq and ar in figure 1.27). These are the transaxial gradients which must be introduced. Combining transaxial gradient aq with axial gradient ab vectorially gives the resultant gradient as such that the wedge angle for the first field is $\tan^{-1}(|G_a| \cot(\theta_h/2)/|G_a|) = 90° − (\theta_h/2)$. By symmetry the wedge angle for the second beam is found in the same way and is identical (because the magnitude of ar is equal to the magnitude of aq).

This result that the wedge angle is $90° − (\theta_h/2)$ with heels together is a well known 'radiotherapy rule' but rather elegantly derived this way. For example, if the two fields enter at right angles ($\theta_h = 90°$) then the wedge angles should be 45°.

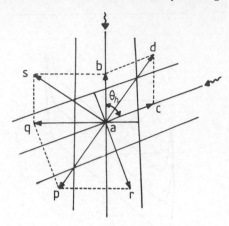

Figure 1.27. *Two wedged beams are combined. The beams subtend an angle θ_h between them. The axial gradients are ab and ac with resultant ad. The addition of vector gradient ap would cancel out ad if ap is equal and opposite. ap is resolved into aq and ar in the two transaxial directions. The resultant of aq and ab gives the wedge gradient as for beam 1 and is at right angles to ad. Equally the resultant of ar and ac (not shown) would have the same wedge angle for beam 2. This result depends on the weights of the two beams being identical and the problem being formulated in 2D only. An alternative way to analyse this problem is shown in figure 1.28. (Adapted from Sherouse 1993.)*

The same vector formulation can be used to work out the effective wedge angle for two separately wedged fields from the same direction. If the separate wedge angles are $\theta_{w,1}$ and $\theta_{w,2}$ and the two fields have weights w_1 and w_2 then the magnitude of the resultant axial dose gradient is $w_1|G_a| + w_2|G_a|$ and the magnitude of the transaxial gradient is $w_1|G_a|\tan(\theta_{w,1}) + w_2|G_a|\tan(\theta_{w,2})$ so the combined effective wedge angle is $\theta_{w,eff}$ where

$$\tan(\theta_{w,eff}) = \frac{w_1|G_a|\tan(\theta_{w,1}) + w_2|G_a|\tan(\theta_{w,2})}{w_1|G_a| + w_2|G_a|}. \quad (1.46)$$

If we arrange for $w_1 + w_2 = 1$ then

$$\tan(\theta_{w,eff}) = w_1\tan(\theta_{w,1}) + (1 - w_1)\tan(\theta_{w,2}) \quad (1.47)$$

from which by rearrangement

$$w_1 = \frac{\tan(\theta_{w,eff}) - \tan(\theta_{w,2})}{\tan(\theta_{w,1}) - \tan(\theta_{w,2})}. \quad (1.48)$$

For example, if the second field is unwedged ($\theta_{w,2} = 0°$) this becomes simply

$$w_1 = \frac{\tan(\theta_{w,eff})}{\tan(\theta_{w,1})}. \quad (1.49)$$

which is a well known rule for combining wedged and unwedged fields to give an arbitrary wedge angle (see the companion Volume, Chapter 7).

Sherouse's method of analysis can also be restated.

(i) For any number of wedged fields, the ith having weight w_i, write down the axial gradient $w_i G_a$ and the transaxial gradient $w_i G_{t_i}$ at the point of intersection of the axes of the beams.

(ii) Resolve these gradients in any three orthogonal directions and set the resolved components to zero (convenient axes may be the axial and two transaxial directions for one of the beams).

(iii) Solve for the unknowns.

For all but the simplest situations the problem will be severely underdetermined. There will be far fewer equations than unknowns. The user must specify all but three variables and solve for the rest. Put another way this means there are many different ways of achieving a uniform field (zero gradient) depending on the choice of which three variables to leave for solution.

For example let us reanalyse the 2D wedged pair this way. The central axes of the two beams have angle θ_h between them. Let the weights be w_1 and w_2. Resolving axial and transaxial components in the direction of the central axis of beam 1 gives:

$$w_1|G_a| + w_2|G_a|\cos(\theta_h) = w_2|G_{t_2}|\sin(\theta_h) \qquad (1.50)$$

and resolving perpendicular to this in the direction transaxial to beam 1 (figure 1.28) gives

$$w_1|G_{t_1}| = w_2|G_a|\sin(\theta_h) + w_2|G_{t_2}|\cos(\theta_h). \qquad (1.51)$$

Here we have two equations in four unknowns. (A third equation cannot be formed for this 2D problem.) However, setting $w_1 = w_2$ by solving the two equations in two unknowns and by simple trigonometry $|G_{t_1}| = |G_{t_2}| = |G_a|\cot(\theta_h/2)$, the same solution as above, from which the wedge angle for each field should be $\tan^{-1}[(|G_a|\cot(\theta_h/2))/|G_a|] = 90° - \theta_h/2$.

Sherouse's (1993) paper gives many more elegant examples of how to combine fields in different geometries. This may not be a panacea for 3D dose planning but it gives a lot of physical insight into what is going on.

Sherouse (1994b) is an advocate of 'geometrical planning' and feels that most problems in conformal radiotherapy can be solved by a suitable arrangement of geometrically shaped and appropriately wedged portals. He has pioneered 'virtual simulation' for this purpose. In a stimulating conference paper entitled 'Is the inverse planning problem a red herring?' he argues (my additions in brackets) that 'battle lines are being drawn in the conformal radiotherapy community between the Axial Arcists (advocates

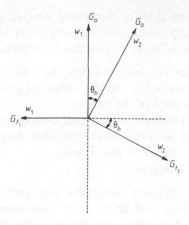

Figure 1.28. *Two wedged beams are combined. The axial and transaxial gradients of each are drawn. The beams subtend an angle θ_h between them. By resolving the gradients (and setting the resolved gradients to zero) along the dotted lines the optimal wedge angles for each beam can be easily calculated (compare with figure 1.27).*

of the use of multiple IMBs) and the Fixed Fielders (who support the evolutionary position that conformal therapy is an exercise in geometry)'. Sherouse (1995) feels the case for IMBs has been overstated, that the advantage is oversold, that the number of clinical cases requiring IMBs is small. He finally concludes that IMBs *are* sometimes needed but are not a panacea for all problems. Indeed (and this is surely sensible) he recognises the large paradigm shift of IMB therapy and when the same problem can be solved with fixed fields it should be. There should be few who argue with that. My comment is that perhaps it is because the inverse-planning method and delivery of IMBs do represent such a quantum leap in conception, that those developing these techniques have perhaps used overstrong statements to introduce them. There is room for both approaches and the true challenge is to develop the wisdom to know when to select which in the clinic (see also Appendix A).

These techniques are really only relevant to the treatment of small tumours. Since the gradient is a local quantity it is clearly not possible to achieve the desired field throughout a large tumour by simply adjusting the weights of open or wedged fields globally. One would have to extend the method to create fields which were some kind of composite of the required wedge weights spatially adjusted over the area of the fields so the method can optimise the solution for all voxels in the tumour and the OAR. As soon as one begins to think like this one is on the road to full intensity modulation once again. The results from Sherouse are presented in some detail in the hope that the challenge to develop such an extended analysis will be taken up.

1.16. CONTROVERSIES IN DETERMINING THE 'BEST PLAN'

Most 3D treatment-planning systems provide several methods to obtain a plan and the question arises of how to determine the 'best'. Tools are provided to assist; for example the DVH, methods to view dose distributions in relation to structures, prediction of TCP and NTCP representing biological response. With increasing use of multiple, possibly non-coplanar, geometrically shaped fields, inverse planning has been tried (Schlegel 1993, Webb 1993b). It is generally recognised that this is the best approach to achieving conformal dose distributions. Controversies include the following.

(i) Should the time allowed for inverse planning be restricted to the same time as would be used in routine forward-planning practice? Since most departments would probably say they are already at capacity time-wise for the known patient throughput, the answer would be 'yes'. However, one would expect that if unbounded time were allowed for investigating inverse planning, the result would be better dose distributions. So it makes sense to answer 'no' to the above question to allow inverse planning to achieve its potential.

(ii) Inverse planning is generally linked with the use of a larger number of beams than conventional forward planning because computer-controlled MLCs are envisaged. Conversely forward planning is generally done for say three or four beams, limited by delivery practicalities. Whatever decision is made time-wise, is it fair to compare the results of inverse planning using fairly large numbers of beams with forward conventional planning using a small, practically manageable, number of beams? It would be argued that forward conventional planning would improve if the number of beams were less bounded. Yet surely it makes little sense to make the comparison for a fixed number of beams? Whilst conventional forward planning has to use a small number of beams, inverse planning is not thus constrained and again can realise its potential if the number of beams can be varied and increased beyond the 'conventional limit'.

A major open question is how many beams are needed? This question has been argued out recently in the context of IMBs (Mackie *et al* 1994a, Brahme 1994 a,b,c). Mackie *et al* (1994a), strong advocates of tomotherapy (see Chapter 2), argue the case for using a large number of IMBs. Firstly they noted that several planning studies showed improved dose distributions and biological outcome for plans with a largish (7–21) number of IMBs (Bortfeld *et al* 1990, Webb 1992a). Secondly, they also invoke the well known prediction that, since $D_{50}(v) = D_{50}(v = 1) \times v^{-n}$, where $D_{50}(v)$ is the dose required to give NTCP $= 50\%$ for an organ when only a fractional volume v is irradiated and n is the empirically determined volume parameter between 0 and 1, for $n < 1$ it is always more favourable to irradiate a larger volume to smaller dose than to irradiate a smaller volume to higher dose, at constant integral dose to the normal organ. (This follows because at constant integral dose, the dose to the normal organ must scale as v^{-1} which, as v decreases, gives a higher dose than that from the v^{-n} law for iso-NTCP.) Thus Mackie *et al* (1994a) argue it is better to 'spread out' the dose over as large a volume as possible using a large number of beams. Thirdly, a large number is also called for if a highly uniform dose is to be delivered to a tumour with a steep fall off at its edge when the tumour has a complicated contour. Fourthly, they argue that equipment which can deliver many beams can easily deliver few whereas the reverse is not true.

Brahme's (1994b) response to Mackie *et al* (1994a) is that, whilst it is true that in certain circumstances the use of a large number of beams is necessary, quite efficacious treatment can be engineered with a few beams if their *direction* is properly chosen (Söderström and Brahme 1992, 1995). Regarding radiobiological considerations, he argues that studying the volume effect of a single normal organ is over-simplistic; it is necessary to consider possible tumour induction in wholly irradiated organs especially paediatric cases. Finally Brahme (1994b) challenges the practicality of implementing multiple-beam techniques and suggests that this may not be possible until into the next century and possibly then in only a limited number of larger centres.

Both Mackie *et al* (1994a) and Brahme (1994b) are agreed that there have not been sufficient studies using IMBs to point to a general answer to the question of how many are needed. Such studies must address not only the question of the number of beams but their disposition. There has recently been further hot debate on this topic (see section 1.10) (Mohan and Ling 1995, Söderström and Brahme 1996, Mohan *et al* 1996). The problem of finding the best non-uniform few-field directions is one of the hardest problems in radiotherapy optimisation. There is also agreement that if the target has a very complex shape it will be necessary to use a large number of beams whose directions are well chosen.

1.17. SUMMARY

Optimising the planning of radiotherapy is a very rapidly developing research field. Analytic and iterative techniques have been proposed and recently a number of independent solutions for the inverse problem of determining intensity-modulated beam profiles have been published. Meanwhile controversy has developed over identifying the class of treatment-planning problems which would benefit from IMBs and optimisation of static-field therapy has also continued to develop. The community is settling to recognise the need for both types of therapy tailored to the problem. In contrast there has been less attention paid to the problem of determining the optimum orientation of beamports although some progress has been made with simplified assumptions. It is important to acknowledge that there is no unique solution to the inverse-planning problem with or without IMBs since the solution depends on the applied constraints. The mid-1990s have seen the incorporation of optimisation techniques into 3D treatment-planning systems which will inevitably lead to their greater use as well as coupling them to the other stages of the treatment-planning process. For example, greater use is being made of 3D medical images to determine the volumes of interest; however, attention is focusing on attendant difficulties in determining these volumes. So whilst the goal is to tailor the high-dose volume to the PTV, this presumes the PTV can be accurately determined. Geometrical misses must be avoided at all costs if the aim is curative therapy.

Comparing the efforts of different research groups is difficult not least because each team inevitably chooses to stress one feature of its optimisation whilst making simplifying assumptions about another. For example, there is no universally accepted dose-calculation algorithm. The constraints applied are also generally different. Some teams concentrate on dose optimisation; the goal for others is to optimise biological outcome. There are even differences in models for TCP and NTCP and a sparcity of supporting data.

However the future is promising in that the wider availability of planning tools should lead to a greater concern for the outcome of therapy and an increased desire to experiment with various planning options working towards a solution which is both efficacious and practical.

Whilst not a panacea for all problems, IMBs are increasingly topical and the next chapter looks in detail at methods by which intensity-modulated radiation might be delivered.

REFERENCES

Austin Seymour M 1994 Inter-institutional comparison of defining planning target volumes (Proc. 36th ASTRO Meeting) *Int. J. Radiat. Oncol. Biol. Phys.* **30** Suppl. 1 117

Balter J, Sandler H M, Lam K, Bree R L, Lichter A S and Ten Haken R K 1993 Measurement of prostate motion over the course of radiotherapy (Proc. 35th ASTRO Meeting) *Int. J. Radiat. Oncol. Biol. Phys.* **27** Suppl. 1 223

Balter J M, Ten Haken R K and Lam K L 1994a Assessment of margins for ventilatory motion during radiotherapy *Med. Phys.* **21** 913

Balter J M, Ten Haken R K, Lam K L, Lawrence T S, Robertson J M and Turrisi A T 1994b The use of spiral CT and spirometry to reduce uncertainties in treatment planning associated with breathing (Proc. 36th ASTRO Meeting) *Int. J. Radiat. Oncol. Biol. Phys.* **30** Suppl. 1 241

Barrett H 1994 private communication

Barrett H, Barber H B, Ervin P A, Myers K J, Paxman R G, Smith W E, Wild W J and Woolfenden J M 1983 New directions in coded-aperture imaging

Information Processing in Medical Imaging ed F Deconinck (Dordrecht: Nijhoff) pp 106–29

Beard C J, Bussiere M R, Plunkett M E, Coleman N and Kijewski P K 1993 Analysis of prostate and seminal vesicle motion (Proc. 35th ASTRO Meeting) *Int. J. Radiat. Oncol. Biol. Phys.* **27** Suppl. 1 136

Bendl R 1995 Virtuelle strahlentherapiesimulation mit VIRTUOS *Dreidimensionale Strahlentherapieplanung* ed W Schlegel, T Bortfeld and J Stein (Heidelberg: DKFZ) pp 31–42

Bendl R, Pross J, Hoess A, Keller M A, Preiser K and Schlegel W 1994 VIRTUOS—A program for virtual radiotherapy simulation and verification *The Use of Computers in Radiation Therapy: Proc. 11th Conf.* ed A R Hounsell *et al* (Manchester: ICCR) pp 226–7

Birkhoff G D 1940 On drawings composed of uniform straight lines *J. Math. Pures Appl.* **19** 221–36

Bohsung J, Nill S, Perelmouter J, Nüsslin F, Kortmann R D and Bamberg M 1996 A software tool for geometric optimisation of radiation beam entries *Proc. CAR 96 (Berlin, 1996)* (Berlin: CAR)

Bortfeld T 1995 Optimierung und inverses problem (1) *Dreidimensionale Strahlentherapieplanung* ed W Schlegel, T Bortfeld and J Stein (Heidelberg: DKFZ) pp 129–35

Bortfeld T R and Boyer A L 1995 The exponential Radon transform and projection filtering in radiotherapy planning *Int. J. Imaging Syst. Technol.* **6** 62–70 (special issue on 'Optimisation of the three-dimensional dose delivery and tomotherapy')

Bortfeld T, Boyer A L, Schlegel W, Kahler D L and Waldron T J 1994 Realisation and verification of three-dimensional conformal radiotherapy with modulated fields *Int. J. Radiat. Oncol. Biol. Phys.* **30** 899–908

Bortfeld T, Burkelbach J, Boesecke R and Schlegel W 1990 Methods of image reconstruction from projections applied to conformation therapy *Phys. Med. Biol.* **35** 1423–34

Bortfeld T, Burkelbach J and Schlegel W 1992 Solution of the inverse problem in 3D *Advanced Radiation Therapy: Tumour Response Monitoring and Treatment Planning* ed A Breit (Berlin: Springer) pp 649–53

Bortfeld T and Schlegel W 1993 Optimisation of beam orientations in radiation therapy: some theoretical considerations *Phys. Med. Biol.* **38** 291–304

Boyer A L, Desobry G E and Wells N H 1991 Potential and limitations of invariant kernel conformal therapy *Med. Phys.* **18** 703–12

Brahme A 1988 Optimisation of stationary and moving beam radiation therapy techniques *Radiother. Oncol.* **12** 129–40

—— 1994a Inverse radiation therapy planning: principles and possibilities *The Use of Computers in Radiation Therapy: Proc. 11th Conf.* ed A R Hounsell *et al* (Manchester: ICCR) pp 6–7

—— 1994b Optimisation of radiation therapy and the development of multileaf collimation *Int. J. Radiat. Oncol. Biol. Phys.* **28** 785–7

—— 1994c The importance of non-uniform dose delivery in treatment planning (Proc. World Congress on Medical Physics and Biomedical Engineering (Rio de Janeiro, 1994)) *Phys. Med. Biol.* **39A** Part 1 498

—— 1995a Treatment optimisation using physical and radiobiological objective functions *Radiation Therapy Physics* ed A Smith (Berlin: Springer) pp 209–46

—— 1995b Development of radiation treatment planning and therapy optimisation *Proc. 19th L H Gray Conf. (Newcastle, 1995)* (Newcastle: L H Gray Trust)

—— 1995c The optimal number of beam portals in radiation therapy (Proc. ESTRO Conf. (Gardone Riviera, 1995)) *Radiother. Oncol.* **37** Suppl. 1 S3

—— 1995d Guest editorial: optimisation of the three dimensional dose delivery and tomotherapy *Int. J. Imaging Syst. Technol.* **6** 1 (special issue on 'Optimisation of the three-dimensional dose delivery and tomotherapy')

—— 1995e Similarities and differences in radiation therapy optimisation and tomographic reconstruction *Int. J. Imaging Syst. Technol.* **6** 6–13 (special issue on 'Optimisation of the three-dimensional dose delivery and tomotherapy')

Breuer H and Wynchank S 1996 Quantitation of brain movement within the skull associated with head position: its relevance to proton therapy planning *Abstracts of the 23rd PTCOG Meeting (Cape Town, 1995)* (Boston: PTCOG) p 31

Brewster L, Mageras G S and Mohan R 1993 Automatic generation of beam apertures *Med. Phys.* **20** 1337–42

Bucciolini M, Banci Buonamici F, Cellai E, Campagnucci A, Fallai C, Magrini S M, Olmi P, Rossi F, Santoni R and Biti G P 1996 The choice of PTV: a crucial point in the treatment plan optimization *Proc. Symp. Principles and Practice of 3-D Radiation Treatment Planning (Munich, 1996)* (Munich: Klinikum rechts der Isar, Technische Universität)

Burman C, Kutcher G J, Emami B and Goitein M 1991 Fitting of normal tissue tolerance data to an analytic function *Int. J. Radiat. Oncol. Biol. Phys.* **21** 123–35

Casamassima F 1994 Studies of the mobility of tumour volume *Hadrontherapy in Oncology* ed U Amaldi and B Larsson (Amsterdam: Elsevier) pp 428–33

Chaney E L and Pizer S M 1992 Defining anatomical structures from medical images *Semin. Radiat. Oncol.* **2** 215–25

Chen G T Y and Pelizzari C A 1995 The role of imaging in tumour localisation and portal design *Radiation Therapy Physics* ed A Smith (Berlin: Springer) pp 1–17

Chen Z, Wang Z, Bortfeld T and Mohan R 1994 Influence of phantom scatter on the design of the intensity modulators *Med. Phys.* **21** 945

Chen Z, Wang X, Bortfeld T, Mohan R and Reinstein L 1995 The influence of scatter on the design of intensity-modulations *Med. Phys.* **22** Part 1 1727–33

Ching-Li H and Volodin V 1993 Opening statement *Three-Dimensional Treatment Planning (Proc. EAR Conf. (WHO, Geneva, 1992))* ed P Minet (Geneva: Minet) pp xix–xxi

Cormack A M 1987 A problem in rotation therapy with x-rays *Int. J. Radiat. Oncol. Biol. Phys.* **13** 623–30

—— 1990 Response to Goitein *Int. J. Radiat. Oncol. Biol. Phys.* **18** 709–10

—— 1995 Some early radiotherapy optimisation work *Int. J. Imaging Syst. Technol.* **6** 2–5 (special issue on 'Optimisation of the three-dimensional dose delivery and tomotherapy')

Cormack A M and Cormack R A 1987 A problem in rotation therapy with x-rays 2: dose distributions with an axis of symmetry *Int. J. Radiat. Oncol. Biol. Phys.* **13** 1921–5

Cormack A M and Quinto E T 1989 On a problem in radiotherapy: questions of non-negativity *Int. J. Imaging Syst. Technol.* **1** 120–4

—— 1990 The mathematics and physics of radiation dose planning using x-rays *Contemp. Math.* **113** 41–55

Deasy J O 1996 General features of multiple local minima in TCP optimisation problems with dose-volume constraints *Med. Phys.* **23** 1112

Deasy J O, De Leone R, Holmes T W and Mackie T R 1994a Beam weight optimisation using the MINOS code *The Use of Computers in Radiation Therapy: Proc. 11th Conf.* ed A R Hounsell *et al* (Manchester: ICCR) pp 64–5

—— 1994b Coplanar beam weight optimisation: electrostatic model and iterative adjustment *The Use of Computers in Radiation Therapy: Proc. 11th Conf.* ed A R Hounsell *et al* (Manchester: ICCR) pp 352–3

Deasy J O, De Leone R, Mackie T R and Holmes T W 1994c A method for tomotherapy treatment planning using an electrostatic analogy *Med. Phys.* **21** 883

—— 1994d Beam weight optimisation using the MINOS code *Med. Phys.* **21** 911

De Vroed M 1994 Optimisation in radiotherapy treatment planning *MSc Thesis* Delft University of Technology, Department of Applied Mathematics and Computer Science

Dunst J 1996 Relation between target volume and chemotherapy *Proc. Symp. Principles and Practice of 3-D Radiation Treatment Planning (Munich, 1996)* (Munich: Klinikum rechts der Isar, Technische Universität)

Elliott P J, Knapman J M and Schlegel W 1992 Interactive image segmentation for radiation treatment planning *IBM Syst. J.* **31** 620–34

Emami B 1994 3D conformal radiation therapy: clinical aspects (Proc. 36th ASTRO Meeting) *Int. J. Radiat. Oncol. Biol. Phys.* **30** Suppl. 1 137

Emami B, Lyman J, Brown A, Coia L, Goitein M, Munzenrider J E, Shank B, Solin L J and Wesson M 1991 Tolerance of normal tissue to therapeutic irradiation *Int. J. Radiat. Oncol. Biol. Phys.* **21** 109–22

Ésik O, Bortfeld T, Bendl R, Németh G and Schlegel W 1995 Inverse radiation treatment planning with dynamic multileaf collimation for a concave–convex target volume in the cervical and upper mediastinal regions (private communication)

Ezzell G A 1996 Genetic and geometric optimization of three-dimensional radiation therapy treatment planning *Med. Phys.* **23** 293–305

Fazio F, Rizzo G, Gilardi M C, Lucignani G and Savi A 1994 Multimodality imaging: a synergistic approach *Diagnostic Imaging* Sept/Oct 43–8

Feldmann H J, Gross M W, Weber W A, Bartenstein P, Schwaiger M and Molls M 1996 *Proc. Symp. Principles and Practice of 3-D Radiation Treatment Planning (Munich, 1996)* (Munich: Klinikum rechts der Isar, Technische Universität)

Finnigan D J, Tanner S F, Dearnaley D P, Edser E, Horwich A, Leach M O and Mayles W P M 1995 Distortion corrected magnetic resonance images for pelvic radiotherapy treatment planning *Proc. 19th L H Gray Conf. (Newcastle, 1995)* (Newcastle: L H Gray Trust)

Forman M D, Mesina C F, He T, Devi S B, Ben-Josef E, Pelizzari C, Vijayakumar S and Chen G T 1993 Evaluation of changes in the location and shape of the prostate and rectum during a seven-week course of conformal radiotherapy (Proc. 35th ASTRO Meeting) *Int. J. Radiat. Oncol. Biol. Phys.* **27** Suppl. 1 222

Fraass B A 1994 Computer-controlled 3D treatment delivery (Proc. 36th ASTRO Meeting) *Int. J. Radiat. Oncol. Biol. Phys.* **30** Suppl. 1 131

—— 1995 The development of conformal radiotherapy *Med. Phys.* **22** 1911–21

Fraass B A and McShan D L 1995 Three dimensional photon beam treatment planning *Radiation Therapy Physics* ed A R Smith (Berlin: Springer) pp 43–93

Gademann G 1996 Definition of target volumes and critical organs *Proc. Symp. Principles and Practice of 3-D Radiation Treatment Planning (Munich, 1996)* (Munich: Klinikum rechts der Isar, Technische Universität)

Gademann G, Schad L R, Schlegel W, Semmler W, Kaick G and Wannenmacher 1993 The definition of target volume in tumours of the brain, base of skull and facial area by means of MRI: its impact on precision radiotherapy *Three-Dimensional Treatment Planning (Proc. EAR Conf. (WHO, Geneva, 1992))* ed P Minet (Geneva: Minet) pp 47–55

Geman S and Geman D 1984 Stochastic relaxation, Gibbs distributions, and

Bayesian resoration of images *IEEE Trans. Patt. Anal. Mach. Int.* **PAMI-6** 721–41

Goitein M 1995 Quantitative target volume definition *Proc. 19th L H Gray Conf. (Newcastle, 1995)* (Newcastle: L H Gray Trust)

Gokhale P, Hussein E M A and Kulkami N 1994 Determination of beam orientation in radiotherapy planning *Med. Phys.* **21** 393–400

Graham M V, Dryzmala R E, Jain N L and Purdy J P 1994 Confirmation of dose-volume histograms and normal tissue complication probability calculations to predict pulmonary complications after radiotherapy for lung cancer (Proc. 36th ASTRO Meeting) *Int. J. Radiat. Oncol. Biol. Phys.* **30** Suppl. 1 198

Grosu A L, Feldmann H J, Albrecht C, Gross M W, Kneschaurek P and Molls M 1996 Three-dimensional conformal external beam irradiation in the treatment of brain gliomas *Proc. Symp. Principles and Practice of 3-D Radiation Treatment Planning (Munich, 1996)* (Munich: Klinikum rechts der Isar, Technische Universität)

Gustafsson A, Lind B K and Brahme A 1994 A generalised pencil beam algorithm for optimisation of radiation therapy *Med. Phys.* **21** 343–56

Haas O C L, Burnham K J, Fisher M H and Mills J A 1995 Genetic algorithm applied to radiotherapy treatment planning *Proc. Int. Conf. Neural Nets and Genetic Algorithms (Alès, France, 1995)* ed D W Pearson and N C Steele (Vienna: Springer) pp 432–5

Halnan K 1995a Landmarks from 100 years of radiotherapy and oncology *Proc. Röntgen Centenary Congress 1995 (Birmingham, 1995)* (London: British Institute of Radiology) p 80

—— 1995b One hundred years of radiotherapy in Britain *The Röntgen Centenary: The Invisible Light: 100 Years of Medical Radiology* ed A M K Thomas, I Isherwood and P N T Wells (Oxford: Blackwell) pp 44–55

Herman G T 1980 *Image Reconstruction from Projections: The Fundamentals of Computed Tomography* (New York: Academic)

Holmes T W 1993 A model for the physical optimisation of external beam radiotherapy *PhD Thesis* University of Wisconsin-Madison

Holmes T and Mackie T R 1992 Simulation studies to characterise the search space of a radiotherapy optimisation algorithm (Proc. AAPM Conf. August 1992) *Med. Phys.* **19** 842

—— 1993a A filtered backprojection dose calculation method useful for inverse treatment planning *Med. Phys.* **21** 303–13

—— 1993b Suppression of artefacts in optimised beam profiles determined by the iterative filtered backprojection inverse treatment planning algorithm *unpublished preprint*

—— 1994 A comparison of three inverse treatment planning algorithms *Phys. Med. Biol.* **39** 91–106

Holmes T, Mackie T R and Jursinic P 1993d Experimental verification of an inverse treatment planning algorithm for tomotherapy *unpublished preprint*

Holmes T W, Mackie T R, Reckwerdt P and Deasy J O 1993a An iterative filtered backprojection inverse treatment planning algorithm for tomotherapy *Int. J. Radiat. Oncol. Biol. Phys.* **32** 1215–25

—— 1994a A prototype inverse treatment planning algorithm for tomotherapy *The Use of Computers in Radiation Therapy: Proc. 11th Conf.* ed A R Hounsell *et al* (Manchester: ICCR) pp 62–3

—— 1994b Optimisation for tomotherapy (Proc. World Congress on Medical Physics and Biomedical Engineering (Rio de Janeiro, 1994)) *Phys. Med. Biol.* **39A** Part 2 675

Horwich A, Wynne C, Nahum A, Swindell W and Dearnaley D P 1994 Conformal radiotherapy at the Royal Marsden Hospital (UK) *Int. J. Radiat. Biol.* **65** 117–22

Jackson A, Wang X H and Mohan R 1994 Optimization of conformal treatment planning and quadratic dose objectives *Med. Phys.* **21** 1006

Jeffrey W and Rosner R 1986 Optimisation algorithms: simulated annealing and neural network processing *Astrophys. J.* **310** 473–81

Jones D, Hafermann M D, Rieke J W and Vermeulen S S 1995 An estimate of the margin required when defining blocks around the prostate *Br. J. Radiol.* **68** 740–6

Källman P, Lind B K and Brahme A 1992 An algorithm for maximising the probability of complication-free tumour control in radiation therapy *Phys. Med. Biol.* **37** 871–90

Kamprad F, Wolf U, Seese A, Wilke W and Otto L 1996 Image correlation and its impact on target volume *Proc. Symp. Principles and Practice of 3-D Radiation Treatment Planning (Munich, 1996)* (Munich: Klinikum rechts der Isar, Technische Universität)

Kessler M L, McShan D L and Fraass B A 1992 Displays for three-dimensional treatment planning *Semin. Radiat. Oncol.* **2** 226–34

Kirkpatrick S, Gelatt C D and Vecci M P 1983 Optimisation by simulated annealing *Science* **220** 671–80

Korin H, Ehman R, Reiderer S J, Felmlee J P and Grimm R C 1992 Respiration kinematics of the upper abdominal organs: a quantitative study *Magn. Reson. Med.* **23** 172–8

Kooy H M, van Herk M, Barnes P D, Alexander E III, Dunbar S F, Tarbell N J, Mulkern R V, Holupka E J and Loeffler J S 1994 Image fusion for stereotactic radiotherapy and radiosurgery treatment planning *Int. J. Radiat. Oncol. Biol. Phys.* **28** 1229–34

Kuhn M 1995 Multimodality medical image analysis for diagnosis and treatment planning: The COVIRA Project (COmputer VIsion in RAdiology) *Report of the AIM Project A 2003* European Union DG 13 pp 1–39

Kutcher G J and Leibel S A 1994 3D conformal radiation therapy—treatment planning (Proc. 36th ASTRO Meeting) *Int. J. Radiat. Oncol. Biol. Phys.* **30** Suppl. 1 124

Kutcher G J, Mageras G G and Leibel S A 1995 Control, correction and modeling of setup errors and organ motion (*Innovations in treatment delivery*) *Semin. Radiat. Oncol.* **5** pp 134–45

Kutcher G J and Mohan R 1995 Introduction: three dimensional treatment delivery (*Innovations in treatment delivery*) *Semin. Radiat. Oncol.* **5** pp 75–6

Langer M, Brown R, Morrill S, Lane R and Lee O 1994 A generic genetic algorithm for generating beam intensities *Med. Phys.* **21** 878

Laughlin J S 1995 The development of medical physics in the USA (Proc. ESTRO Conf. (Gardone Riviera, 1995)) *Radiother. Oncol.* **37** Suppl. 1 S28

Lawrence T S 1994 The influence of target definition on NTCP—a clinical perspective (Proc. 36th ASTRO Meeting) *Int. J. Radiat. Oncol. Biol. Phys.* **30** Suppl. 1 117

Leibel S A, Kutcher G J, Zelefsky M J and Fuks Z 1994 Coping with prostate and seminal vesicle motion in three-dimensional conformal radiation therapy *Int. J. Radiat. Oncol. Biol. Phys.* **28** 327–8

Levegrün S, Waschek T, van Kampen M, Engenhart R and Schlegel W 1995 A new approach of target volume definition based on fuzzy logic *Medizinische Physik 95 Röntgen Gedächtnis-Kongress* ed J Richter (Würzburg: Kongress) pp 144–5

Levegrün S, Waschek T, van Kampen M, Engenhart-Cabillic R and Schlegel W 1996 Biological scoring of different target volume extensions in 3D radiotherapy treatment planning *Proc. Symp. Principles and Practice of 3-D Radiation Treatment Planning (Munich, 1996)* (Munich: Klinikum rechts der Isar, Technische Universität)

Levy R P, Schulte R W M, Frankel K A, Steinberg G K, Marks M P, Lane B, Slater J D and Slater J M 1994 CT slice-by-slice target volume delineation for stereotactic proton irradiation of large intracranial arteriovenous malformations: an iterative approach using angiographic, CT and MRI data (Proc. 36th ASTRO Meeting) *Int. J. Radiat. Oncol. Biol. Phys.* **30** Suppl. 1 162

Lichter A S 1994a The influence of dose and volume on normal tissue tolerance to radiation (Proc. 36th ASTRO Meeting) *Int. J. Radiat. Oncol. Biol. Phys.* **30** Suppl. 1 99

—— 1994b Potential for three-dimensional treatment planning and conformal dose delivery *A Categorical Course in Physics: Three-Dimensional Radiation Therapy Treatment Planning* ed J A Purdy and B A Fraass (Chicago: RSNA) pp 21–5

—— 1996 The future of technology in radiation oncology *Med. Phys.* **23** 1050

Lichter A S 1990 private communication

Lichter A S, Sandler H M, Robertson J M and Lawrence T S 1993 The role of 3D treatment planning in radiation oncology *Three-Dimensional Treatment Planning (Proc. EAR Conf. (WHO, Geneva, 1992))* ed P Minet (Geneva: Minet) pp 1–6

Lichter A S, Sandler H M, Robertson J M, Lawrence T S, Ten Haken R K, McShan D L and Fraass B A 1992 Clinical experience with three-dimensional treatment planning *Semin. Radiat. Oncol.* **2** 257–66

Lind B K and Källman P 1990 Experimental verification of an algorithm for inverse radiation therapy planning *Radiother. Oncol.* **17** 359–68

Liu S, Lind B K and Brahme A 1993 Two accurate algorithms for calculating the energy fluence profile in inverse radiation therapy planning *Phys. Med. Biol.* **38** 1809–24

Lu H M, Kooy H M, Leber Z H and Ledoux R J 1994 Optimization of treatment planning for stereotactic radiosurgery *Med. Phys.* **21** 874

Lukas P 1996 Relation between target volume and surgical procedure *Proc. Symp. Principles and Practice of 3-D Radiation Treatment Planning (Munich, 1996)* (Munich: Klinikum rechts der Isar, Technische Universität)

Mackie T R 1995 Status of Monte Carlo dose planning (Proc. ESTRO Conf. (Gardone Riviera, 1995)) *Radiother. Oncol.* **37** Suppl. 1 S28

Mackie R, Deasy J, Holmes T and Fowler J 1994a Letter in response to 'optimisation of radiation therapy and the development of multileaf collimation' by Anders Brahme *Int. J. Radiat. Oncol. Biol. Phys.* **28** 784–5

Mackie T R, Holmes T W, Deasy J O and Reckwert P J 1994b New trends in treatment planning (Proc. World Congress on Medical Physics and Biomedical Engineering (Rio de Janeiro, 1994)) *Phys. Med. Biol.* **39A** Part 1 480

Mackie T R, Reckwerdt P J, Wells C M, Yang J N, Deasy J O, Podgorsak M, Holmes M A, Rogers D W O, Ding G X, Faddegon B A, Ma C, Bielajew A F and Cygler J 1994c The OMEGA project: comparison among EGS4 electron beam simulations, 3D Fermi-Eyges calculations and dose measurements *The Use of Computers in Radiation Therapy: Proc. 11th Conf.* ed A R Hounsell *et al* (Manchester: ICCR) pp 152–3

Mageras G S and Mohan R 1992 Application of fast simulated annealing to optimisation of conformal radiation treatments *Med. Phys.* **20** 639–47

Manfredotti C, Nastasi U and Zanini A 1995 Monte Carlo calculation of dose delivered to patients in radiotherapy treatments (Proc. ESTRO Conf. (Gardone Riviera, 1995)) *Radiother. Oncol.* **37** Suppl. 1 S29

Marks L B, Sherouse G W, Spencer D P, Bentel G, Clough R, Vann K, Jaszczak R, Coleman E, Anscher M A and Prosnitz L R 1994 The role of 3-dimensional functional lung imaging in treatment planning: the functional DVH (Proc. 36th ASTRO Meeting) *Int. J. Radiat. Oncol. Biol. Phys.* **30** Suppl. 1 197

Mayles W P M, Chow M, Dyer J, Fernandez E M, Heisig S, Knight R T, Moore I, Nahum A E, Shentall G S and Tait D M 1993 The Royal Marsden Hospital pelvic radiotherapy trial: technical aspects and quality assurance *Radiother. Oncol.* **29** 184–91

McShan D 1990 private communication

Melian E, Kutcher G, Leibel S, Zelefsky M, Baldwin B and Fuks Z 1993 Variation in prostate position: quantitation and implications for three-dimensional conformal radiation therapy (Proc. 35th ASTRO Meeting) *Int. J. Radiat. Oncol. Biol. Phys.* **27** Suppl. 1 137

Metropolis N, Rosenbluth A W, Rosenbluth M N, Teller A H, and Teller E 1953 Equations of state calculations by fast computing machines *J. Chem. Phys.* **21** 1087–92

Moerland M A, Beersma R, Bhagwandien R and Bakker C J G 1995 Analysis and correction of geometric distortions in 1.5 T magnetic resonance head images *Proc. 19th L H Gray Conf. (Newcastle, 1995)* (Newcastle: L H Gray Trust)

Mohan R 1990 Clinically relevant optimisation of 3D conformal treatments *The Use of Computers in Radiation Therapy: Proc. 10th Int. Conf. Use of Computers in Radiation Therapy* ed S Hukku and P S Iyer (Lucknow: ICCR) pp 36–9

—— 1994 Computer-aided optimisation of radiation treatment plans (Proc. World Congress on Medical Physics and Biomedical Engineering (Rio de Janeiro, 1994)) *Phys. Med. Biol.* **39A** Part 2 p 676

Mohan R and Ling C C 1995 When becometh less more? *Int. J. Radiat. Oncol. Biol. Phys.* **33** 235–7

Mohan R, Ling C C, Stein J and Wang X H 1996 The number of beams in intensity-modulated treatments: in response to Drs Söderström and Brahme *Int. J. Radiat. Oncol. Biol. Phys.* **34** 758–9

Mohan R, Mageras G S, Baldwin B, Brewster L J, Kutcher G J, Leibel S, Burman C M, Ling C C and Fuks Z 1992 Clinically relevant optimisation of 3D conformal treatments *Med. Phys.* **19** 933–44

Mohan R, Wang X, Jackson A, Bortfeld T, Boyer A L, Kutcher G J, Leibel S A, Fuks Z and Ling C C 1994 The potential and limitations of the inverse radiotherapy technique *Radiother. Oncol.* **32** 232–48

Mould R 1993 *A Century of X-rays and Radioactivity in Medicine with Emphasis on Photographic Records of the Early Years* (Bristol: Institute of Physics) pp 108–47

Murtagh B A and Saunders M A 1983 *MINOS 5.1 Users guide, Technical report SOL 83-20R* Systems Optimisation Laboratory, Stanford University, CA (revised 1987)

Nahum A E 1994 Conformal radiotherapy (Proc. World Congress on Medical Physics and Biomedical Engineering (Rio de Janeiro, 1994)) *Phys. Med. Biol.* **39A** Part 1 494

—— 1995 Clinical trials of conformal therapy—physics aspects (Proc. ESTRO Conf. (Gardone Riviera, 1995)) *Radiother. Oncol.* **37** Suppl. 1 S16

Neal A J, Sivewright G and Bentley R 1994a Evaluation of a region growing algorithm for segmenting pelvic CT images during radiotherapy planning *Br. J. Radiol.* **67** 392–5

—— 1994b Clinical evaluation of a computer segmentation algorithm for radiotherapy treatment planning *The Use of Computers in Radiation Therapy: Proc. 11th Conf.* ed A R Hounsell *et al* (Manchester: ICCR) pp 198–9

Newton C M 1970 What next in radiation treatment optimisation? Computers in radiotherapy (Proc. 3rd Int. Conf. on computers in radiotherapy (Glasgow, 1970)) *Br. J. Radiol.* Special Report **5** 83–9

Niemierko 1996a Optimisation of intensity modulated beams: local or global optimum *Med. Phys.* **23** 1072

—— 1996b Selection of objective functions for optimisation of intensity modulated beams *Med. Phys.* **23** 1172

Niemierko A and Goitein M 1991 Calculation of normal tissue complication probability and dose-volume histogram reduction schemes for tissues with a critical element architecture *Radiother. Oncol.* **20** 166–76

Nitz W R, Bradley W G Jr, Watanabe A S, Lee R R, Burgoyne B, O'Sullivan R M and Herbst M D 1992 Flow dynamics of cerebrospinal fluid: assessment with phase-contrast velocity MR imaging performed with retrospective cardiac gating *Radiology* **183** 395–405

Oldham M and Webb S 1993 Clinical implementation of inverse treatment planning *Br. J. Radiol.* **66** Congress Suppl. 162

—— 1994 Inverse planning and the optimisation of radiotherapy plans by simulated annealing incorporating dual weighting *The Use of Computers in Radiation Therapy: Proc. 11th Conf.* ed A R Hounsell *et al* (Manchester: ICCR) pp 60–1

—— 1995a The optimisation and inherent limitations of 3D conformal radiotherapy treatment plans of the prostate *Br. J. Radiol.* **68** 882–93

—— 1995b Optimisation by fast simulated annealing of 3D conformal radiotherapy treatment plans of the prostate and theoretical limits of improvement *Proc. Röntgen Centenary Congress 1995 (Birmingham, 1995)* (London: British Institute of Radiology) p 474

—— 1995c The optimisation and inherent limitations of 3D conformal radiotherapy of the prostate (Proc. ESTRO Conf. (Gardone Riviera, 1995)) *Radiother. Oncol.* **37** Suppl. 1 S14

Oldham M, Neal A and Webb S 1995a A comparison of conventional forward planning with inverse planning for 3D conformal radiotherapy of the prostate *Proc. Röntgen Centenary Congress 1995 (Birmingham, 1995)* (London: British Institute of Radiology) p 229

—— 1995b A comparison of forward planning and optimised inverse planning (Proc. ESTRO Conf. (Gardone Riviera, 1995)) *Radiother. Oncol.* **37** Suppl. 1 S5

—— 1995c Custom optimisation of wedge angles in prostate radiotherapy (Proc. ESTRO Conf. (Gardone Riviera, 1995)) *Radiother. Oncol.* **37** Suppl. 1 S60

—— 1995d A comparison of conventional forward planning with inverse planning for 3D conformal radiotherapy of the prostate *Radiother. Oncol.* **35** 248–62

—— 1995e The optimisation of wedge filters in radiotherapy of the prostate *Radiother. Oncol.* **37** 209–20

Paxman R G, Barrett H H, Smith W E and Milster T D 1985 Image reconstruction from coded data: 2: code design *J. Opt. Soc. Am.* A **2** 501–9

Pickett B, Roach M III, Horine P, Verhey L and Phillips T L 1994 Optimisation of the oblique angles in the treatment of prostate cancer during six-field conformal radiotherapy *Med. Dosimetry* **19** 237–54

Pickett B, Roach M, Rosenthal S, Horine P and Phillips T 1993 Defining 'ideal margins' for six field conformal irradiation of localized prostate cancer (Proc. 35th ASTRO Meeting) *Int. J. Radiat. Oncol. Biol. Phys.* **27** Suppl. 1 223–4

Poncelet B P, Wedeen V J, Weisskoff R M and Cohen M S 1992 Brain parenchyma motion: measurement with cine echo-planar MR imaging *Radiology* **185** 645–51

Pötter R, Wachter S, Sperveslage P, Prott F J and Dieckmann K 1996 The value of MRI assisted planning of radiotherapy to the posterior fossa in medulloblastoma *Proc. Symp. Principles and Practice of 3-D Radiation Treatment Planning (Munich, 1996)* (Munich: Klinikum rechts der Isar, Technische Universität)

Press W H, Flannery B P, Teukolsky S A and Vetterling W T 1986 *Numerical Recipes: The Art of Scientific Computing* (Cambridge: Cambridge University Press) pp 326–34

Press W H and Teukolsky S A 1991 Simulated annealing optimisation over continuous spaces *Comput. Phys.* **5** 426–9

Pross J 1993 Tool for manual segmentation of volumes in multi modality 3D imaging *Three-Dimensional Treatment Planning (Proc. EAR Conf. (WHO, Geneva, 1992))* ed P Minet (Geneva: Minet) pp 245–51

—— 1995 3D imaging bestimmung von zielvolumen und riskoorganen *Dreidimensionale Strahlentherapieplanung* ed W Schlegel, T Bortfeld and J Stein (Heidelberg: DKFZ) pp 23–30

Pross J, Bendl R and Schlegel W 1994 TOMAS, A tool for manual segmentation based on multiple image data sets *The Use of Computers in Radiation Therapy: Proc. 11th Conf.* ed A R Hounsell *et al* (Manchester: ICCR) pp 192–3

Prott F J, Haverkamp U, Willich N, Resch A, Stöber U and Pötter R 1995 Comparison of imaging accuracy at different MRI units based on phantom measurements *Radiother. Oncol.* **37** 221–4

Purdy J A 1994a Consistent volume and dose specification: a prerequisite for 3D conformal radiation therapy quality assurance (Proc. World Congress on Medical Physics and Biomedical Engineering (Rio de Janeiro, 1994)) *Phys. Med. Biol.* **39A** Part 1 495

—— 1994b ICRU report on prescribing, recording, and reporting photon beam therapy (Proc. 36th ASTRO Meeting) *Int. J. Radiat. Oncol. Biol. Phys.* **30** Suppl. 1 117

Radcliffe N and Wilson G 1990 Natural solutions give their best *New Scientist* April pp 47–50

Rampling R, McNee S G, Khalifa A and Hadlee D 1993 Integrated use of CT and MRI for 3D planning of radiotherapy for brain tumours *Proc. IPSM Meeting 3D Treatment Planning and Conformal Therapy* (London: British Institute of Radiology) p 13

Reinstein L, Wang X, Burman C, Chen Z, Leibel S, Fuks Z and Mohan R 1994 3D treatment plan optimisation for cancer of the prostate using intensity modulated beams *Med. Phys.* **21** 927

Roach M, Pickett B, Rosenthal S A, Verhey L and Phillips T L 1994 Defining treatment margins for six field conformal irradiation of localized prostate cancer *Int. J. Radiat. Oncol. Biol. Phys.* **28** 267–75

Robison R F 1994 The race for megavoltage therapy machines: 1911 to 1951 (Proc. 36th ASTRO Meeting) *Int. J. Radiat. Oncol. Biol. Phys.* **30** Suppl. 1 99

Rosen I I 1994 Use of simulated annealing for conformal therapy plan optimization *Med. Phys.* **21** 1005

Rosen I I, Lam K S, Lane R G, Langer M and Morrill S M 1995 Comparison of simulated annealing algorithms for conformal therapy treatment planning *Int. J. Radiat. Oncol. Biol. Phys.* **33** 1091–9

Rosen I I, Morrill S M and Lane R G 1992 Optimised dynamic rotation with wedges *Med. Phys.* **19** 971–7

Rosenman J, Soltys M, Cullip T and Chen J 1994 Improving treatment planning accuracy through multiple modality imaging (Proc. 36th ASTRO Meeting) *Int. J. Radiat. Oncol. Biol. Phys.* **30** Suppl. 1 241

Rosenwald J C, Belshi R, Gaboriaud G, Pontvert D, Mazal A, Ferrand R and Drouard J 1995 Useful software tools for conformal radiotherapy *Medizinische Physik 95 Röntgen Gedächtnis-Kongress* ed J Richter (Würzburg: Kongress) pp 146–7

Sailer S L, Chaney E L, Rosenman J G, Sherouse G W and Tepper J E 1992 Treatment planning at the University of North Carolina at Chapel Hill *Semin. Radiat. Oncol.* **2** 267–73

Sailer S L, Rosenman J G, Symon J R, Cullip T J and Chaney E L 1993 The tetrad and hexad: maximum beam separation as a starting point for non-coplanar 3D treatment planning (Proc. 35th ASTRO Meeting) *Int. J. Radiat. Oncol. Biol. Phys.* **27** Suppl. 1 138

—— 1994 The tetrad and hexad: maximum beam separation as a starting point for non-coplanar 3D treatment planning: prostate cancer as a test case *Int. J. Radiat. Oncol. Biol. Phys.* **30** 439–46

Sailer S L and Tepper J E 1995 Clinical implications of new innovations in radiation treatment delivery (*Innovations in treatment delivery*) *Semin. Radiat. Oncol.* **5** 166–71

Sandham W A, Yuan Y and Durrani T S 1995 Conformal therapy using maximum entropy optimisation *Int. J. Imaging Syst. Technol.* **6** 80–90 (special issue on 'Optimisation of the three-dimensional dose delivery and tomotherapy')

Sandler H, McLaughlin P W, Ten Haken R, Addison H, Forman J and Lichter A 1993 3D conformal radiotherapy for the treatment of prostate cancer: low risk of chronic rectal morbidity observed in a large series of patients (Proc. 35th ASTRO Meeting) *Int. J. Radiat. Oncol. Biol. Phys.* **27** Suppl. 1 135

Sandler H M, Radany E H, Greenberg H S, Junck L and Lichter A S 1994 Dose escalation using 3D conformal radiotherapy for high grade astrocytomas (Proc. 36th ASTRO Meeting) *Int. J. Radiat. Oncol. Biol. Phys.* **30** Suppl. 1 214

Sauer O A and Linton N 1995 The transition from 2D to 3D treatment planning *Activity: Int. Nucletron-Oldelft Radiother. J.* (Special report No 6 Treatment planning, external beam, stereotactic radiosurgery, brachytherapy and hyperthermia 3–11)

Schlegel W 1993 Impact of 3D treatment planning on treatment techniques *Three-Dimensional Treatment Planning (Proc. EAR Conf. (WHO, Geneva, 1992))* ed P Minet (Geneva: Minet) pp 131–42

Schmidt B F, Carduck H P, Janas R, Kraft R and Gustorf-Aeckerle R 1996 Individually angled sectional NMR imaging for 3D radiotherapy planning in brain tumours *Proc. Symp. Principles and Practice of 3-D Radiation Treatment Planning (Munich, 1996)* (Munich: Klinikum rechts der Isar, Technische Universität)

Schultheiss T E, Hanks G F, Hunt M A, Epstein B and Peter R 1993 Factors influencing incidence of acute grade 2 morbidity in conformal and standard radiation treatment of prostate cancer: univariate and multivariate analysis (Proc. 35th ASTRO Meeting) *Int. J. Radiat. Oncol. Biol. Phys.* **27** Suppl. 1 134

Schweikard A, Rodduluri M, Tombropoulos R and Adler J R 1996 Planning, calibration and collision-avoidance for image-guided radiosurgery *Proc. Symp. Principles and Practice of 3-D Radiation Treatment Planning (Munich, 1996)* (Munich: Klinikum rechts der Isar, Technische Universität)

Sherouse G W 1993 A mathematical basis for selection of wedge angle and orientation *Med. Phys.* **20** 1211–8

—— 1994a Dose homogenisation using vector analysis of dose gradients *The Use of Computers in Radiation Therapy: Proc. 11th Conf.* ed A R Hounsell *et al* (Manchester: ICCR) pp 248–9

—— 1994b Is the inverse problem a red herring? *The Use of Computers in Radiation Therapy: Proc. 11th Conf.* ed A R Hounsell *et al* (Manchester: ICCR) pp 66–7

—— 1994c A simple method for achieving uniform dose from arbitrary arrangements of possibly non-coplanar fixed fields (Proc. 36th ASTRO Meeting) *Int. J. Radiat. Oncol. Biol. Phys.* **30** Suppl. 1 240

—— 1995 When does treatment plan optimisation require inverse planning? (Proc. ESTRO Conf. (Gardone Riviera, 1995)) *Radiother. Oncol.* **37** Suppl. 1 S4

Silverman A and Adler J 1992 Animated simulated annealing *Comput. Phys.* **6** 277–81

Smit S 1993 Optimisation in radiotherapy treatment planning *MSc Thesis* Delft University of Technology, Department of Applied Mathematics and Computer Science

Smith W E, Paxman R G and Barrett H H 1985 Image reconstruction from coded data: 1: reconstruction algorithms and experimental results *J. Opt. Soc. Am.* A **2** 491–500

Söderström S and Brahme A 1992 Selection of suitable beam orientations in radiation therapy using entropy and Fourier transform measures *Phys. Med. Biol.* **37** 911–24

—— 1993 Optimisation of the dose delivery in a few field techniques using radiobiological objective functions *Med. Phys.* **20** 1201–9

—— 1995 Which is the most suitable number of photon beam portals in coplanar radiation therapy? *Int. J. Radiat. Oncol. Biol. Phys.* **33** 151–9

—— 1996 Small is beautiful—and often enough: in response to Drs Mohan and Ling *Int. J. Radiat. Oncol. Biol. Phys.* **33** 235–7 *Int. J. Radiat. Oncol. Biol. Phys.* **34** 757–9

Söderström S, Gustafsson A and Brahme A 1995 Few-field radiation therapy optimisation in the phase space of complication-free tumour control *Int. J. Imaging Syst. Technol.* **6** 91–103 (special issue on 'Optimisation of the three-dimensional dose delivery and tomotherapy')

Sonntag A 1975a Ein Weg zur räumlichen Bestrahlungsplanung: 1. Das Problem der Dosishomogenität innerhalb des Überschneidungbereichs nichtkomplanar angeordneter Strahlenbündel, Basisuntersuchungen zu einem Optimierungsverfahren *Strahlentherapie* **150** 507–20

—— 1975b Ein Weg zur räumlichen Bestrahlungsplanung: 2. Vektorielle Darstellung der räumlichen Dosisverteilung im einzelnen Strahlenbündel und erst Folgerungen für die Praxis *Strahlentherapie* **150** 569–78

—— 1976a Ein Weg zur räumlichen Bestrahlungsplanung: 3. Näheres über das Verfahren zur Erzielung von Dosishomogenität bei nichtkomplanaren Strahlenbündelanordnungen und dessen experimentelle Überprüfung *Strahlentherapie* **151** 10–25

—— 1976b Ein Weg zur räumlichen Bestrahlungsplanung: 4. Beispiele aus der strahlentherapeutischen Praxis und Schlussfolgerungen *Strahlentherapie* **151** 26–46

Stein J, Mohan R, Wang X H, Bortfeld T, Wu Q, Preiser K, Schlegel W and Ling C C 1996 Optimum number and orientations of beams for intensity-modulated treatments *Med. Phys.* **23** 1063

Suit H and du Bois W 1991 The importance of optimal treatment planning in radiation therapy *Int. J. Radiat. Oncol. Biol. Phys.* **21** 1471–8

Surridge M and Scielzo G 1995 EUROPORT RAPT ESPRIT European porting project: Focus on radiotherapy (St Augusta: GMD)

Szu H 1987 Fast simulated annealing *AIP Conf. Proc. Neural Networks for Computing (Snowbird, UT, 1986) (AIP Conf. Series 151)* (New York: AIP) pp 420–5

Szu H and Hartley R 1987 Fast simulated annealing *Phys. Lett.* **122A** 157–62

Tait D M, Nahum A E, Meyer L, Law M, Dearnaley D P, Horwich A, Mayles W P and Yarnold J R 1997 A randomized trial on the effect of reducing the volume of irradiated normal tissue on acute toxicity in pelvic radiotherapy *Radiother. Oncol.* at press

Tait D M, Nahum A E, Rigby L, Chow M, Mayles W P M, Dearnaley D P and Horwich A 1993 Conformal radiotherapy of the pelvis: assessment of acute toxicity *Radiother. Oncol.* **29** 117–26

Ten Haken R K, Balter J M, Lam K L and Robertson J M 1994 Effects of patient breathing on CT-based 3D planning for lung irradiation *Med. Phys.* **21** 914

Thames H D, Schultheiss T E, Hendry J H, Tucker S L, Dubray B M and Brock W A 1992 Can modest escalations of dose be detected as increased tumour control *Int. J. Radiat. Oncol. Biol. Phys.* **22** 241–6

Thornton A F, Ten Haken R K, Gerhardsson A and Correll M 1991 Three dimensional motion analysis of an improved head immobilisation system for simulation, CT, MRI and PET imaging *Radiother. Oncol.* **20** 224–8

Tsien K C 1955 The application of automatic computing machines to radiation therapy planning *Br. J. Radiol.* **28** 432–9

Turner S L, Swindell R, Bowl N, Read G and Cowan R A 1994 Bladder movement during radiation therapy for bladder cancer: implications for treatment planning (Proc. 36th ASTRO Meeting) *Int. J. Radiat. Oncol. Biol. Phys.* **30** Suppl. 1 199

Vance R, Sandham W A and Durrani T S 1994 Optimisation of beam profiles in conformal therapy using genetic algorithms (Proc. World Congress on Medical Physics and Biomedical Engineering (Rio de Janeiro, 1994)) *Phys. Med. Biol.* **39A** Part 1 518

Van de Geijn J 1965 The computation of two and three dimensional dose distributions in cobalt-60 teletherapy *Br. J. Radiol.* **38** 369–7

Van Kampen M, Levegrün S, Waschek T, Engenhart-Cabillic R and Schlegel W 1996 Quantification of subjective judgement of imaging diagnostics and its impact on target volume definition *Proc. Symp. Principles and Practice of 3-D Radiation Treatment Planning (Munich, 1996)* (Munich: Klinikum rechts der Isar, Technische Universität)

Verhey L, Goitein M, McNulty P, Munzenrider J E and Suit H 1982 Precise positioning of patients for radiation therapy *Int. J. Radiat. Oncol. Biol. Phys.* **8** 289–94

Wachter S, Dieckmann K and Pötter R 1996 Conformal radiotherapy of localized prostate cancer *Proc. Symp. Principles and Practice of 3-D Radiation Treatment Planning (Munich, 1996)* (Munich: Klinikum rechts der Isar, Technische Universität)

Walstam R 1995 Medical radiation physics in Europe: a historical review (Proc. ESTRO Conf. (Gardone Riviera, 1995)) *Radiother. Oncol.* **37** Suppl. 1 S15

Warmelink C, Sharma R and Forman J 1994 Three-dimensional treatment planning in a dose escalation study for locally advanced prostate cancer (Proc. World Congress on Medical Physics and Biomedical Engineering (Rio de Janeiro, 1994)) *Phys. Med. Biol.* **39A** Part 1 497

Wang X H, Jackson A, Phillips T and Mohan R 1994 Optimisation of intensity distributions for conformal treatment for prostate, brain and lung *Med. Phys.* **21** 874

Webb S 1988 *The Physics of Medical Imaging* (Bristol: Institute of Physics)

—— 1989a SPECT reconstruction by simulated annealing *Phys. Med. Biol.* **34** 259–81

—— 1989b Optimisation of conformal radiotherapy dose distributions by simulated annealing *Phys. Med. Biol.* **34** 1349–69

—— 1990a Inverse tomograph *Nature* **344** 284

—— 1990b A new dose optimising technique using simulated annealing *Proc. 9th Annual Meeting of ESTRO (Montecatini, 1990)* p 263

—— 1991a Optimisation by simulated annealing of three-dimensional conformal treatment planning for radiation fields defined by a multileaf collimator *Phys. Med. Biol.* **36** 1201–26

—— 1991b Optimisation of conformal radiotherapy dose distributions by simulated annealing 2: inclusion of scatter in the 2D technique *Phys. Med. Biol.* **36** 1227–37

—— 1992a Optimisation by simulated annealing of three-dimensional conformal treatment planning for radiation fields defined by a multileaf collimator: 2. Inclusion of two-dimensional modulation of the X-ray intensity *Phys. Med. Biol.* **37** 1689–704

—— 1992b Optimised three dimensional treatment planning for volumes with concave outlines, using a multileaf collimator (Proc. ART91

(Munich, 1991)) (abstract book p 66) *Advanced Radiation Therapy: Tumour Response Monitoring and Treatment Planning* ed A Breit (Berlin: Springer) pp 495–502

—— 1992c Optimising dose with a multileaf collimator for conformal radiotherapy *Proc. 50th Ann. Congress of the British Institute of Radiology (Birmingham, 1992)* (London: British Institute of Radiology) p 15

—— 1993a Techniques for optimisation of dose with a multileaf collimator for conformal radiotherapy of target volumes with concave outlines *Three-Dimensional Treatment Planning (Proc. EAR Conf. (WHO, Geneva, 1992))* ed P Minet (Geneva: Minet) pp 163–72

—— 1993b Beam geometry and beam shaping *Three-Dimensional Treatment Planning (Proc. EAR Conf. (WHO, Geneva, 1992))* ed P Minet (Geneva: Minet) pp 75–88

—— 1993c *The Physics of Three-dimensional Radiation Therapy: Conformal Radiotherapy, Radiosurgery and Treatment Planning* (Bristol: Institute of Physics)

—— 1994a Tomotherapy and beamweight stratification *The Use of Computers in Radiation Therapy: Proc. 11th Conf.* ed A R Hounsell *et al* (Manchester: ICCR) pp 58–9

—— 1994b Optimising the planning of intensity-modulated radiotherapy *Phys. Med. Biol.* **39** 2229–46

—— 1995a A note on the problem of isotropically orienting N vectors in space with application to radiotherapy planning *Phys. Med. Biol.* **40** 945–54

—— 1995b Optimising radiation therapy inverse treatment planning using the simulated annealing technique *Dreidimensionale Strahlentherapieplanung* ed W Schlegel, T Bortfeld and J Stein (Heidelberg: DKFZ) pp 137–56

—— 1995c Optimising radiation therapy inverse treatment planning using the simulated annealing technique *Int. J. Imaging Syst. Technol.* **6** 71–9 (special issue on 'Optimisation of the three-dimensional dose delivery and tomotherapy')

—— 1995d Optimising intensity modulated therapy planning (Proc. ESTRO Conf. (Gardone Riviera, 1995)) *Radiother. Oncol.* **37** Suppl. 1 S3

—— 1996 Inverse planning for intensity-modulated radiation therapy: the role of simulated annealing *Proc. Intensity-Modulated Radiation Therapy (ICRT) Workshop, sponsored by the NOMOS Corporation (Durango, CO, 1996)* (Sewickley, PA: NOMOS) pp 6–8

Webb S and Oldham M 1993 Optimisation of conformal 3D dose distributions created by a multileaf collimator; options including modelling biological response *Proc. Meeting on 3D Treatment Planning and Conformal Radiotherapy (British Institute of Radiology, London, March 1993)* (London: British Institute of Radiology)

Willie L T 1986 Searching potential energy surfaces by simulated annealing *Nature* **324** 46–8

Wong W L, Hawkes D and Maisey M N 1994 Computerised combination and display of PET, MR and CT images *Rad. Mag.* **20** (228) 21–3

Yuan Y, Sandham W A and Durrani T S 1994a Towards beam-based dose optimisation for conformation RTP *The Use of Computers in Radiation Therapy: Proc. 11th Conf.* ed A R Hounsell *et al* (Manchester: ICCR) pp 250–1

—— 1994b Optimisation of beam offset for conformal therapy inverse planning (Proc. World Congress on Medical Physics and Biomedical Engineering (Rio de Janeiro, 1994)) *Phys. Med. Biol.* **39A** Part 1 p 497

Yuan Y, Sandham W A, Durrani T S, Mills J A and Deehan C 1994c Application of Bayesian and Maximum Entropy optimisation to conformation radiotherapy treatment planning *Appl. Signal Processing* **1** 20–34

Zimmermann F B, Kaisig D, Wehrmann R and Kneschaurek P 1996 The influence of organ movements on calculated NTCP in the radiotherapy of prostate cancer *Proc. Symp. Principles and Practice of 3-D Radiation Treatment Planning (Munich, 1996)* (Munich: Klinikum rechts der Isar, Technische Universität)

CHAPTER 2

METHODS TO CREATE INTENSITY-MODULATED BEAMS (IMBS)

In the companion Volume (Chapter 2) and in the first chapter of this Volume it has been seen that, by combining intensity-modulated radiation fields, a conformal distribution of dose can be obtained even for PTVs which do not have a convex surface outline. Such volumes are sometimes described as being invaginated or having a bifurcated shape. To obtain a distribution which is conformal in a 2D slice, a set of 1D beams are combined, each collimated to a slit defining the slice, all lying in that slice and each with a potentially different intensity modulation.

One way of obtaining a 3D conformal distribution of dose is to repeat this for each slice spanning the target volume. This is described as a 'coplanar' treatment because all the slices are parallel and the gantry rotates about a single axis normal to the slices. (There are other, specifically 'non-coplanar' treatment techniques in which the intensity-modulated 2D beams cannot be represented as a set of parallel 1D IMBs.)

Provided we are considering the former (coplanar) category, the 3D planning problem can be regarded as a set of formally equivalent 2D planning problems whose solution gives the set of 1D intensity-modulated profiles for that slice. So here we restrict ourselves to this problem and specifically consider how to *deliver* such 1D fields in practice. There have been six practical proposals which define classes of IMB delivery technique. Within these classes there are some subdivisions. These are summarised in table 2.1. The field of radiotherapy with IMBs has been reviewed by Mackie *et al* (1994b). It is in the field of intensity-modulated therapy that the biggest potential for a major improvement in treatment outcome lies. However, sadly, this development calls for significant changes in treatment technology and the solution of a number of attendant problems which are either new to this form of therapy or are worse (e.g. coping with tissue movement) than for therapy without intensity modulation. Most techniques

96

require significant developments in the electrotechnical industries. Purdy (1996) has emphasised that the use of IMBs is a major paradigm shift for clinical planning which will only come to fruition after four 'phases': resistance, confusion, integration and commitment.

2.1.   CLASS 1: 2D COMPENSATORS

A 2D intensity modulation can be achieved by placing a compensator in the shadow tray of the accelerator. A compensator is so called because the first devices so constructed were made to compensate for 'missing tissue', due to changes in anatomical outline of the patient and internal tissue inhomogeneities. In the present context this would be regarded as a special case of creating a modulation of intensity. Methods to construct compensators have been discussed by Djordjevich *et al* (1990) and Mageras *et al* (1991).

Compensators are no more than blocks of metal in which the local thickness varies with position to achieve differential attenuation of the beam. They were the only means to achieve this before there were computer-controlled linac jaws, multileaf collimators (MLC) or special-purpose devices of the kind described in section 2.4. Their principal disadvantage is the need to fabricate compensators for each field, a time-consuming process requiring special machinery. As computer-controlled linac collimation becomes more common, we may expect to see the gradual demise of individually fabricated compensators.

Two issues of importance in the design of compensators are (i) the minimum transmissiuon which determines the maximum thickness—thick compensators can be difficult to make—and (ii) the lateral spatial resolution—if this can be quite coarse then thick drills can be used to speed up the fabrication process. Bortfeld *et al* (1996) have developed a method of casting in which a material called MCP96 is poured into a mould made of a material called MDF. The mould material MDF is joined to metal plates of good heat conductivity so that cavity formation is avoided and the surface of the resulting compensator stays faithful to the design.

Another concept is the multileaf compensator (see Chapter 3) in which several planes of thin aluminium leaves are separately set to different fieldshapes. These planes lie one above the other in the accelerator blocking tray creating a 2D variable thickness map. This technique can only work for a limited small amount of intensity modulation of the type needed in breast compensation.

As well as compensators, one could think of simple blocking as providing the simplest method of creating an intensity modulation, i.e. that in which part of the field receives virtually zero primary radiation. For different gantry orientations the blocking must correspond with the projection of the OARs

Table 2.1. *Methods to create a 1D modulation of fluence (intensity) across a radiation field.*

(By combining a set of such 1D fields at different gantry orientations a 2D conformal distribution of dose in the common plane of the 1D fields can result. To obtain a 3D conformal distribution of dose the set must be different from plane to plane.) All methods allow local as well as global extrema in the IMB profile.

Method[f]	Advantages	Disadvantages	Requirements	Relative efficiency per slice[a]
Class 1				
1. Individual 2D compensators	All planes treated simultaneously; no electromechanical movements; probably only method to have been routinely used	Time consuming	Compensator cutting machine; software to design compensator	1
Class 2				
2. Dynamic collimation[b] (Convery and Rosenbloom, Svensson *et al*, Stein *et al*, Spirou and Chui, Yu *et al*)	X-rays on all the time so stable; unidirectional jaw/leaf motion; jaw/leaf in continuous movement; no beam-off periods	Must consider penumbra, scatter, leakage; no unique solution for jaw/leaf velocities—must solve optimisation problem subject to constraints on velocity and minimising total irradiation time	Jaw/leaf must be computer controlled; must cross central axis	$< 1^c$
3. Intensity-modulated-arc-therapy (IMAT)	Shorter treatment times than dynamic leaf movement; wide field (static for each arc) is less error prone	Few but requires multiple gantry rotations	Needs an 'interpreter'	$< 1^d$
4. Scanning photon beam	When combined with method 2 can give shorter treatment times	High width of scanning beam	Only available on Racetrack Microtron	< 1
Class 3				
5. MLC 'leaf-sweep' (Bortfeld *et al*)	Intuitive relation of leaf positions to intensity profile; coinciding beam-off times allow all slices to be modulated simultaneously	X-rays may not be stable switched on and off repeatedly; beam-off time reduces efficiency	Set of irradiations with static MLC leaves	$\leqslant 1^d$
6. MLC 'close-in' (Bortfeld *et al*)		as (3) 'leaf-sweep'		$\leqslant 1^d$
Class 4				
7. Tomotherapy (Carol *et al*)	Add-on linac attachment; prototype exists; reliable electropneumatic control; gantry continuously rotates	Resolution limited by 20 stubby vanes	Requires special software to link planning and setting the vanes	1

Table 2.1. *(Continued)*

Method[f]	Advantages	Disadvantages	Requirements	Relative efficiency per slice[a]
8. Tomotherapy (Mackie *et al*)	Treatment analogue of spiral CT so in principle no local hotspots; gantry continuously rotates	Not yet built; complex treatment planning	Special purpose device—does not exist; principle patented including concept of 2 'layers' of vanes	1
Class 5 9. Scanning attenuating bar	Simple to construct	Each transaxial plane must receive the same modulation; some IMB patterns cannot be generated	Simple engineering	< 1
10. Scanning narrow slit at variable speed	Easy to arrange	Incredibly inefficient—no one would do this!	Moving jaws	≪1

[a]Methods 1,2,3,4,5,6 can be used to treat all slices simultaneously whereas methods 7,8 treat just one or two slices at a time. Thus the true efficiency may end up worse eventually for methods 7,8.
[b]Different 1D profiles at each gantry orientation required for different tomographic slices demand the use of an MLC. If variations in fluence perpendicular to the tomographic slices are ignored the modulation can be achieved by jaw movement only.
[c]Depends on maximum leaf speed and complexity of profile.
[d]The efficiency must account also for the existence of beam-off times whilst the leaves move.
[e]The dynamic wedge is a special case with just one jaw moving.
[f]Main scientific developers mentioned; refer to text for detailed references.

to be shielded. Consequently in dynamic therapy there is a need for a continuously varying shield. Fiorino *et al* (1992, 1993) have presented an elegant extension of the synchronous shielding technique first proposed by Proimos (see the companion Volume, Chapter 2). The spinal cord is shielded by a hinged two-part block in which the hinge angle varies as the gantry rotates. The synchronous shielding is under motor control (figure 2.1).

2.2. CLASS 2: DYNAMIC-LEAF COLLIMATION

Dynamic collimation is the method by which the two corresponding leaves of a MLC, defining a slice, move unidirectionally, each with a different velocity as a function of time (figure 2.2) (Bortfeld 1995). The method has also been called the 'camera-shutter technique', 'leaf-chasing' and 'sliding window'. In recent years it has received great theoretical attention and some preliminary practical irradiations have been made. Hence we review this topic in some depth.

(If variations in anatomy normal to the slice are ignored, the same philosophy applies to a pair of linac jaws. The dynamic wedge (Leavitt *et al*

Figure 2.1.   *The hinged-block synchronous-shielding apparatus of Fiorino. The blocks shield the spinal cord as the gantry rotates, continuously adjusting the angle of one block with respect to the other to take care of the changing projection of the cord. (From Fiorino et al 1992.)*

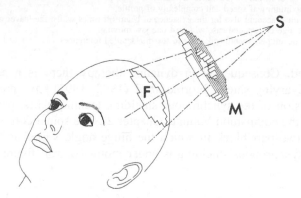

Figure 2.2.   *X-rays emitted from a source (S) pass through an MLC (M) which creates a geometrically shaped field (F) matched to the view of the tumour from this source position. The new technique is to vary the* intensity *of the rays across the field by* moving *the leaves in the direction of their length at variable speed. In practice many fields are combined (only one is shown). The figure is not to scale.*

1990) is a special case of dynamic collimation with only one jaw moving.) For either jaws or leaves the motion is continuous with the radiation on. The leaves must be able to cross the central axis ('over-running') and be under computer control (figure 2.3). Bortfeld (1995) reviewed the hardware

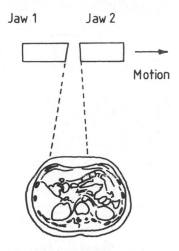

Figure 2.3. *The configuration for dynamic therapy. Two jaws (or pairs of corresponding MLC leaves) move in a direction parallel to the slice to be treated and with a differential velocity profile. The beam is on all the time.*

and software requirements to achieve dynamic therapy. All leaves must be motor driven, capable of greater than 2 cm s^{-1} leaf-speed and have a method to accurately monitor the leaf positions. There is no unique solution for the two velocity profiles which lead to a given intensity modulation. Instead the problem is posed as an optimisation problem: determine the velocity profiles subject to the constraints of the maximum practical leaf velocity and aim to minimise the treatment time.

To introduce the problems let us consider just one leaf-pair creating a single 1D IMB. The direction of movement of the leaf-pair corresponds to the definition of a particular slice of the patient. Everything that follows may be applied separately to each of the many leaf-pairs giving a 2D IMB, although some problems related to coupling will be highlighted.

2.2.1. The solution of Convery and Rosenbloom

The problem of determining the leaf velocity profiles has been solved by the Simplex method by Convery and Rosenbloom (1992). The output is in the form of a diagram showing the leaf position $x_i(t)$, for each leaf ($i = 1$ (trailing leaf), 2 (leading leaf)), as a function of cumulative monitor units (MUs) (representing cumulative time t). Cumulative time or cumulative MUs is an important concept to grasp. It is the total elapsed time or number of MUs since the start of irradiation. Conversely the diagram shows the cumulative time or MUs $t(x_i)$ at which the leaves are at each location. By definition, the horizontal separation on such diagrams is the intensity profile $I(x)$ making up the IMB as a function of position (figure 2.4), i.e.

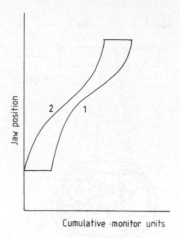

Figure 2.4. *The jaw or leaf positions as a function of cumulative MUs for two jaws or leaves (the trailing jaw or leaf is number 1) in dynamic therapy. The horizontal axis is a measure of cumulative MUs representing time t. The vertical axis is a measure of position x. At any position x, the horizontal width between the two curves gives the intensity of the IMB in MU.*

considering the rate of output of the accelerator to be constant, and with time and intensity both measured in MUs,

$$t_1(x) - t_2(x) = I(x). \tag{2.1}$$

With x representing leaf position and with t now representing only cumulative time, and $I(x)$ representing the intensity profile in MUs as a function of distance, then, since

$$\frac{dx}{dt} = \frac{dI}{dt} \bigg/ \frac{dI}{dx} \tag{2.2}$$

and when dI/dt is a constant rate of output of the accelerator, it follows that the leaf velocity dx/dt is *slowest* where the rate of change of the intensity profile dI/dx is *largest* and vice versa, a result which at first sight appears counter-intuitive, but is nevertheless true. It is more difficult to create slowly varying intensity profiles than quickly varying ones because the leaves have to move fast and may be required to move faster than the maximum allowed speed.

Convery and Rosenbloom (1992) found that, as the limiting maximum leaf velocity increased, the overall total treatment times decreased but not in inverse proportion. The algorithm is *not* equivalent to scanning a narrow slit at variable velocity (which would be very inefficient) but gains its efficiency because the aperture is both variable and large. By calculating the ratio of the total cumulative MUs required to deliver a single profile to the maximum in the modulated profile (a figure always greater than unity but decreasing with increasing limiting maximum leaf velocity), they showed that the efficiency of the dynamic collimation technique was always lower than that of a static collimator but could be typically only 50% worse (depending on the shape of the required modulation). Because there are no beam-off periods this efficiency is not further reduced (compare with the multiple-segment (static) field technique; section 2.3).

The optimisation algorithm has to be adjusted slightly when it is desired to use this method for sets of leaf-pairs to deliver different modulations to different slices at the same gantry angle, since in general the cumulative irradiation time will be different for each 1D profile. This is achieved by demanding that the irradiation finishes with the leaf-pairs closed for all except the profile with the longest irradiation time. A small disadvantage is that this means all the profiles, except that taking the longest, take a little longer than they would if the unadjusted algorithm were used, but the increased inefficiency was shown by Convery and Rosenbloom (1992) to be $\leqslant 4\%$ (depending on limiting leaf velocity). This is one of the principal advantages of dynamic therapy, that the 1D profiles for all slices can be delivered simultaneously even though the cumulative time will vary between them. In practice this is achieved using a MLC.

Convery and Rosenbloom (1995) studied, for one plan, the effects of three types of potential inaccuracies in delivering IMBs, namely: (i) gantry angle uncertainty; (ii) patient position uncertainty; (iii) IMB element inaccuracy. A detailed analysis of the effects of changes in two of seven beams was presented on the resulting total dose distribution using dose statistics and TCP, NTCP calculations. The conclusion is that expected practical uncertainties hardly affect the outcome.

Following the work of Convery and Rosenbloom (1992) a number of groups independently provided new solutions to the optimisation process with rapid computational times. The features of dose delivery were studied in detail and three groups independently discovered and published the same algorithm for minimising the treatment time. We now consider each of these in turn.

2.2.2. The solution of Svensson et al

From equation (2.1)

$$\frac{\mathrm{d}I}{\mathrm{d}x} = \frac{\mathrm{d}t_1(x)}{\mathrm{d}x} - \frac{\mathrm{d}t_2(x)}{\mathrm{d}x}. \tag{2.3}$$

If $v_i(x)$ represents the velocity of the ith leaf at location x

$$\frac{\mathrm{d}I}{\mathrm{d}x} = \frac{1}{v_1(x)} - \frac{1}{v_2(x)}. \tag{2.4}$$

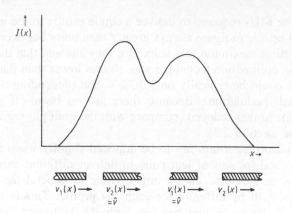

Figure 2.5. *When the gradient dI/dx of the intensity profile $I(x)$ is positive, the leading leaf 2 should move at the maximum velocity $\hat{v}$; conversely when the gradient dI/dx of the intensity profile $I(x)$ is negative, the trailing leaf 1 should move at the maximum velocity $\hat{v}$. Equations (2.5) and (2.6) are illustrated by showing a schematic of the pair of leaves in two separate locations delivering the IMB profile shown in the upper part of the figure.*

Svensson *et al* (1994a,b) have considered the implication of this equation. Suppose that at the end, x_{end}, of delivering the 1D profile the two leaves come together to close (as required if many 1D profiles are to be delivered by multiple leaf-pairs), the aim is then to minimise the cumulative time $t_2(x_{end})$. When the gradient dI/dx of the intensity profile is positive, the optimal solution is always to move the leading leaf ($i = 2$) at maximum speed $\hat{v}$ to make the negative part of equation (2.4) minimal and open up the field as fast as possible, modulating the field with the trailing leaf ($i = 1$), and thus satisfying equation (2.4). Conversely, if the gradient dI/dx of the intensity profile is negative, the optimal solution is always to move the trailing leaf ($i = 1$) at maximum speed $\hat{v}$ to make the positive part of equation (2.4) minimal, and modulate the field with the leading leaf ($i = 2$), satisfying equation (2.4), i.e.

$$v_2(x) = \hat{v}$$
$$v_1(x) = \frac{\hat{v}}{1 + \hat{v}(dI/dx)} \qquad \text{when } \frac{dI}{dx} \geqslant 0 \qquad (2.5)$$

and

$$v_1(x) = \hat{v}$$
$$v_2(x) = \frac{\hat{v}}{1 - \hat{v}(dI/dx)} \qquad \text{when } \frac{dI}{dx} < 0. \qquad (2.6)$$

The formal proof is given in section 2.2.4 and the principle is illustrated in figure 2.5.

Since acceleration a is given by

$$a = \frac{\mathrm{d}v}{\mathrm{d}t} = \frac{\mathrm{d}x}{\mathrm{d}t}\frac{\mathrm{d}v}{\mathrm{d}x} = v\frac{\mathrm{d}v}{\mathrm{d}x} \tag{2.7}$$

by differentiating equation (2.4)

$$\frac{\mathrm{d}^2 I}{\mathrm{d}x^2} = \frac{a_2(x)}{v_2^3(x)} - \frac{a_1(x)}{v_1^3(x)}. \tag{2.8}$$

For example, when the leaves are modulating a positive slope $\mathrm{d}I/\mathrm{d}x$, the leading leaf moves at maximum speed $\hat{v}$ and zero acceleration $a_2(x) = 0$, so the acceleration of the trailing leaf is

$$a_1(x) = -\frac{\mathrm{d}^2 I}{\mathrm{d}x^2}(x)v_1^3(x) = -\frac{\mathrm{d}^2 I}{\mathrm{d}x^2}(x)\bigg/\left[\frac{1}{\hat{v}} + \frac{\mathrm{d}I}{\mathrm{d}x}(x)\right]^3. \tag{2.9}$$

These equations are analytic forms when the velocities and accelerations can take any (including infinite) values. In practice, if the equations are applied strictly to compute the required leaf movements to give a desired intensity profile, unrealistic velocities and accelerations may be required. For example, at an extremum in a profile, a leaf will be required to move from some small speed to the maximum speed or vice versa (depending on which leaf and whether the extremum is a maximum or minimum). Also in regions where $\mathrm{d}I/\mathrm{d}x$ is small, the required leaf speed may be more than can be achieved in practice (see equation (2.2)). Svensson *et al* (1994a) discussed these problems at length and provided solutions for how to overcome them by arranging for the modulation close to extrema to be controlled by not just one, but by both leaves.

Svensson *et al* (1994a) also analysed the close-in or shrinking-field method of intensity modulation, comparing it with the dynamic-leaf-sweep technique. They concluded that similar treatment times ensue.

These equations for computing leaf movements are based on assuming a spatially and temporally invariant fluence upstream of the collimation. Some new machines are capable of modulating the beam intensity in both time and space. Temporal modulation is no real help for this problem because the temporal modulation for one leaf-pair will not in general be the same as that required for some other leaf-pair modulating the profile for a different CT slice. However, spatial modulation upstream of the collimator could be useful because it decreases the added modulation needed from the leaves and so decreases the overall treatment time (see section 2.2.7). Svensson *et al* (1994a) give examples of how this helps. In this respect their work considerably extends that of Convery and Rosenbloom (1992) who considered the case where the upstream fluence was constant in both space and time. It is emphasised that in this section, time $t(x)$ and intensity $I(x)$ are in the same dimensions of MUs. Corresponding velocities have dimensions (distance MU^{-1}). To obtain $t(x)$ in seconds or velocity in (distance s^{-1}) requires conversion by the accelerator output dI/dt in MUs s^{-1}.

2.2.3. The solution of Stein et al

Stein (1993) and Stein *et al* (1994a,b) at DKFZ, Heidelberg have also recently extended the work of Convery and Rosenbloom (1992) and provided a new iterative algorithm for calculating the trajectories of the leading and trailing leaves when the upstream fluence is constant. The algorithm is based on exactly the same feature as in the Svensson *et al* (1994a,b) approach, deduced independently, that:

(i) when the spatial gradient of the fluence profile is positive, the leading leaf should move at maximum speed and the modulation be provided by the trailing leaf;

(ii) when the spatial gradient of the fluence profile is negative, the trailing leaf should move at maximum speed and the modulation be provided by the leading leaf. The formal proof is given in section 2.2.4.

Stein's algorithm iteratively fits a 'virtual profile' (VP), initially set equal to the desired intensity profile (DP), and optimises the leaf motion such that the overall treatment time is minimised subject to the above modulation rules. The profiles are specified by N points $DP(x_i)$, $VP(x_i)$, $i = 1, 2, \ldots, N$. Initially $I(x_i)$ is set equal to $VP(x_i)$ which has been initially set to $DP(x_i)$. Using the above rules the algorithm for the time profiles $t_1(x_i)$ and $t_2(x_i)$ becomes:

if $I(x_i) \geqslant I(x_{i-1})$ then for $i = 2, 3, \ldots, N$

$$t_2(x_i) = t_2(x_{i-1}) + t_{min} \qquad (2.10)$$

and

$$t_1(x_i) = I(x_i)/I_0 + t_2(x_i), \qquad (2.11)$$

where I_0 is the constant upstream fluence rate (dimensions [MU T^{-1}]), and t_{min} is the minimum time taken for a leaf to move from position x_i to x_{i+1}. This is equal to the distance between the ith and the $(i + 1)$th data point divided by the maximum leaf velocity. Note in Stein's notation, reproduced in this section, time is in units [T] rather than MU, hence the division by I_0 (compare with equation (2.1)).

Conversely if $I(x_i) \leqslant I(x_{i-1})$ then for $i = 2, 3, \ldots, N$

$$t_1(x_i) = t_1(x_{i-1}) + t_{min} \qquad (2.12)$$

and

$$t_2(x_i) = -I(x_i)/I_0 + t_1(x_i). \qquad (2.13)$$

It may be seen that equations (2.10)–(2.13) are formally exactly the same as equations (2.5) and (2.6). The starting values were taken to be

$$t_1(x_1) = I(x_1)/I_0 \qquad (2.14)$$

and

$$t_2(x_1) = 0. \qquad (2.15)$$

This algorithm minimises the total time taken for the trailing leaf 1 to cross the field aperture. The total time for creating a profile is the sum of the time increments taken for leaf 1 to move through all its locations x_i. Using the above equations it follows that this total time T is

$$T = \sum_{x_i} dt_1(x_i) = N t_{min} + \sum_{l} \left[\frac{\Delta I_{min \to max}(l)}{I_0} \right] \qquad (2.16)$$

where l is the index for sections of the profile where the intensity is increasing and $\Delta I_{min \to max}(l)$ is the fluence difference between a local minimum and the succeeding lth local maximum. For a field of total width W and, representing the maximum leaf velocity by $\hat{v}$ $(= dx_i/t_{min})$, Stein *et al* (1994a,b) were able to write, since $W = N dx_i$,

$$T = \frac{W}{\hat{v}} + \sum_{l} \left[\frac{\Delta I_{min \to max}(l)}{I_0} \right]. \qquad (2.17)$$

This very useful equation allows us to estimate the total treatment time in terms of the complexity of the profile.

After this first step to create a first estimate of the leaf trajectories, the 'realised intensity profile' (RP), the profile which would actually be delivered by these leaf movements, was created by folding in the leaf penumbra and the in-plane scatter. Functions representing these were analytic forms fitted to measured data, i.e. if $P_1(x_1(t), x_i)$ represents the penumbra profile generated by the first leaf when it is situated at point $x_1(t)$ and the fluence is measured at position x_i and similarly $P_2(x_2(t), x_i)$ represents the penumbra profile generated by the second leaf when it is situated at point $x_2(t)$ and the fluence is measured at position x_i, then RP is given by

$$RP(x_i) = I_0 \int P_1(x_1(t), x_i) \times P_2(x_2(t), x_i) \times S(|x_1(t) - x_2(t)|) \, dt \quad (2.18)$$

where $S(|x_1(t) - x_2(t)|)$ is unity for large distances and includes a correction when the leaves are very close together. The realised profile may not exactly be the desired profile after this first fitting stage and so the process was repeated iteratively as follows.

The desired profile was subtracted from the realised profile, point by point, to create a 'difference profile' $D(x_i)$, i.e.

$$D(x_i) = \text{RP}(x_i) - \text{DP}(x_i). \tag{2.19}$$

If the difference $D(x_i)$ were less than some specified percentage (e.g. 2%) for all i then the optimisation of leaf velocities terminated. If not, a new virtual profile

$$\text{VP}_k(x_i) = \text{VP}_{k-1}(x_i) - D(x_i) \tag{2.20}$$

was created for all points x_i where k represents the iteration loop number. Then the leaf trajectories were re-computed to fit this new virtual profile, i.e. $I(x_i)$ was reset to $\text{VP}(x_i)$ in equations (2.11) and (2.13). This process was cyclically continued to convergence. In the ideal situation, when finally the virtual profile is no longer changing on iteration, i.e. $\text{VP}_k(x_i) = \text{VP}_{k-1}(x_i)$, the difference profile $D(x_i)$ is zero and the realised profile $\text{RP}(x_i)$ is equal to the desired profile $\text{DP}(x_i)$.

The main improvements of the work of Stein *et al* (1994a,b) over that of Convery and Rosenbloom (1992) were:

(i) the inclusion of penumbra and scatter;

(ii) the imposition of the condition for maximum leaf speed depending on the sign of the gradient of the intensity profile (like Svensson *et al* 1994a,b); and

(iii) the computations were faster than the Simplex algorithm.

Stein *et al* (1994a,b) found that the algorithm converged in just a couple of iterations and that very good fitting of the desired intensity profile could be obtained (figure 2.6). They worked out the ratio of the total treatment time to the fluence maximum (which is a measure of the efficiency of the method of creating intensity-modulated profiles) and obtained typical values of the order three, which are a little higher than obtained by the Convery and Rosenbloom method for the same maximum leaf speed. The total treatment time increases with the complexity of the profile (equation (2.17) shows this directly) and also as the number of iterations increases (since this leads to less smooth profiles) (figure 2.7).

As with other MLC-based techniques, several slices can be treated simultaneously at each gantry orientation by using several leaf-pairs, and like the Convery and Rosenbloom technique, this demands that all but one profile be delivered ending with the leaves totally closed, a feature not in the early work by Stein but easily added. Also the work of Stein has yet to consider the effects of scatter into adjacent planes and the effect of finite acceleration.

Presently DKFZ are constructing an MLC with a motor on each leaf-pair and sufficient over-running capability that the MLC should be capable of delivering dynamic therapy in this way (see section 3.3.3). Of course

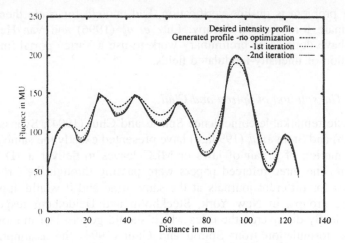

Figure 2.6. *An example of the degree of fit between a realised fluence profile generated by the iterative technique of Stein and the desired fluence profile. The realised profile includes the effect of penumbra and transmission. Note that even the unoptimised profile is quite a good fit and that after two iterations the fit is almost perfect. (Reprinted from Stein et al 1994b, with kind permission from Elsevier Science Ireland Ltd, Bay 15K, Shannon Industrial Estate, Co. Clare, Ireland.)*

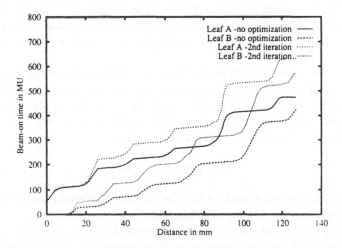

Figure 2.7. *Leaf trajectories generated by the iterative technique of Stein showing graphically the increase in treatment time required to deliver the realised profiles after iterative optimisation compared with the unoptimised profiles. This is because the optimised profiles are sharper. (Reprinted from Stein et al 1994b, with kind permission from Elsevier Science Ireland Ltd, Bay 15K, Shannon Industrial Estate, Co. Clare, Ireland.)*

a major problem is on-line verification, bad enough for static therapy and a nightmare for dynamic therapy. Lutz *et al* (1996) and Van Herk *et al* (1996) have done some preliminary work to use a Varian portal imager for verification of intensity-modulated fields.

2.2.4. The solution of Spirou and Chui

By a quite remarkable coincidence Spirou and Chui (1994), Svensson *et al* (1994a,b) and Stein *et al* (1994a,b) have presented exactly the same solution for the motion of a pair of jaws or MLC leaves to deliver a 1D IMB of radiation. The three refereed papers were passing through their refereeing processes for different journals at the same time and it would appear that the three groups, in New York, Stockholm and Heidelberg respectively, independently discovered the rules which should govern leaf motion.

In the formulation from Spirou and Chui (1994) the assumption was made that penumbra, collimator scatter and leakage between leaves could be ignored. The 1D intensity profiles were also assumed to be piecewise linear, i.e. consist of straight lines between closely spaced locations x_i. IMBs were realised by a 'sliding window' technique in which leaf number 2 leads leaf number 1. Initially it was assumed that each leaf totally blocks the beam below it. The algorithm was then extended to account for leaf penetration. Consider first the simplest case with perfect shielding. The algorithm consists of two parts: (i) the equations for the leaf motion; (ii) the start and finish positions of the leaves. The algorithm aims to maximise efficiency in all segments $dx_i = x_{i+1} - x_i$ so that overall the most efficient pattern of dynamic motion is obtained.

Using the same notation as Svensson *et al* (1994a,b) (to ease comparison) let the 1D intensity profile be represented by $I(x_i)$. Let the velocities of the leading and trailing leaves be $v_2(x_i)$ and $v_1(x_i)$ respectively. Let the cumulative beam-on time when leaves 1 and 2 are at x_i be $t_1(x_i)$ and $t_2(x_i)$ respectively. Then the intensity profile satisfies

$$I(x_i) = t_1(x_i) - t_2(x_i) \tag{2.21}$$

because the irradiation starts when the leading leaf unblocks the location x_i and stops when the trailing leaf blocks it.

The algorithm follows by calculating and then inspecting the velocity of the trailing and leading leaves over the segment dx_i. From equation (2.21) it follows that

$$I(x_{i+1}) = t_1(x_{i+1}) - t_2(x_{i+1}). \tag{2.22}$$

The speed of the trailing leaf is

$$v_1(x_{i+1}) = \frac{dx_i}{t_1(x_{i+1}) - t_1(x_i)}. \tag{2.23}$$

Similarly the speed of the leading leaf is

$$v_2(x_{i+1}) = \frac{dx_i}{t_2(x_{i+1}) - t_2(x_i)}. \tag{2.24}$$

Substituting equations (2.21) and (2.22) in equation (2.24) we have

$$v_2(x_{i+1}) = \frac{dx_i}{t_1(x_{i+1}) - t_1(x_i) - I(x_{i+1}) + I(x_i)}. \tag{2.25}$$

From equations (2.23) and (2.25) we may see that where the IMB profile is *increasing* with x_i, i.e. $I(x_{i+1}) \geqslant I(x_i)$ or dI/dx is positive, $v_2 \geqslant v_1$. In this case set $v_2 = \hat{v}$, the maximum speed. Hence the movement of the leading leaf is

$$t_2(x_{i+1}) = t_2(x_i) + dx_i/\hat{v}. \tag{2.26}$$

Equation (2.22) gives the movement of the trailing leaf

$$t_1(x_{i+1}) = t_2(x_{i+1}) + I(x_{i+1}). \tag{2.27}$$

From equations (2.26) and (2.27)

$$t_1(x_{i+1}) = t_2(x_i) + dx_i/\hat{v} + I(x_{i+1}). \tag{2.28}$$

Substituting from equation (2.21)

$$t_1(x_{i+1}) = t_1(x_i) - I(x_i) + dx_i/\hat{v} + I(x_{i+1}). \tag{2.29}$$

So

$$dt_1(x_i) = dx_i/\hat{v} + dI(x_i) \tag{2.30}$$

since $dt_1(x_i) = t_1(x_{i+1}) - t_1(x_i)$ and $dI(x_i) = I(x_{i+1}) - I(x_i)$ by definition. With

$$v_1(x_i) = \frac{dx_i}{dt_1}(x_i) \tag{2.31}$$

we obtain

$$\frac{1}{v_1(x_i)} = \frac{1}{\hat{v}} + \frac{dI(x_i)}{dx_i} \tag{2.32}$$

which, rearranged, gives

$$v_1(x_i) = \frac{\hat{v}}{1 + \hat{v}(dI(x_i)/dx_i)}. \tag{2.33}$$

This is the leaf speed of the trailing leaf.

On the other hand if the IMB profile is *decreasing* with x_i, i.e. $I(x_{i+1}) < I(x_i)$ or dI/dx is negative, from equations (2.23) and (2.25) $v_2 < v_1$. In this case set $v_1 = \hat{v}$, the maximum speed. Hence the movement of the trailing leaf is

$$t_1(x_{i+1}) = t_1(x_i) + dx_i/\hat{v}. \tag{2.34}$$

Equation (2.22) gives the movement of the leading leaf

$$t_2(x_{i+1}) = t_1(x_{i+1}) - I(x_{i+1}). \tag{2.35}$$

From equations (2.34) and (2.35)

$$t_2(x_{i+1}) = t_1(x_i) + dx_i/\hat{v} - I(x_{i+1}). \tag{2.36}$$

Substituting from equation (2.21)

$$t_2(x_{i+1}) = t_2(x_i) + I(x_i) + dx_i/\hat{v} - I(x_{i+1}). \tag{2.37}$$

So

$$dt_2(x_i) = dx_i/\hat{v} - dI(x_i). \tag{2.38}$$

With

$$v_2(x_i) = \frac{dx_i}{dt_2}(x_i) \tag{2.39}$$

we obtain

$$\frac{1}{v_2(x_i)} = \frac{1}{\hat{v}} - \frac{dI(x_i)}{dx_i} \tag{2.40}$$

which, rearranged, gives

$$v_2(x_i) = \frac{\hat{v}}{1 - \hat{v}(dI(x_i)/dx_i)}. \tag{2.41}$$

This is the leaf speed of the leading leaf.

This completes the specification of the algorithm for the leaf movements and one may observe that it is the same as presented by Svensson *et al* (1994a,b) and Stein *et al* (1994a,b), i.e. when the profile is increasing, the leading leaf should travel at maximum velocity and the modulation be provided by the trailing leaf (equation (2.33)) and vice versa (equation (2.41)).

When $dI/dx \geqslant 0$, equations (2.26) and (2.27) are formally the same as equations (2.10) and (2.11) of Stein *et al* (1994a,b) and equation (2.33) is formally the same as equation (2.5) of Svensson *et al* (1994a,b). Conversely when $dI/dx < 0$, equations (2.34) and (2.35) are formally the same as equations (2.12) and (2.13) of Stein *et al* (1994a,b) and equation (2.41) is formally the same as equation (2.6) of Svensson *et al* (1994a,b).

Some further mathematics allows us to see that this algorithm does in fact generate the *most efficient* leaf movement and this analysis from Spirou and

Chui (1994) is the easiest way to visualise the formal proof. Supposing there were some other set of trajectories through time (which we shall denote with a prime) which produce the same IMB. Then, from equation (2.21)

$$I(x_i) = t_1(x_i) - t_2(x_i) = t'_1(x_i) - t'_2(x_i). \qquad (2.42)$$

Consider the segment-to-segment differences $t_1(x_{i+1}) - t_1(x_i)$ and $t'_1(x_{i+1}) - t'_1(x_i)$. As before we consider the two cases separately depending on the slope of the profile.

When $I(x_{i+1}) < I(x_i)$, i.e. the IMB profile is *decreasing,* then

$$t_1(x_{i+1}) - t_1(x_i) = dx_i/\hat{v}. \qquad (2.34)$$

and

$$t'_1(x_{i+1}) - t'_1(x_i) = dx_i/v'_1 \qquad (2.43)$$

where

$$\hat{v} \geqslant v'_1 \qquad (2.44)$$

and v'_1 is the velocity of the trailing leaf in the primed path. It follows that

$$t_1(x_{i+1}) - t_1(x_i) \leqslant t'_1(x_{i+1}) - t'_1(x_i). \qquad (2.45)$$

Now, alternatively when $I(x_{i+1}) \geqslant I(x_i)$, i.e. the IMB profile is *increasing,*

$$t_1(x_{i+1}) = t_2(x_{i+1}) + I(x_{i+1}). \qquad (2.27)$$

So substituting from equation (2.26)

$$t_1(x_{i+1}) = t_2(x_i) + dx_i/\hat{v} + I(x_{i+1}) \qquad (2.28)$$

and further substituting from equation (2.21)

$$t_1(x_{i+1}) = t_1(x_i) - I(x_i) + dx_i/\hat{v} + I(x_{i+1}). \qquad (2.29)$$

Rearranging

$$t_1(x_{i+1}) - t_1(x_i) = dx_i/\hat{v} + I(x_{i+1}) - I(x_i). \qquad (2.46)$$

Also, by analogy with equation (2.26),

$$t'_2(x_{i+1}) - t'_2(x_i) = dx_i/v'_2 \qquad (2.47)$$

where

$$\hat{v} \geqslant v'_2 \qquad (2.48)$$

and v_2' is the velocity of the leading leaf in the primed path. It follows that since $dx_i/\hat{v} < dx_i/v_2'$,

$$t_1(x_{i+1}) - t_1(x_i) \leqslant t_2'(x_{i+1}) - t_2'(x_i) + I(x_{i+1}) - I(x_i) \qquad (2.49)$$

i.e. using equation (2.27) rewritten in the primed path

$$t_1(x_{i+1}) - t_1(x_i) \leqslant t_1'(x_{i+1}) - t_1'(x_i) \qquad (2.50)$$

which is identical to the inequality (2.45) for the decreasing IMB segments.

Thus, taking the two cases $dI/dx < 0$ (inequality (2.45)) and $dI/dx \geqslant 0$ (inequality (2.50)), for *all* segments of the profile

$$t_1(x_{i+1}) - t_1(x_i) \leqslant t_1'(x_{i+1}) - t_1'(x_i). \qquad (2.51)$$

Since this is true for each and all segments it means the total beam-on time will be least for the unprimed path, i.e. for the path determined by the algorithm (equations (2.33) and (2.41)) of Spirou and Chui (1994). This completes the formal proof that use of this rule gives the minimum treatment time.

The second part of the algorithm concerns the specification of the start and stop positions of the leaves. Clearly the trailing leaf 1 starts at the leftmost point x_1 and the leading leaf 2 finishes at the rightmost point x_N. If the initial rate of change of the profile dI/dx is greater than $1/\hat{v}$ then there is no need for the leading leaf 2 to start at x_1 since all the modulation can be accomplished by the trailing leaf 1 alone. Similarly if the closing part of the profile decreases at a sufficient rate, the trailing leaf 1 can stop short of x_N and all the modulation can be provided by the leading leaf 2.

As others have done, Spirou and Chui (1994) discussed the extension of the algorithm to the MLC where the total irradiation time may be different for each leaf-pair. Instead of the suggestion that all leaf-pairs be forced to close, they suggest instead that the maximum velocity $\hat{v}$ in the last segment only is changed to make all the beam-on times the same, since this is easy to implement.

Spirou and Chui (1994) presented the extension of the algorithm to account for non-trivial transmission τ through the leaves. If the total beam-on time is T the desired intensity $I(x_i)$ is related to a modified intensity $i(x_i)$, which represents the time when x_i is not covered by either leaf, via

$$I(x_i) = \{[T - i(x_i)] \times \tau\} + i(x_i) \qquad (2.52)$$

then the desired intensity is the sum of the leaking transmitted radiation (first term in equation (2.52)) and the direct exposure (second term in equation (2.52)) (figure 2.8). Rearranging

$$i(x_i) = \frac{I(x_i) - T \times \tau}{1 - \tau}. \qquad (2.53)$$

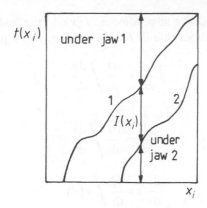

Figure 2.8. *This figure illustrates how the contribution to the radiation intensity $I(x_i)$ does not only depend on the leaf (or jaw) positions but also on the leakage radiation. The diagram shows the leaf (or jaw) trajectories for two leaves 1 and 2 as a function of time. The vertical axis is time in cumulative MUs. Location x_i thus receives open field radiation for time $t_1(x_i) - t_2(x_i)$ but also leakage radiation proportional to the time for which location x_i is shielded by either leaf. (Adapted from Spirou and Chui 1994.)*

Further equations in the paper by Spirou and Chui (1994) develop the expression for the total beam-on time. Hence the problem reduces to using $i(x_i)$ instead of $I(x_i)$ in the equations to deduce the velocity profiles. There is one small caveat that some profiles may not be achievable if the minimum intensity at any point falls below the leaked radiation intensity $T\tau$ (since this would require impossible negative $i(x_i)$), i.e. it is required that $I(x_i) > T\tau$ for all i.

Like Stein *et al* (1994a,b), Spirou and Chui (1994) show that scattered radiation can be accounted for by computing the velocity profiles without it, calculating the delivered IMB with scatter and iteratively adjusting the motion to compute a slightly modified velocity sequence. Interestingly they do not consider the effects of finite acceleration (which Svensson *et al* (1994a,b) studied in depth) to be important. They note the need to use portal imaging for verification of delivered IMBs. Lutz *et al* (1996) are working on this. They have shown several examples illustrating the accuracy of the algorithm and demonstrating the fast speed which they claim is an advantage over the computational technique of Convery and Rosenbloom (1992). Chui *et al* (1994) have implemented dynamic MLC movement to achieve intensity modulation for breast radiotherapy and Wang *et al* (1996) have implemented the method for prostate therapy (see also section 2.2.8).

2.2.5. *The solution of Yu and Wong*

2.2.5.1. *Dynamic-leaf movement.*   As we have seen, considerable insight may be gained regarding the creation of an IMB using the equation which

relates the required fluence at position x, $I(x)$, to the leaf or jaw movement and the accelerator output rate dI/dt. Repeating, this equation is

$$\frac{dx}{dt} = \frac{dI}{dt} \bigg/ \frac{dI}{dx}. \qquad (2.2)$$

Rearranging

$$\frac{dI}{dx} = \frac{dI}{dt} \bigg/ \frac{dx}{dt}. \qquad (2.54)$$

or

$$\frac{dI}{dx} = \frac{dI}{dt} \bigg/ v(x). \qquad (2.55)$$

From this equation it is clear that the shape of the IMB profile dI/dx may be adjusted either by varying the leaf velocity $v(x)$ at constant accelerator output rate dI/dt or by varying the accelerator output rate dI/dt at constant leaf velocity $v(x)$ or by varying *both*. If the profile monotonically increases or decreases it is only necessary to move one leaf or jaw whilst the other remains fixed. However, as we have seen, if the IMB has both rising and falling components it is necessary to execute movements for both leaves or jaws.

In practice a linear accelerator has constraints on the maximum leaf or jaw speed and on the rate of change of machine dose-rate so that it is sometimes necessary to control both to ensure each remains within its limits even when the IMB is monotonic. However, as we saw in section 2.2.2, varying the output is only useful if all leaf IMB profiles are to be the same.

Yu *et al* (1995c) have shown that, in order to deliver intensity $I(x)$ at position x, which is given by the difference of two terms, the cumulative times (expressed in MUs) for the transit of the left (1) and right (2) leaves are

$$t_1(x) - t_2(x) = I(x). \qquad (2.1)$$

They did not explicitly give the rule derived by Stein and others concerning the relation between the leaf which should move fastest and the sign of the gradient of the IMB. Instead they separate the delivery into left and right components which can be delivered in one of two distinct ways. The first of these is 'step-and-shoot'. Leaf-pairs take up a series of finite positions in which a quantum of beam intensity is delivered. Between these quanta, while the leaf-pairs are moving the radiation is switched off. The second way is called 'dynamic-step-and-shoot'. Once again the IMB profile is separated into a series of components. However, the leaves take up these positions and move between them while the radiation is switched on (with a small proviso discussed below).

Yu *et al* (1995c) generated a wedged isodose distribution using equation (2.2). First note that the equation for the IMB for a wedge can be derived as follows. Let $D(0, 10)$ be the dose profile at a depth of 10 cm at $x = 0$, i.e. on the central axis, and $D(x, 10)$ be the dose at the same depth at position x. If the wedge angle is α and μ_m is the mean linear-attenuation coefficient of the spectral beam, we have that

$$D(x, 10) = D(0, 10) \exp[-\mu_m x \tan(\alpha)]. \tag{2.56}$$

Since the IMB fluence is proportional to $D(x, 10)$ we may write

$$I(x) = k \exp[-\mu_m x \tan(\alpha)]. \tag{2.57}$$

When Yu *et al* (1995c) implemented this equation (2.57) with equation (2.55) they kept the leaf *speed* constant and varied the accelerator output rate. The method was implemented experimentally on a Siemens KD-2 accelerator using just the main jaws. The dose distributions were measured for a field size of 15×15 cm^2 and 10 MV x-rays using film. A jaw speed of 5 mm s^{-1} was used and wedges with $\alpha = 30°$ and $\alpha = 60°$ were created. No comparison of absolute dosimetry was made but the isodoses were qualitatively correct (figure 2.9). Indeed Yu *et al* (1995a) comment that by varying the attenuation constant μ_m the experimental profiles may be brought into agreement with the prescription and this gives a method to determine the effective attenuation coefficient μ_m.

Du *et al* (1995) have shown how the general method of dynamic beam-intensity modulation has been used to create a wedged field in which the orientation of the wedge is at an arbitrary angle with respect to the MLC leaves. They created a 60° dynamic wedge with relative collimator angles of 0°, 15°, 30° and 45°. This removes one of the significant limitations of the use of an MLC with internal wedge at fixed orientation. It also demonstrates that the problem of inter-leaf leakage may not be as great as initially feared.

Yu and Wong (1994) studied a 2D fluence profile which had both rising and falling parts; hence both leaves had to move. The movement profiles for a *bank of* leaves were calculated corresponding to a 2D horse-saddle distribution of radiation intensity. Then using a step-and-shoot mode, similar to that of Bortfeld *et al* (1994c) (see section 2.4), they irradiated a phantom containing plastic scintillation dosimeters (Chawla *et al* 1995) and showed the ability to realise the modulation.

In a later development the leaves of the Philips' MLC were moved with the beam on to achieve these IMBs without step-and-shoot (Yu *et al* 1994, 1995c). This method of creating IMBs with the beam on can be implemented in either of two distinct ways. Either the leaves can be made to move dynamically continuously with the beam on (true dynamic delivery) or a 'dynamic-step-and-shoot' method can be used. In this latter, the radiation is on all the time, the profile comprises essentially static field elements, but the movement from one to the next static field position is

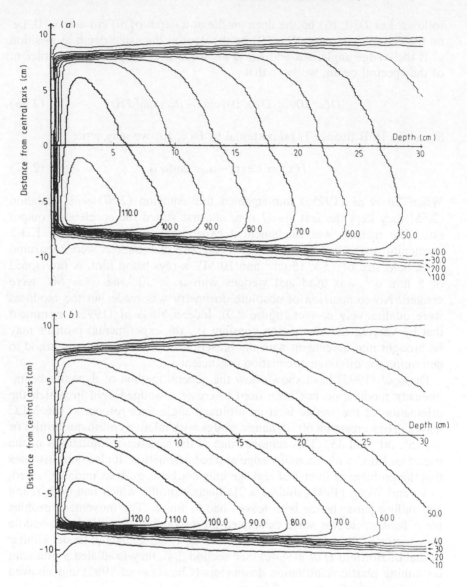

Figure 2.9. *Isodose contours derived from film measurements of (a) a 30°
and (b) a 60° wedge implementation on a Siemens KD-2 linear accelerator
equipped with independent jaws. The wedge distributions were made by
dynamically scanning the jaws with a differential velocity profile. (From Yu
et al 1995c.)*

achieved by moving the leaves with the beam still on. This has the effect
of blurring the incremental steps. Yu *et al* (1995c) used the latter method.
Each step was given 2% of the total MUs and the accelerator output rate

was kept constant. Hence all the modulation was achieved by varying the leaf positions. Three 2D distributions were generated, the saddle again, a 'Mexican hat' sinc function (figure 2.10) and a random superposition of four Gaussians (figure 2.11) . Again the experimental distributions were measured using film, digitised and the resulting dose distributions displayed for comparison with the prescriptions. The qualitative feasibility was clearly shown.

The advantage of the two dynamic methods is that there is little dependence on beam start-up whereas step-and-shoot requires a fast turn-on time for the beam. All methods require accurate verification of leaf position. For both dynamic methods the dose-rate is limited by the leaf speed. For the step-and-shoot method the beam-off time extends the overall treatment time leading to an overall lowering of the efficiency of treatment. The dynamic-step-and-shoot method requires quick response of the MLC control system. This method of controlling the MLC on the Philips SL20 accelerator requires no hardware modifications as all control is software engineered. The development at the William Beaumont Hospital, Royal Oak, USA, can accept prescriptions from the NOMOS PEACOCKPLAN planning system for IMBs (see section 2.3) (Philips Medical Systems 1994).

The dynamic-step-and-shoot method has an additional feature which is now discussed. Whilst the leaves are moving at their maximum speed between positions the MLC controller compares the number of delivered MUs with the MU prescribed for the step. If the leaf reaches the next location just as the MU for this step is delivered all well and good. However, if the leaf reaches this position *before* the quantum of MU is delivered it will stop at this position and wait there. On the other hand if the quantum of MU is delivered before the leaf has reached its next position, the MLC controller generates a pause signal which halts the radiation by temporarily switching off the microwave power to the accelerating guide whilst the leaves 'catch up'. Yu *et al* (1995c) arrange that this happens infrequently.

There are outstanding issues to be addressed concerning the accuracy of treatment. Beam penumbra was ignored. The effects of beam non-uniformity and photon leakage through the leaves need to be studied (the solution of Spirou and Chui (1994) could for example be adopted). Photon scatter has been ignored. For absolute quantitation a method of relating photon dose to intensity is required. In its more general form when both the accelerator output and the leaf speeds can be varied, attention must be paid to adjusting these so as not to exceed constraints.

Throughout these discussions of all these theoretical studies the concept of maximum leaf speed has played a prominent role. Mohan (1995) has given this speed for three commercial MLCs. It is 1, 2 and 5 cm s^{-1} for Scanditronix, Philips and Varian respectively.

In concluding this discussion of techniques to create intensity-modulated fields by dynamic-leaf movement it must be pointed out that, with the present

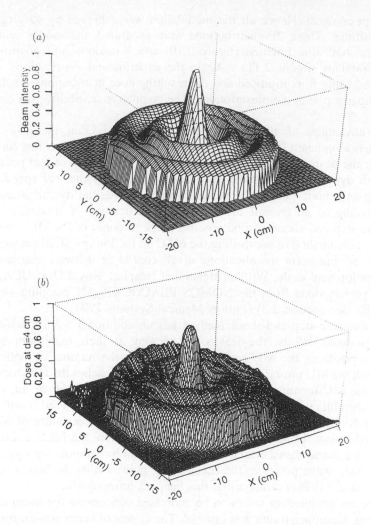

Figure 2.10. *(a) A 3D perspective plot of the desired intensity distribution of a model 2D sinc function. (b) A similar perspective of the actual dose distribution, delivered by an IMB, determined from measurements on a film, sandwiched in a water-equivalent phantom at a depth of 4 cm. A total of 200 MU of 6 MV x-rays were delivered in approximately 2 min using a machine dose-rate of 100 MU min⁻¹. 50 segments were used, each delivering 2% of the total number of MUs. (From Yu et al 1995c.)*

state-of-the-art, solutions are generally considered leaf-pair by leaf-pair. That is, there is considered to be no coupling between the leaf-pairs. The dose distribution developed in some particular slice would be related to the sets of 1D fluence profiles which are generated at different gantry angles by each leaf-pair. However, a problem arises due to the stepped nature of the long

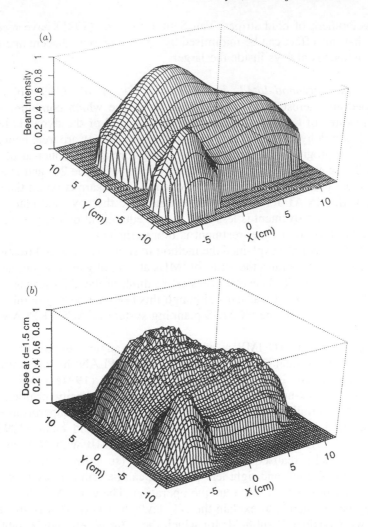

Figure 2.11. *(a) A 3D perspective plot of the desired intensity distribution of a model 2D function comprising a superposition of four Gaussian functions. (b) A similar perspective of the actual dose distribution, delivered by an IMB, determined from measurements on a film, sandwiched in a water-equivalent phantom at a depth of 1.5 cm. A total of 100 MU of 6 MV x-rays were delivered in approximately 1 min. 50 segments were used. (From Yu et al 1995c.)*

edges of the leaves (see also Chapter 3). Because adjacent leaves slide over each other in some kind of tongue-and-groove arrangement, there will be times when some leaves are ahead or behind their neighbours and the leaf side will be visible as a partial attenuator. The effect of this is complex depending on the adjacent leaf velocities but it can be shown that it leads to the development of cold stripes. Van Santvoort *et al* (1995) have recently shown that this effect can be minimised by synchronising adjacent apertures so the smaller is always inside the larger.

2.2.5.2. Intensity-modulated arc therapy. Yu *et al* (1995a,b) have developed an interesting IMB-generating technique which combines some of the features of tomotherapy (see section 2.4) and of the dynamic MLC. During beam delivery the linac is programmed to deliver arc treatments and the MLC is programmed to dynamically step through a sequence of field shapes. The gantry makes many arc traverses (for example 20) and at every 5° of arc the position of the leaves can change. The beam is on all the time. They call this IMAT for 'intensity-modulated arc therapy' and claim that, because the leaf movement is less than for the equivalent delivery at a series of static field locations, the treatment is more efficient.

Yu (1995, 1996) has explained the method in some detail. A 3D treatment-planning system generates the series of IMBs at a set of gantry angles needed to treat a series of CT slices. (It is easiest to think of the CT slices as in the same plane as the set of leaf-pairs although this need not be so.) In this work these are generated by the NOMOS planning system PEACOCKPLAN (see section 2.4).

Consider any one 1D IMB at any one gantry angle. Suppose it can be discretised into N intensity quanta. (In PEACOCKPLAN this discretisation is an optimisation feature. In the work of Webb (1994a,b,c) it is an *a posteriori* break-up of an otherwise continuous intensity profile.) The discrete IMB profile can be experimentally realised by superposition of a set of subfields in the manner described by Bortfeld *et al* (1994b,c) (see section 2.3). Yu (1995) nicely shows that an IMB with N levels can be experimentally realised in any of $N!^2$ ways (so called 'decomposition patterns'), i.e. the left and right leaf positions can be arranged as a set of N subfields but this is far from a unique operation. There are $N!^2$ choices.

It may be instructive to explain the $N!^2$ law (Yu 1995). This is arrived at as follows. Consider N subfields for which there are N left and N right leaf locations required (see figure 2.15 illustrating Bortfeld and Boyer's static-field method). The first left leaf has N choices of right leaf. For *each choice* made by the first left leaf, there are $(N - 1)$ choices of right leaf for the second left leaf. So for the first two left leaves there are $N(N - 1)$ choices of right leaf. For each of these there are $(N - 2)$ choices to pair a right leaf with a third left leaf. So by induction, for an IMB with N subfields there are $N(N - 1)(N - 2) \ldots 1 = N!$ ways to pair N left leaves with N right leaves. For each of these matches they can be arranged in $N!$ different orders. So the total number of independent decomposition patterns is $N!^2$.

Yu (1995, 1996) then seeks those choices for which the leaf-pair movement (for each slice) is minimised as the gantry angle is changed. Then N arcs of therapy are delivered with the leaves moving to a new

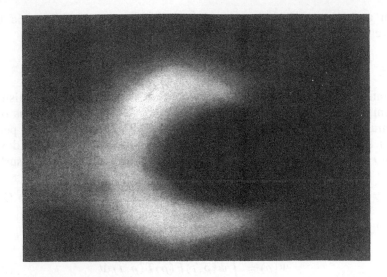

Figure 2.12. *A C-shaped high dose distribution delivered by the IMAT technique. (From Yu 1995.)*

position every 5°. Each arc generates one subfield for each gantry angle (for each slice, identified with a leaf-pair). Then the next arc generates the next subfield and so on until N arcs are completed.

There are some refinements. Since not all IMBs will have the same number of intensity quanta, the number of arcs is set by the most complex profile and other more simple ones will be broken down into further subfields. Special attention has to be paid to closed leaf-pairs.

Yu (1995) has implemented this technique on a Philips SL20 accelerator with the Philips' MLC. Code has been written to synchronise the MLC controller and the linear accelerator. The sequences of subfields are programmed just as the MLC is programmed for static fields (except there are more of them). Yu planned a C-shaped target with an OAR in the cusp of the 'C' for a phantom (a familiar planning problem) and showed that the delivered distribution qualitatively matched the planned distribution (figure 2.12). It was argued that for arc treatments only five levels of intensity may be enough. This is debatable.

It was argued that IMAT has the following advantages over other slice-based treatment methods:

(i) it can be implemented on existing accelerators;
(ii) it does not collimate the beam to a slit and so photons are efficiently utilised;
(iii) there is no need to abut treatment slices, so there are no potential matchline problems or difficulties with patient movement;
(iv) the 'along the leaf' spatial resolution is 'continuous'.

Yu and Wong (1996) have compared IMAT with multiple-static IMB delivery and they favoured the IMAT technique.

2.2.6. *Iterative optimisation of scanning-leaf configurations*

Gustafsson *et al* (1995) have presented a very general formalism of optimisation of 3D treatment planning. Dose is formulated in terms of an integration over the energy fluence $\Psi(\rho)$ (units Joules m^{-2}) and the pencil-beam kernel $p(r, \rho)$ (units kg^{-1}), representing the contribution to dose at point r in the patient from a beam entering at point ρ on the patient surface, as:

$$D(r) = \iint_S p(r, \rho)\Psi(\rho)\, \mathrm{d}S. \tag{2.58}$$

The fluence is further expressed as the product

$$\Psi(\rho) = \int_t \psi(\rho, t)F(\rho)T(\rho, t)\, \mathrm{d}t \tag{2.59}$$

where $\psi(\rho, t)$ is the energy fluence rate at time t produced by an accelerator, $F(\rho)$ is a transmission factor for a filter in the beam and $T(\rho, t)$ represents the general transmission properties of MLC leaves. F, for example, could be provided by a fixed wedge or a variable-angle wedge. $\psi(\rho, t)$ could, for example, represent the time-varying fluence produced by a scanning beam.

Gustafsson *et al* (1995) show how in practice this formalism can be reduced to a matrix operation. They consider a further simplification with the kernel dependent only on the relative position of r and ρ, i.e. $p(r, \rho) = p(r - \rho)$. Photon beam elements are discretised into M bixels and 3D dose-space discretised into N voxels. Thus equation (2.58) can be cast as a discrete matrix operation

$$\mathrm{d} = \mathbf{P}\Psi \tag{2.60}$$

where $\mathbf{P}$ is an $M \times N$ dimensional pencil-beam matrix. Gustafsson *et al* (1995) optimise Ψ by an iterative algorithm similar in functional form to that previously presented by this group in Sweden (see the companion Volume) namely

$$\Psi^{k+1} = C[\Psi^k + \mathbf{A}\Delta_\Psi P_+(d^k)] \tag{2.61}$$

where k represents iteration number, C represents the application of constraints, P_+ is the probability of uncomplicated tumour control (which is the basis adopted for the optimisation) and Δ is a gradient operator with respect to the fluence vector. The matrix $\mathbf{A}$ controls the speed of convergence.

The novel feature of this formulation is that at each step of the optimisation procedure each component is individually corrected according to the objective function gradient for the single component. The optimisation thus focuses directly on the parameters which define the energy fluence (see equation (2.59) above). It does *not* determine the dynamic multileaf collimation, for example, for a *previously determined* IMB profile. Instead it finds the optimal physically realisable dynamic collimation directly.

The constraint operator C obviously takes account of the fact that beam fluences must be positive, but it also plays a more sophisticated role. For example, it includes the limits within which MLC leaves may move, that opposing leaves may not overlap, that velocity and acceleration constraints are not exceeded etc.

To demonstrate the efficacy of the optimisation a single treatment case was considered of optimising a case of female pelvic radiotherapy with just two beams at right angles to each other. However, the optimisation was performed for *ten* different levels of treatment complexity! The simplest was the case of two open unit-weight fields which produced $P_+ = 0.129$. Optimising fixed collimation geometry and beamweights immediately raised this to $P_+ = 0.685$. Designing two metal compensating filters in this way led to $P_+ = 0.854$, a very acceptable result but of course requiring individual block fabrication. Designing the velocity profiles of MLC leaves with a fixed beam-scan pattern led to $P_+ = 0.812$ and with a variable scanning pattern led to $P_+ = 0.813$. Only by allowing full pencil-beam delivery could they obtain $P_+ = 0.860$, comparable to the use of a fixed 2D metal compensator. The conclusions are striking that: (i) very significant gains can be had from including intensity modulation; (ii) the optimum scanning patterns for leaves could be designed *directly* by this method; (iii) that dynamic MLC therapy gave significant increase in P_+ compared with the use of static MLC fields.

2.2.7. The solution of Spirou and Chui including upstream fluence modulation

In section 2.2.4 the solution of Spirou and Chui (1994) was given for the problem of determining the optimum leaf velocities to deliver a known intensity modulation. From equations (2.21), (2.23) and (2.24) we have that

$$\frac{1}{v_1(x_{i+1})} - \frac{1}{v_2(x_{i+1})} = \frac{I(x_{i+1}) - I(x_i)}{dx_i} \qquad (2.62)$$

which is the discrete form of equation (2.4). From this the fundamental rule was established that if the right-hand side (rhs) is negative, the trailing leaf number 1 is set to have maximum velocity but if, conversely, the rhs is positive then the leading leaf number 2 is set to have maximum velocity instead. In the two cases equation (2.62) reduces to equations (2.40) and (2.32) respectively.

This theory assumes that the only degrees of freedom in the optimisation are the setting of the leaf velocities. It was assumed that the accelerator delivered a uniform upstream rate of fluence. Spirou and Chui (1996) have given the modified theory appropriate to the situation when the upstream fluence can itself be *spatially* modified. Suppose, without any collimation at all, the upstream fluence varies across the field such that the relative intensity distribution at the point x_i is $S(x_i)$, normalised to some location. Equation (2.21), the discrete form of equation (2.1), now takes the form

$$I(x_i) = S(x_i)[t_1(x_i) - t_2(x_i)] \tag{2.63}$$

since the time interval for which the point x_i should be open simply scales by $1/S(x_i)$. Thus the quantity $I(x_i)/S(x_i)$ takes the place of the simple expression $I(x_i)$ in all the equations in section 2.2.4. In particular equation (2.62) now becomes

$$\frac{1}{v_1(x_{i+1})} - \frac{1}{v_2(x_{i+1})} = \frac{[I(x_{i+1})/S(x_{i+1})] - [I(x_i)/S(x_i)]}{dx_i}. \tag{2.64}$$

From inspection of equation (2.64) it is now apparent that there are extra degrees of freedom in the solution. For any given intensity profile, from which, for an elementary section, the values $I(x_{i+1})$ and $I(x_i)$ are extracted, the scaling of the two rhs terms means that whichever leaf is to become the fastest is *not* solely determined by the sign of the intensity change. In fact one can change the sign of the rhs by changing the upstream modulation. Spirou and Chui (1996) discuss the options.

Firstly to minimise the treatment time one would like the trailing leaf to move at maximum speed. This makes the rhs either negative or zero. One would also like the leading leaf to travel at maximum speed (thus making both sides zero) since this will lead to minimum errors. However, this may not be possible because there is, in addition to the constraints on the maximum leaf speed, an extra constraint on the rate of change of upstream fluence that

$$\frac{|S(x_{i+1}) - S(x_i)|}{dx_{i+1}} \leqslant \left(\frac{dS}{dx}\right)_{max} \tag{2.65}$$

because of the Gaussian-like shape of the upstream fluence kernel. Whilst for any given S profile one can always choose the leaf speeds to satisfy equation (2.64) the converse is not true. For a given set of leaf speeds one cannot always satisfy equation (2.64) within the constraint of equation (2.65).

The solution is as follows. Firstly, given $t_1(x_i)$, $t_2(x_i)$, $I(x_i)$, $I(x_{i+1})$ and $S(x_i)$ the rhs of equation (2.64) is set to zero. This determines $S(x_{i+1})$. If equation (2.65) is satisfied then both leaf speeds are set to the maximum and this becomes the solution. If, however, equation (2.65) is not satisfied $S(x_{i+1})$ is recalculated to satisfy it and the rhs becomes non-zero. Depending on the sign of the rhs the maximum leaf speed is assigned to one or the other leaf by the same rule as discussed before (section 2.2.4) for the case with uniform upstream fluence.

Spirou and Chui (1996) discuss the full case of multiple moving leaves with upstream fluence modulation which is more complex. They also state that the problems of leaf transmission and penumbra can be accommodated in much the same way as discussed in section 2.2.4.

The advantage of using upstream modulation is that the overall treatment time can be reduced. They show an example of how a wedged field can be delivered via the sliding window technique both with and without upstream intensity modulation, the former leading to 40% reduction in time for the case considered. The method is also needed when the overtravel of the leaves is limited for a particular MLC, for example to 5 cm on the Scanditronix MM50 machine, which fortunately has the capability for upstream modulation.

The analysis represented by these equations is somewhat idealised because it is assumed that the upstream fluence profile can be generated with perfect spatial resolution. In practice of course the upstream profile can only be generated with a finite spatial resolution. Once the optimum scanning profile has been determined this must be converted to a practically realisable scan pattern taking this into account and then the leaf trajectories must be recalculated. So the practical algorithm requires a measure of iteration.

Spirou and Chui (1996) also indicate two further practical constraints. The first is that the verification of the scanning beam is limited by the finite spatial resolution of the monitor chamber attached to the MM50. In practice this corresponds to a 5×4 cm^2 area at 100 cm source-to-skin distance (SSD). The second constraint is that the leaf motion must be synchronised with the scan pattern for the upstream fluence. It must be said that, elegant though this concept may be, the practical realities are somewhat daunting. At the time of writing no patient irradiations using this method have been reported.

With the description of this theoretically elegant but possibly somewhat futuristic method we are near to concluding this moderately lengthy review of dynamic-leaf collimation for the production of IMBs.

2.2.8. *Experimental delivery of IMBs with dynamic-leaf movement*

Whilst there has been much progress in the development of the theory of the delivery of IMBs with dynamic-leaf movement, the limitations of the capabilities of commercial accelerators has resulted in few experimental studies so far. Over the next few years this situation will change.

Wang *et al* (1996) have used an MLC attached to a Varian 2100C accelerator to deliver IMBs by dynamic-leaf movement. They used the inverse-planning technique developed by Bortfeld *et al* (1994c) (see section 1.9) inplemented integrally within the Memorial Sloan Kettering Cancer Centre (MSKCC) 3D treatment-planning system. The case studied was the treatment of a prostate. Nine coplanar gantry orientations were selected equispaced in 360° and the beams were geometrically shaped to the beam's-eye-view (BEV) of the planning target volume (PTV). The inverse-planning technique minimised the quadratic differences between the desired and computed dose distributions with the dose in the OAR only playing a part when the dose rose above tolerance (controlled by a penalty function). That is, so long as the dose in a normal tissue does not exceed the tolerance limit, the voxel does not contribute to the quadratic score function. The target was 7 cm wide in the direction normal to the leaves so seven discrete line IMB profiles were generated for each gantry angle. The profiles at any particular gantry angle were computed independently of the others.

The IMB profiles were then converted into a sequence of leaf patterns by the technique discussed by Spirou and Chui (1994) (including the effect of 2% leaf transmission as discussed in section 2.2.4). The leaves are of course required to move at variable speeds. Since the Varian leaves can be in only one of two binary states, either stationary or moving at a speed of 5 cm s^{-1}, the variable speed required was simulated by dividing each transition into a number of substeps of smaller size in which the leaves were either stationary or moving at 5 cm s^{-1}. In the rare event that one or more pairs of leaves could not move fast enough, the MLC control computer inserted a pause in the irradiation. During the leaf movement the jaws were held open at the envelope of the irregular field.

The dose distribution corresponding to each fluence pattern was computed using the pencil-beam-convolution method. The dose distribution was also *measured* by calibrated film, in a cubic polystyrene phantom, the measurement plane being 100 cm from the source, the direction of incidence of the beam normal to the phantom surface and with 5 cm of polystyrene above the film plane.

Wang et al (1996) found that in general the two sets of isodose curves from calculation and experimental measurement had similar shapes but that there were discrepancies in excess of 5–10% which were unacceptable. The MSKCC criterion for acceptability was 3% in dose or 3 mm in displacement of isodose lines, whichever is lower. They concluded that the discrepancies were due to: (i) the rounded shape of the leading edges of the leaves; (ii) scattered radiation originating from the components of the treatment machine. They found that source (i) was inadequate to account for all the difference but that by adding a small but constant (2.4 mm) widening of the sweeping gap, residual discrepancies could be reduced to 2% (figure 2.13).

Wang et al (1996) also investigated the 'tongue-and-groove' effect, namely that the small narrow (1 mm wide) strip between adjacent leaf sides, can be underdosed. This arises because this region can be 'seen' by half-depth shielding for parts of the sweep. Although the fluence reduction in the strip can be as much as 60–70% the dose reduction is by only some 10–15% because of electron transport, although this can widen the region in which the effect is noticed. The solution was found to be synchronising the movements of leaf-pairs so that all leaf-pairs finish closed at the same

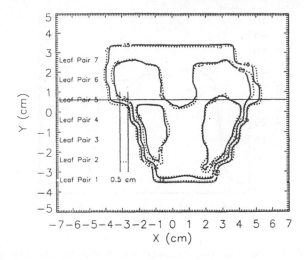

Figure 2.13. *An example of a 2D intensity-modulated field. This field was delivered with the gantry at 0° and with seven pairs of moving leaves. It is one of nine fields (at different gantry angles) contributing to a prostate treatment. The distribution shows the dose in a plane normal to the beam direction measured and calculated for a phantom geometry. The solid lines show the measured isodose distributions and the dotted lines show the calculated dose distributions. An effective constant gap-widening (see text) has been incorporated to improve the agreement between calculation and measurement. (From Wang et al 1996.)*

time. This is achieved by a reduction in the speed of some leaf-pairs (i.e. some profiles are delivered individually suboptimally) with correspondingly reduced gap widths between a left and right leaf-pair. This does not *eliminate* the problem but it increases the time for which leaf-pairs travel together and makes the effect essentially inconsequential (figure 2.14).

Wang *et al* (1996) also observed that the gaps between a left and right leaf-pair can sometimes become quite small which is associated with a reduced output factor. However, they concluded that features of the Clinac 2100C beam optics reduce this effect (the variation with field size being only a few per cent from the smallest to the largest field sizes.) Also, for IMBs in which the variation in fluence is small the effect can be neglected.

It is interesting to note that the approach taken analysed the component parts of the full nine-orientation delivery separately. The main conclusion was that the delivery of IMBs by dynamic-leaf movement was certainly possible and that some problems such as the 'tongue-and-groove' effect and the variation of output factor with field size may not be as important as first thought provided synchronisation is invoked at least for this particular accelerator and set of IMBs. The work is a landmark study in the development of IMB therapy.

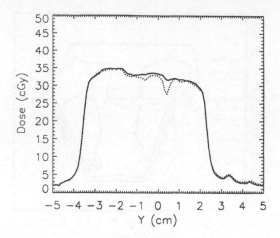

Figure 2.14. *Line dose profiles sampled normal to the direction of leaf movement showing the tongue-and-groove effect for the 120° gantry-angle beam (one of a set of nine) for experiments before (dotted line) and after (solid line) synchronisation of the leaves. The dips in the dotted profile at −0.5 cm and +0.5 cm are due to the tongue-and-groove effect. (From Wang et al 1996.)*

2.3. CLASS 3: LEAF-SWEEP AND CLOSE-IN; THE BORTFELD AND BOYER METHOD OF MULTIPLE-SEGMENTED (STATIC) IMBS

'Leaf-sweep' and 'close-in' are alternative techniques in which the MLC leaf-pair takes up a number of static locations. The radiation from each static field combines to give the modulated beam profile (figure 2.15) (Bortfeld *et al* 1994c, Boyer 1994, Boyer *et al* 1994). One conceptual advantage of the methods is that the leaf locations bear an intuitively obvious relation to the intensity profile (figures 2.16 and 2.17). A possible disadvantage is that the x-rays are switched off while the leaves move and it may be necessary to allow time for the radiation to stabilise each time it is switched on. However, Williams (1993) suggests this wait time is probably no more than a small fraction of an MU. The use of a gridded gun could overcome the problem but not all manufacturers have this on their linear accelerators (Bortfeld *et al* 1994a,b). However, the total patient treatment time must include both the beam-on and the beam-off periods and this reduces efficiency.

The determination of the static locations is performed as follows:

(i) convert from fluence as a function of the discrete positions in the 1D profile to a continuous function;
(ii) select a discrete interval of fluence;
(iii) resample to obtain the set of discrete positions at which the continuous profile crosses the fluence intervals. This is *not* a monotonic inversion. The

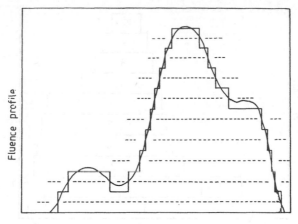

Distance along central axis

Figure 2.15. *Shows how a 1D intensity modulation may be created for a radiotherapy beam profile. The horizontal axis is the distance along the direction of travel of the leaf, measured at the isocentre of the beam (called a central axis in the transaxial cross section of the patient). The vertical axis is x-ray fluence. The solid line is the intensity modulation expressed as a continuous function of distance, interpolated from the discrete modulation resulting from some method of inverse planning. The horizontal dotted lines are the discrete intervals of fluence. Vertical lines are created where the dotted lines intersect the continuous profile thus giving a set of discrete distances at which discrete fluence increments or decrements take place. These are realised by setting the left and right leaves of an MLC leaf-pair at these distances in either 'close-in' or 'leaf-sweep' technique. Note all left leaf settings occur at positions where the fluence is increasing and all right leaf settings occur at positions where the fluence is decreasing.*

result is an even number of crossings per boundary and a discretely sampled 1D profile (discrete now in *fluence* not *position*);

(iv) determine the positions of the leaf ends to realise this profile experimentally—this comprises lists of pairs of right and left leaf positions corresponding to each fluence interval;

(v) from these lists follow the leaf-setting trajectories, either by *leaf-sweep* or *close-in*—the two are entirely equivalent (see the proof in the appendix of Bortfeld *et al* (1994c) and figures 2.16 and 2.17 where the equivalence can be immediately seen);

(vi) fluence profiles are converted to dose distributions at depth using convolution dosimetry.

The method was verified experimentally using film. Obviously the accuracy depends on the fluence interval chosen (Boyer *et al* 1994).

The conclusion is that even complex profiles can be realised with a small number of leaf positions. Bortfeld *et al* (1994c) claim this is better than the

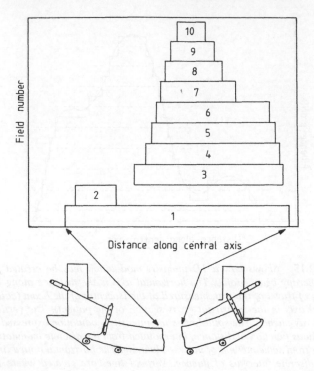

Figure 2.16. *The ten separate fields which when combined would give the distribution of fluence shown in figure 2.15. Each rectangle represents a field and the left vertical edge is the position of the left leaf and the right vertical edge is the position of the right leaf. This method of setting the leaves is known as the 'close-in' technique. A schematic diagram of a pair of MLC leaves is shown below the fields with arrows indicating the correspondence with the field edges.*

method of dynamic delivery first discussed by Convery and Rosenbloom (1992). In practice the two methods offer complementary approaches with different features. For example, Bortfeld *et al* (1994c) disregard the dead-time in which the beam is off and do not determine the practical arrangement in terms of the limiting velocities of the leaves. Bortfeld *et al* (1994b) do discuss the predicted treatment times including 'beam-off'.

If $x_L(i, j)$ and $x_R(i, j)$ $(i = 1, \ldots, N)$ represent the set of x locations for left and right leaves for fluence level j where i labels the locations and there is an even number N of these for each of the two leaves at that level of fluence, two schemes for practical delivery may be adopted.

For 'close-in' the leaves take up locations by pairing left and right leaves with the same i and j values, i.e. the leaves are either side of a local maximum intensity (figure 2.16).

To implement 'leaf-sweep', firstly, the locations for each leaf are

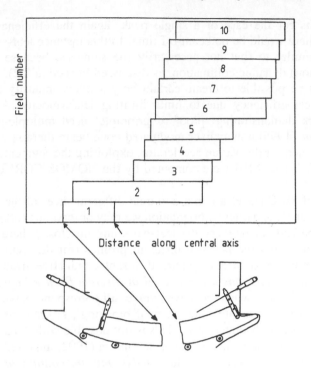

Figure 2.17. *The ten separate fields which when combined would give the distribution of fluence shown in figure 2.15. Each rectangle represents a field and the left vertical edge is the position of the left leaf and the right vertical edge is the position of the right leaf. This method of setting the leaves is known as the 'leaf-sweep' technique. A schematic diagram of a pair of MLC leaves is shown below the fields with arrows indicating the correspondence with the field edges.*

separately sorted by the magnitude of location x with no regard for the j value labelling the intensity intervals. Then the leaves are paired in order of magnitude of location (figure 2.17), i.e. firstly smallest x_L value with smallest x_R value and so on up to largest x_L value with largest x_R value.

By comparing figures 2.16 and 2.17 it may be seen that these two schemes are entirely equivalent. Also they both take exactly the same time (total number of cumulative MUs). As we have already seen in section 2.2.5.2 there are actually $N!^2$ possible arrangements of which these are just two. Bortfeld *et al* (1994c) prove the equivalence mathematically but it is just as easy, for this example, to see by diagrams. In terms of beam-on time only, both methods are in general more inefficient than the use of a compensator unless the intensity modulation had just a single maximum and no minimum, in which case they would be equal in efficiency to the use of a compensator. This argument ignores the reduction in efficiency due to beam-off time. (If close-in and leaf-sweep were alternatively realised with continuous movements for the case of a single peak, again the efficiency would be unity as there would be no beam-off time. In this instance leaf-sweep would become similar to (but not necessarily the same as, because there is no optimisation) dynamic collimation, as discussed in section 2.2). Of course it might not be possible to create certain very shallow intensity modulations with perfect efficiency due to finite limiting leaf velocity. Although the early work demonstrating 'proof of principle' used multiple-static fields, Boyer *et al* (1996) now deliver modulated cone-beam therapy via dynamic-leaf movement of the Varian accelerator, exploiting the segmented treatment table (STT). The IMBs were computed by the NOMOS CORVUS planning system.

If sets of MLC leaves are in use to treat different slices of the patient with the gantry at some particular orientation we may observe the following. Since in both methods of delivery the leaf-pairs end up closed there is no extra complication from the fact that different profiles (for different slices) may have different maxima and require different cumulative irradiation times *provided the irradiations are delivered in discrete fluence intervals* so that the beam-off times, when the leaves move, all correspond between slices. It is even possible to create 'internal field blocking', i.e. regions in the field where the dose is absolutely zero. This is not possible with dynamic therapy.

On the other hand Convery and Rosenbloom (1992) analysed the close-in technique in terms of a series of close-in *dynamic and continuous* movements with the beam on, separated by beam-off periods while the leaves moved to start different close-in sequences. They made the observation that whilst this would indeed achieve the same result as their method of unidirectional dynamic collimation with no beam-off periods (but with different efficiency), it would be impossible to apply the close-in technique simultaneously, with an MLC and with optimum efficiency, for the profiles required for multiple slices, because the beam-off periods would not correspond. One possible solution would be to divide up each of the profiles into peaks and let each pair of leaves close-in on one peak (this process will take the time for the largest peak); then the beam could be switched off, all leaves repositioned to the base of a new peak and the process repeated and so on until each profile has been completed. It would be hard to estimate the efficiency of this total process because this would depend not only on the complexity of each profile but also on their *relative complexity*.

The use of discrete fluence intervals overcomes this problem, except that the beam-off time is determined by the time it takes for the biggest changes of leaf position to occur, so some leaf-pairs will be ready and idle whilst others are still moving into position. Moreover, some profiles will be finished whilst others are continuing. The concept of delivery by discrete fluence intervals is a very important contribution made by Bortfeld *et al* (1994c).

Bortfeld *et al* (1994a,b) report an experiment to verify the method for just one 'patient'. Using the method described in Chapter 1, section 1.9, they prepared the intensity profiles for nine fields with the number of leaf settings per leaf-pair per field varying between 21 and 30. A polystyrene multi-slice phantom was constructed based on the outlines of the CT data for this patient for whom a concave-surface, high-dose volume was required to the prostate, sparing rectum and bladder. The phantom comprised nine slices, 1 cm thick with films sandwiched between the slices. The treatment was delivered with a Varian accelerator fitted with a Varian MLC using the 'sweep' technique. This one-off experiment confirmed the conformal nature of the 3D dose distribution although some positional inaccuracies were noted. The 80% contour was a very accurate match between plan and measurement. The experiment took many hours to perform because the MLC was not under fully automatic computer control, but this requirement is being addressed by Varian.

Figure 2.18 shows that the experiment was very successful in generating a high-dose volume with a concave outline. The necessity for this is clear from figure 2.19 which shows that the prostate wraps around the rectum and is in turn itself wrapped around by part of the bladder, creating the requirement for IMB therapy.

Other workers have also developed and discussed the multiple-segmented (static) method. Mageras *et al* (1992) describe the features of a computer system which is able to control the treatment machine automatically without manual setup. The controlled features include the gantry orientation, the couch rotation and translation, the head twist, the beam energy and modality, the dose, dose-rate, wedges and *the position of the leaves of an MLC*. The implementation is for a Scanditronix MM50 Medical Microtron and so an additional computer-controlled possibility is the creation of *IMBs* via continuously varying MLC-shaped fields or scanning the beam. The essential feature is that all the control sequences are down-loaded from a control computer which communicates with a host computer which, for example, accommodates the treatment-planning package. Mageras *et al* (1992) are concerned with how to ensure that such a system performs reliably and they give details of how to 'dry-run' the movements (without the beam on) to test for potential collisions etc. Software security is fundamental. The design can also be realised on a Varian Clinac 2100C with an MLC. The software is written for a VAXstation host. Typically ten fields each separately shaped with an MLC delivering 2 Gy take only 4 min, some three times shorter than the setup time. This is quicker than their 'conventional' manual setup of a few fields.

Similar work using an MLC with multiple-segmented fields to create intensity modulation has been reported by Fraass *et al* (1994), De Neve *et al* (1995) and De Jaeger *et al* (1995). De Neve *et al* (1995) and De Jaeger *et al* (1995) used the GRATIS treatment-planning system and adapted it to create simple segmented fields. Using 6–8 beam incidences they created 30–

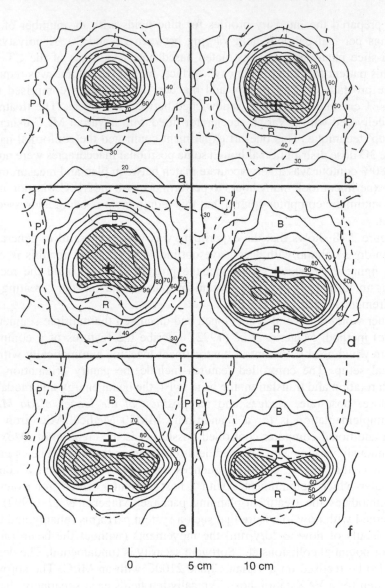

Figure 2.18. *An evaluation of the film scans for the experiment in which Bortfeld delivered IMB therapy to a phantom. Isodose lines are superposed on the anatomy for slices at distances of (a) −3 cm, (b) −2 cm, (c) −1 cm, (d) 1 cm, (e) 2 cm and (f) 3 cm from the central slice. Isodose values are in per cent in increments of 10%. The 95% isodose line is also shown. The 100% isodose line only appears in slice (d) which contains the dose maximum. Anatomical structures are rectum (R), bladder (B) and pelvis (P). The target is displayed by the shaded areas. The cross marks the axis of rotation. (Reprinted by permission of the publisher from Bortfeld et al 1994b; © 1994 Elsevier Science Inc.)*

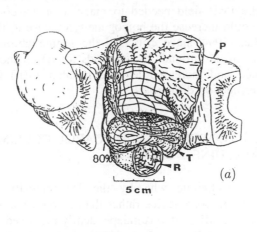

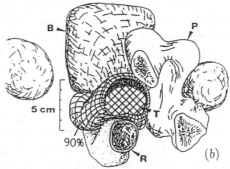

Figure 2.19. *Three-dimensional display of isodose surfaces and the anatomical structures. R is the rectum, B is the bladder, P is the pelvis and T is the target. (a) Direction of view: superior to inferior. The isodose value is 80% of the maximum dose. The dome of the bladder is shown cutaway for better appreciation of the isodose surface with the inner wall of the bladder. The seminal vesicles are enclosed in the invaginating target volume. (b) Visualisation from inferior to superior, showing the 90% isodose. The pubis synphysis and right pelvic bones are removed to show the juxtaposition of the isodose surface between the bladder and the rectum. (Reprinted by permission of the publisher from Bortfeld et al 1994b; © 1994 Elsevier Science Inc.)*

50 segments in total and were able to generate 80–85% dose homogeneity in the concave target.

Evans *et al* (1997) have developed a technique to set the optimum intensity levels at which an IMB profile could be 'chopped up' into a small finite number of fields. This technique relied on minimising the least-squares difference between the real histogram of the intensity levels and the rearranged histogram with just a few levels. They showed that applying the method to the IMB field needed to make a breast compensator (see Chapter 4) they could increase the homogeneity of dose to the breast.

Galvin and Han (1994) have criticised the segmentation technique for beam modulation on the grounds that it leads to: (i) an increased number of photons needed and increased leakage problems; (ii) field mismatch problems; (iii) verification difficulties.

2.4.   CLASS 4: TOMOTHERAPY; THE CAROL COLLIMATOR AND THE MACKIE MACHINE

Tomotherapy is a technique whereby the 1D intensity modulation is created by a special-purpose device rather than using a compensator or the accelerator jaws or the MLC. It was independently proposed by two groups, one at the University of Wisconsin and the other at Pittsburg (Mackie *et al* 1993, 1994a, Holmes *et al* 1993, Carol 1992a,b, 1993, 1994a,b, 1995, 1996a,b, Carol *et al* 1993, Curran 1996, Grant 1996) and gains its name from being the therapy analogue of computed tomography. At any one instant in time one or two slices (Greek 'tomos' = slice) are being irradiated. A special collimator, to be described next, generates the intensity profile. At the same time the gantry rotates about the long axis of the patient. The two proposals are slightly different. In one (Carol) the patient is translated longitudinally *between* gantry locations by the slit width of the collimator to treat sequential transaxial slices. In the other (Mackie), the patient is translated longitudinally slowly and continuously *during* gantry rotation. Strictly the term tomotherapy refers only to the second arrangement but the first is sufficiently similar to be a useful generic title.

Carol's collimator (called the MIMiC (Multivane Intensity Modulating Collimator) and manufactured by the NOMOS Corporation) comprises a slit aperture which defines two slices to be treated. The collimator and source combination rotate about an axis normal to the slice planes defined by this slit. The aperture is provided with 40 stubby leaves (or vanes) in two banks, each with 20 leaves of thickness approximately 8 cm which can be activated electropneumatically to move at right angles to the slit aperture. Each bank can be independently controlled. Two slices can be treated simultaneously. When a valve is turned on, this causes high-pressure air to flow to the front side of the piston and drive the vane out of the field; the device thus fails safe (closed) if the air pressure were to drop or fail completely. When the valve is turned off, constant lower air pressure at the backside of the piston drives it back into the field. The pistons operate in 40–60 ms in either direction. These stubby leaves or vanes can be in the aperture for all, none, or part of the time under computer control and thus a 1D intensity-modulated field is constructed (figures 2.20, 2.21 and 2.22). Hence the collimator has unit

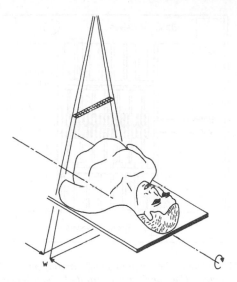

Figure 2.20. *Shows a method to obtain a 1D intensity profile of radiation without the use of an MLC. The beam is collimated to a narrow slit of width w, across which the intensity is modulated by a series of stubby vanes or leaves which move at right angles to this slit. The vanes can be in the radiation field either all the time, none of the time or for part of the time, thus constructing an intensity profile. The experimental means to achieve this was developed by Carol (1992a,b), who originally proposed two sets of 20 vanes so that two different intensity profiles could be simultaneously created for two adjacent transaxial slices of the patient. The apparatus (the MIMiC) rotates around the patient so that different profiles can be created at each gantry orientation. The vanes are shown schematically. In practice they comprise approximately 8 cm of tungsten in the direction of the beam. The maximum field size at the isocentre depends on to which accelerator the MIMiC is mounted.*

efficiency *per slice*, the same as a compensator. Of course because only two slices are treated at the same time, whereas for dynamic therapy and for the multiple-segmented static method all CT slices may be simultaneously treated, the true efficiency of the MIMiC may end up poorer for PTVs of large longitudinal extent. Only a metal compensator has truly unit efficiency. A prototype has single-focused vanes or leaves. The vanes are focused in the transaxial direction and are 6 mm wide on the patient side and 5 mm wide on the source side. The adjacent sides of the vanes have a five-element 'tongue-and-groove' shape to prevent radiation leakage between them. The aperture of travel can be set to either 5 mm or 10 mm. The field-of-view depends on the exact placement of the MIMiC in the blocking tray. On a Philips SL25 accelerator the diameter of the field-of-view is 182 mm. It is claimed to be able to fit any manufacturer's linac (figures 2.23, 2.24 and 2.25). The machine includes a set of sensors which monitor the location and

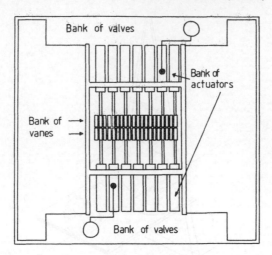

Figure 2.21. *Shows the experimental arrangement of the MIMiC stubby-vane device. The vanes are controlled electropneumatically. Each vane has an associated actuator and valve. Not all the actuators and valves are shown for clarity. Two back-to-back collimators are shown treating adjacent slices of the patient. The diagram is adapted from one by Carol (1992b).*

speed of movement of the vanes.

The MIMiC collimator comes with a system for inverse planning which is based (Carol 1993, 1994a) on the simulated-annealing approach described by Webb (1989, 1991) but which was independently designed and constructed. The 'aggression' of the inverse planning can be controlled by weighting the cost function differently for the target volume and for the OAR and by specifying dose limits. This has a big effect on the resulting plan (Carol and Targovnik 1994). Planning is very computer intensive and can take many hours but is implemented off-line in batch mode, once the user specifications have been entered. The treatment-planning package is known as 'PEACOCKPLAN' (Woo *et al* 1994, Kalnicki *et al* 1994, Wu *et al* 1995).

The treatment delivery is monitored by computer and light emitting diodes (LEDs), which give the patient position, and can interupt the treatment delivery should there be any error. When first proposed there was no facility for executing couch translation but this has now been incorporated. A special indexing table called the CRANE controls longitudinal movement of the couch to 0.1–0.2 mm. This CRANE consists of a large vertical column supporting two ball-bearing driven arms each equipped with a digital scale and clamped to rails on the treatment couch. By releasing the locks on the couch and adjusting the arms the position of the couch is determined and then relocked. The three-hundred-pound (~136 kg) weight of the indexing system prevents backlash and ensures precise positioning. The first patient was treated in Houston, Texas in March 1994. The leakage radiation of

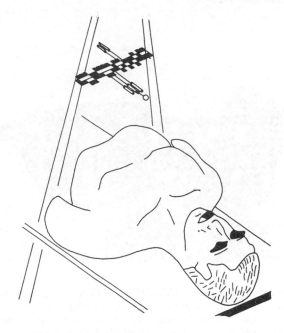

Figure 2.22. *Shows the location of the MIMiC stubby-vane collimator in relation to the patient with some of the vanes closed and others open. The two back-to-back slit apertures each comprise 20 vanes.*

the NOMOS machine was measured by Grant *et al* (1994a) and shown to be about 1% of the delivered dose. Grant *et al* (1994b) have also verified that the MIMiC collimator delivers dose distributions as planned by PEACOCKPLAN with an accuracy of within 5 mm for isodoses above 50%. Kania *et al* (1995) have discussed the calibration of PEACOCKPLAN. The NOMOS products MIMiC and PEACOCKPLAN are together known as the PEACOCK system. PEACOCKPLAN only computes IMBs and is 'close-linked' to the MIMiC. A new planning system from NOMOS, known as CORVUS (the collective name for ravens; all NOMOS products are named after birds) will also compute plans for uniform MLC fields as well as having the PEACOCKPLAN facility. Sternick and Sussman (1996) have summarised the quite complex legal requirements which had to be met for the MIMiC and PEACOCKPLAN before receiving 'clearance' by the US Food and Drug Administration (FDA) in 1996.

A potential problem with the use of the MIMiC was considered to be the possibility of matchline difficulties when more than one pair of slices is required to fully irradiate the PTV. The NOMOS Corporation have always been aware of this concern and recently Carol *et al* (1996) have studied the potential dose inhomogeneity. First a set of three cylindrical targets of diameters 4, 10 and 18 cm and length 15 mm were planned for IMB irradiation. In this arrangement there is no requirement for indexing the

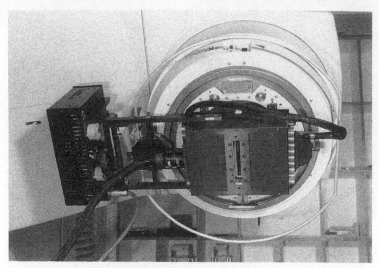

Figure 2.23.   *Shows the NOMOS MIMiC collimator attached to a Philips' SL25 linear accelerator at the Royal Marsden NHS Trust, Sutton, UK. The two sets (banks) of leaves (or vanes) can be seen with some open and some shut. (Loan of MIMiC courtesy of the NOMOS Corporation.)*

couch since the target fits within the field-of-view of the pair of slices. Target inhomogeneity, defined as the range of dose, was computed as 11, 14 and 16% respectively. This arises purely as a result of the use of IMBs. It was not stated what the requirements for conformality were but this was claimed to be a similar dose homogeneity to that reported by others.

Then cylinders of the same diameters but requiring multiple slices were planned. If it were assumed that the slices matchlined correctly with no error this only introduced a further 2–3% dose inhomogeneity due to the way in which longitudinal dose profiles overlapped. However, even small uncertainties in the indexing of the sequential slices led to very large unacceptable increases in dose inhomogeneity. For example, a 2 mm error led to a 41% dose inhomogeneity, emphasising the need for very accurate indexing. It is claimed that the CRANE can perform indexing to 0.1–0.2 mm accuracy which would introduce only a 3% increase in dose inhomogeneity.

Another issue of concern is the dosimetry of small fields generated by the MIMiC as modelled by PEACOCKPLAN. PEACOCKPLAN models each bixel as if it had closed walls whereas, at the time of treatment delivery, some walls are open. The differences between planning and delivery have been studied by Webb and Oldham (1996). They found that the way in which PEACOCKPLAN models the elemental bixel dose distribution via a stretched kernel actually minimises the difference between planning and delivery.

Mackie's device is at an earlier stage of development (figure 2.26). It was

Figure 2.24. *A more general view than figure 2.23 showing the NOMOS MIMiC collimator plus the computer which controls the movement of the leaves. The computer is mounted on two arms which protrude from the MIMiC. The computer is controlled via a touch-screen and has a floppy disk drive for the disk which contains the information for changing the leaf positions, determined by the PEACOCKPLAN planning system. The two cables which can be seen are the airline and the electrical links to the equipment. These two cables rotate with the equipment. (Loan of MIMiC courtesy of the NOMOS Corporation.)*

proposed first in the summer of 1989 by Stuart Swerdloff, working with Rock Mackie (Mackie 1994, 1996, Deasy 1995). No working prototype yet exists but one is under construction (Mackie 1995, Mackie *et al* 1996). Mackie *et al* (1993, 1994a,b) proposed a new radiotherapy treatment machine of dream-like conception. The treatment accelerator will be mounted on a CT-like gantry and will rotate through a full circle. The beam will be collimated to a narrow slit, at right angles to which will be a set of (say 64) leaves of an MLC called a dynamic collimator. The leaves will be able to fly in and out of the slit aperture under computer control to define a continuously changing 1D beam profile (like the MIMiC). At the same time the treatment head and gantry will rotate and the patient will translate through the aperture in a manner analogous to a spiral CT scanner. This is the strict definition of *tomotherapy*. This will achieve conformal radiotherapy and will be the practical implementation of the technique, planning for which has been solved by many authors addressing the inverse problem.

If the gantry were to remain stationary but the patient were still to translate through the aperture with the dynamic collimator activated, the same equipment would achieve *topotherapy*.

More than this, the gantry will carry a diagnostic CT scanner for planning,

Figure 2.25. *Another view of the NOMOS MIMiC collimator showing a leaf pattern. (Loan of MIMiC courtesy of the NOMOS Corporation.)*

for verifying the treatment position during therapy, and much more. It will also carry a digital portal imager to verify the beam positions. It has been described (Holmes 1993) as a 'radiotherapy department in a box' since the machine will replace so many 'conventional' tasks including CT-simulator, block and compensator-making mill, and simulator.

Mackie *et al* (1993) worked out the engineering specification for such a machine. This is clearly futuristic radiotherapy but it is a very serious proposal. A spiral CT gantry has been acquired for mounting the tomotherapy collimator and construction is underway (Mackie 1995, Mackie *et al* 1996). Since this is a quite revolutionary concept for delivering radiation therapy we shall concentrate on some of the details. The proposals from the group at the University of Wisconsin are embodied in two US patents (Swerdloff *et al* 1994a,b). Swerdloff *et al* (1994a) show the technique by which individual elements of a fan-beam of radiation may be differentially attenuated. Referring to figure 2.27 a fan-beam of radiation (14) from a point source (18) of an x-ray generator (12) is primarily collimated to a fan by collimator (16) so that (20) represents the central section through the fan. A housing mechanism (22) allows a plurality of wedge-shaped attenuators (30), focused back to the source, to be moved into and out of the beam, separately, by a set of actuators (32), coupled to the attenuators by flexible rods (34). The actuators are driven by compressed air through tubes (35).

Figure 2.28 shows a cross section through the collimator at (2) in figure 2.27 showing how the vanes (24) slide on rails (36) which in turn are securely anchored to a backplate (26). The rails or sleeves are radiotranslucent in the fan of radiation but can be more substantial outside

(a) **Tomotherapy Unit**

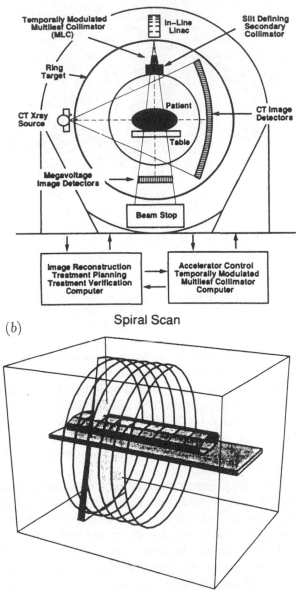

(b) Spiral Scan

Figure 2.26. *A schematic diagram of Mackie's proposal for tomotherapy. (a) An in-line linac with a temporally modulated MLC is mounted on a gantry along with other components which will perform megavoltage CT and diagnostic CT. The equipment rotates continually as the patient is translated slowly through the beam, hence (b) the motion is spiral with respect to the patient. (From Mackie et al 1993.)*

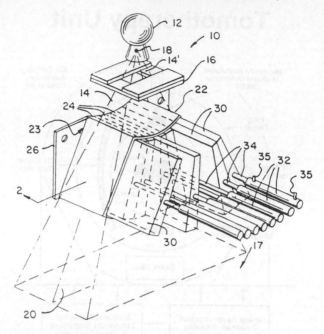

Figure 2.27. *A general schematic diagram of the collimator proposed by Swerdloff et al (1994a) for creating intensity-modulated fan-beams of radiation (see text for details.) (From Swerdloff et al 1994a.)*

the irradiated area. The backplate can also be substantial since it is not in the beam. The rails are of sufficient length to support the vanes even when they are outside the x-ray field, the so called open position, as well as when inside, the so called closed position. The relative radiation fluence within each beam (28) is controlled by the duty cycle of each vane. Figure 2.29 shows how the vanes are captured and supported by rigid collars (42) when in the fully open or closed positions, giving them more support than can be obtained by the guiderails alone.

Swerdloff *et al* (1994b) present a proposal for a modified collimator which overcomes certain problems associated with the one just described. Two problems are that with a single bank of attenuating vanes there is friction between adjacent vanes which can slow down the rate at which the vanes may be moved. If the tolerance between vanes is adjusted to reduce this friction to assist overcoming this problem, this would lead to a second problem of leakage radiation between the vanes.

They solve this rather neatly with the proposal shown in figure 2.30. The basic principle of collimation to a fan is the same as shown in figure 2.27 and will not be commented on further. The new idea is to have two banks of attenuating vanes, examples of which are shown at (28) and (34). Each plurality of vanes is again focused at the source but one bank lies closer to

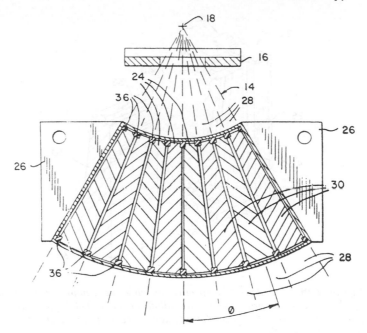

Figure 2.28. *A cross section of the collimator proposed by Swerdloff et al (1994a) for creating intensity-modulated fan-beams of radiation (see text for details.) (From Swerdloff et al 1994a.)*

the source than the other. Each bank contains alternate vanes and vane-width spaces. The banks are arranged so that where there is a vane in the upper bank there is a space in the lower bank and vice versa. Figure 2.31 makes this very clear. Vanes are shown with shaded circles such as at (28) and spaces as (30). Once again the attenuating vanes move on rails or sleeves and are driven by air-operated pistons. Figure 2.32 shows a perspective view of the arrangement by which each bank is captured by supporting collars (26). The vanes thus do not have any surface bearing on an adjacent vane since the 'adjacent' vane is in the other bank. Also the vanes are constructed just a little wider than the spaces so there is no leakage radiation between adjacent rays. Thus this arrangement simultaneously overcomes both of the problems described above.

A second new proposal in the patent by Swerdloff *et al* (1994b) is that each vane should be able to move out of the radiation fan on either side of the fan. To achieve this the guiderails are of sufficient length each side of the collimated beam longitudinally. The reason for this is as follows. Imagine a vane in the closed position. Let it move one side of the radiation fan to the fully open position and then return back to the fully closed position. This is the movement needed to achieve a certain attenuation with the attenuation being determined by the duty cycle, the ratio of the open to closed times.

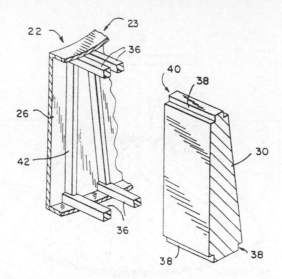

Figure 2.29. *A cutaway perspective showing the capture of the vane of the collimator proposed by Swerdloff et al (1994a) for creating intensity-modulated fan-beams of radiation (see text for details.) (From Swerdloff et al 1994a.)*

However, in making this movement a *gradient* of radiation fluence arises longitudinally due to the finite time it takes for the vane to execute the movements. This is a small but nevertheless important perturbation. Now, however, imagine that the vane can execute the same movements—fully open to fully closed to fully open—but *with the vane moving to the other side, longitudinally, of the fan.* In this case the longitudinal gradient will be exactly opposite to that in the first case. However, provided the magnitudes of all accelerations and velocities are exactly the same in the two cases, the fluence profile created by *summing both movements* will be exactly flat, a very neat solution. Swerdloff *et al* (1994b) provide for this degree of control in the associated circuitry and microswitches which indicate where the vanes are and lock them into positions. Mackie *et al* (1995) suggest that an optimal tomotherapy design will be able to deliver a wider field than the MIMiC and also will only need one bank of vanes thus reducing the noise (which the MIMiC makes) because there will be no colliding banks.

It should be noted that there is a fundamental difference between the two approaches, although many similarities. Carol's implementation has the patient couch stationary as the dynamic collimator executes one gantry revolution. One (or, as discussed, two) slices are treated. The couch is then incremented and the rotation repeated to treat one or two further slices. This process then repeats for sequential slices. In Mackie's proposal the couch is in continuous slow translational motion as the gantry with its dynamic collimator rotates. The dynamic collimator thus executes a spiral

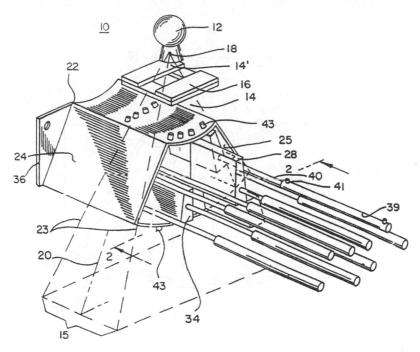

Figure 2.30. *A general schematic diagram of a second collimator proposed by Swerdloff et al (1994b) for creating intensity-modulated fan-beams of radiation. This collimator has two banks of attenuating vanes which overcomes some difficulties with the collimator shown in figure 2.27 (see text for details.) (From Swerdloff et al 1994b.)*

movement with respect to the patient (as in spiral CT). The ratio of the pitch of the spiral or screw motion to the slice width could be important in ensuring there are no matchline difficulties, a potential difficulty with the Carol-type approach (although minimised with the CRANE). To investigate this, Mackie (1993) conducted experiments with phantoms on a horizontal platter mounted on a rotating vertical screw thread with a horizontal slit of radiation collimated by a prototype NOMOS MIMiC dynamic collimator. This arrangement simulated spiral tomotherapy (figure 2.33).

Mackie *et al* (1995) showed some results of this experimental simulation of their tomotherapy machine using the NOMOS MIMiC collimator. They compared the delivery in sequential slice-by-slice mode with that of a continuous helical motion. More than this, they added the facility to allow the phantom to execute a periodic motion with variable amplitude and periodicity. It was discovered that with an amplitude of 0.5–1 cm and a frequency of twice the rotation frequency, the simulated breathing did not significantly affect the dose delivery but it will make some perturbations. Figure 2.34 shows a radiograph of the distribution of dose in a transaxial

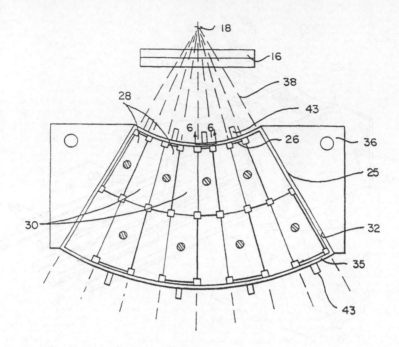

Figure 2.31. *A cross section of a second collimator proposed by Swerdloff et al (1994b) for creating intensity-modulated fan-beams of radiation (see text for details.) (From Swerdloff et al 1994b.)*

slice achieved with the NOMOS MIMiC collimator set to deliver a C-shaped distribution and using the setup described above. There are some ring artefacts evident at about the ±3% level probably due to the imprecise setup of the collimators (i.e. not exactly $\frac{1}{4}$ leaf spacing offset as is arranged with CT detectors).

Mackie *et al* (1993, 1995) suggest using an inverse filtered-backprojection method of computing the required intensity-modulated profiles. An iterative stage subsequently refines the profiles. They draw the analogy with SPECT reconstruction and claim this method is better than an analytic technique (see section 1.8). They have compared the results of a calculation of dose for a C-shaped high-dose region enclosing a small square OAR with the experimental dose delivered by a MIMiC collimator programmed to receive the 1440 beam intensities for 20 vanes at 72 angular orientations. The agreement was within 5% or isodose lines agreed to within 0.5 mm (Mackie *et al* 1995).

Webb (1994a,b,c) has compared the dosimetric outcomes of the Bortfeld/Boyer (BB) and the Carol/Mackie (CM) techniques of collimation. In essence CM allows a continuum of gantry orientations with a continuous intensity modulation at each angle whereas the BB method is only practical

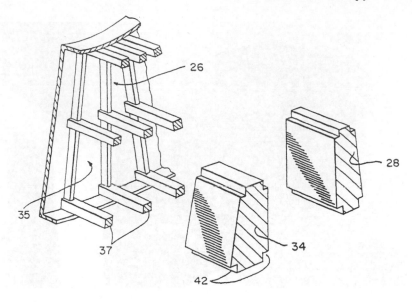

Figure 2.32. *A cutaway perspective showing the capture of the vanes in both banks of the collimator proposed by Swerdloff et al (1994b) for creating intensity-modulated fan-beams of radiation (see text for details.) (From Swerdloff et al 1994b.)*

with a finite number of orientations and gantry angles. Webb (1994a,b) represented the CM technique with 120 angles at 3° intervals with continuous modulation and the BB technique with 9 angles at 40° intervals and either 10 or 30 discrete strata of intensities in each profile. For some model problems, in which the PTV and OAR were close and partially overlapped, treatment plans were prepared under the two conditions. The DVHs in the PTV and OAR were computed and converted to TCP and NTCP measurements using the Webb and Nahum (1993) model for computing TCP when the dose distribution in the PTV was inhomogeneous and the Lyman (1985) four-parameter model with the Emami *et al* (1991) data for NTCP. It was found that, when the number of strata was small, the stratification and smaller number of gantry angles in the BB technique led to increased inhomogeneity of dose in the PTV and hence decreased TCP compared with the CM method. If for the BB method the mean target dose were increased so that the TCPs were the same for the two methods then the NTCP for the BB method was significantly larger than for the CM method. This conclusion was reached for clinical problems planning the prostate as PTV when the OAR were rectum and bladder. Part of each OAR was in the PTV and this explains the high rectal NTCPs obtained because part of the rectal OAR receives a tumourcidal dose. The rectum only showed these significant NTCPs and was dose limiting. (The bladder has such a large volume effect that although

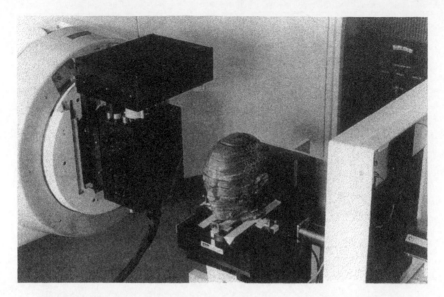

Figure 2.33. *The computer-controlled phantom rotator/translator system developed by the group at Wisconsin to simulate tomotherapy. The NOMOS MIMiC collimator is attached to a Varian 2100C linac which also supports the PortalVision portal imaging system shown to the lower right of the picture. (From Mackie et al 1995; reprinted by permission of John Wiley and Sons.)*

parts of it too receive a tumourcidal dose, the value of $D_{50}(v_{eff})$, the dose for which the NTCP is 0.5 at partial volume v_{eff}, is so large that the NTCP of the bladder was negligible.) However, when the number of strata was increased to 30 there was very little difference between the BB and the CM methods (Webb 1994b,c). We may expect to see a growth in the number of studies designed to decide under what circumstances tomotherapy is needed and when a smaller number of IMBs can be delivered from fixed gantry orientations (see also Appendix A).

2.4.1. The NOMOS 2D intensity-modulating collimator

The NOMOS MIMiC described in section 2.4 is a 1D intensity-modulating collimator. The movable vanes create a 20-bixel IMB (strictly the two banks create two such IMBs).

The NOMOS Corporation is also developing a 2D intensity-modulating collimator. This comprises ($N \times N$) cells (bixels) arranged in a chess-board pattern. Alternate cells are permanently blocked with tungsten (imagine the black squares of a chess-board). The remaining cells (the white squares) each contain a latex balloon. The shape of the cells is focused at the source. The cells connect to a reservoir of mercury under a pressure of 5 psi ($\sim$34.5 kPa).

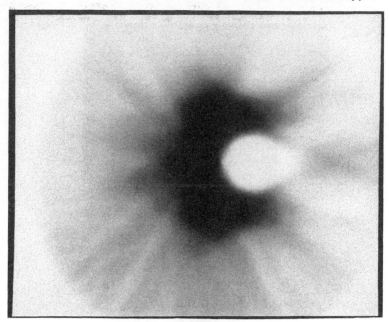

Figure 2.34.   *A radiograph of the distribution of dose in a transaxial slice achieved with the NOMOS MIMiC collimator set to deliver a C-shaped distribution. (Courtesy of Professor R Mackie.)*

Each balloon is equipped with a switchable 20 psi (~137.9 kPa) airline. When the air to a balloon is switched off, the pressure on the mercury forces it to fill the cell so blocking radiation. When the air to a balloon is switched on, the mercury is forced out of the cell allowing radiation to pass. The white squares thus act as binary shutters and a transition between states can take place in 100 ms. As with the MIMiC the intensity of radiation transmitted by a cell is determined by the dwell time of the attenuating material (in this case mercury) in that cell. The filling status of each balloon can be monitored electrically by providing contacts whereby a circuit is completed when a balloon contains mercury. In this way the computer 'knows' about the filling state.

This arrangement takes care of creating an arbitrarily specified IMB on the white squares. The IMB on the black squares is similarly created after rotating the collimator by 90° about the normal to its centre or by laterally shifting the housing of the collimator by one square's distance. If the collimator were instead rectangular, it must be laterally shifted. To deliver a 3D conformal dose distribution to a *volume* requires two arcs of the gantry with the collimator rotated between them. There is no requirement for the cells to be of uniform size. A finer chess-board cell size could be situated, for example, at the centre. The collimator can, like the MIMiC, but unlike a conventional MLC, create internal blocking. It can act as a field

(a)

(b)

Figure 2.35. *The prototype 2D intensity-modulating collimator from NOMOS, (a) components; (b) assembled. (Courtesy of Dr M P Carol.)*

shaper and also as a missing-tissue compensator. This design improves on an earlier concept in which all squares of the chess-board contained such a variable-status balloon arranged within a wire-frame mesh.

A 4×4 prototype has been constructed (figure 2.35). The 'volume box' 2D IMB-creating collimator has been patented (Carol 1996b). One other useful feature is that each elemental beam profile is the same and independent of the opening time, unlike the MIMiC where the beam profile through each bixel depends on the state of opening of its neighbours (Webb 1996, Webb and Oldham 1996). This feature simplifies dosimetry. The progress of this development will be watched with interest.

2.4.2. *A diagnostic analogue*

The tomotherapy devices have a hardware analogue in diagnostic radiology. In conventional diagnostic radiology a problem occurs when film is used as detector as the dynamic range is not wide enough to cope with the extremes of exposure which arise. For example, in chest radiology the lung regions can be overexposed whilst the mediastinum may be simultaneously underexposed. The problem is characterised by the Hurter–Driffield response curve for film density plotted against film exposure.

One solution has been advanced multiple beam equalisation radiography (AMBER) (figure 2.36) (Kool *et al* 1988, Vlasbloem and Kool 1988). The AMBER system scans a fan beam of radiation collimated to a slit aperture across the field. The aperture is divided into 20 subfields of size 1.6×0.6 in^2 ($\sim 4.06 \times 1.52$ cm^2) measured at the detector. During scanning these subfields can be partially blocked by a radiation-opaque absorber to modulate the beam intensity. Each absorbing modulator is connected in a feedback loop to a corresponding almost-transparent radiation detector placed in front of the film. The detector produces a voltage depending on the radiation intensity it receives and this voltage controls a piezoelectric actuator which bends to offer the modulator to the beam. The idea is that the more radiation exposure the detector receives, the more the absorbing modulator moves into the field and reduces the radiation intensity incident on the patient. The detector providing feedback transmits more than 99% of radiation at 70 keV to the film.

The feedback can be varied. Clearly it cannot be so total as to ensure that all parts of the film receive *the same* exposure or nothing will be visible! In practice the feedback is governed by a control curve (figure 2.37) which distorts the conventional linear relationship between log film exposure and radiological absorption thickness. By reducing the exposure when the absorption thickness is low to that which would be obtained conventionally from a thicker path, the film density for these low absorption paths is reduced to the 'well-exposed' part of the Hurter–Driffield curve.

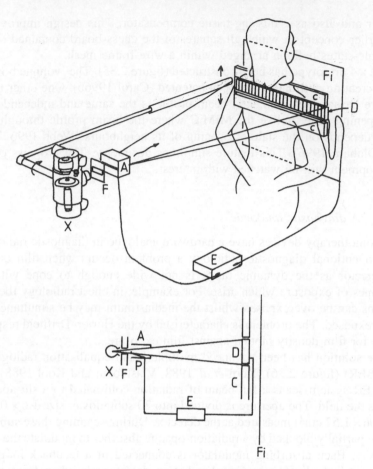

Figure 2.36. *The components of the AMBER system for scanning-multiple-beam-equalisation radiography, the diagnostic analogue of to-motherapy. The beam is collimated to a fan and passes through a slit aperture (A) and fore (F) and distal (C) collimators, falling onto a film (Fi) via an almost perfectly transmitting detector bank (D). The aperture (A) is divided into 20 segments which can each be separately modulated by an absorber which can be introduced into the beam under feedback control (via electronics (E)) from a radiation detector (D). The fan beam is scanned mechanically as shown by arrows. The absorber is a radiation-opaque strip mounted on a piezoelectric strip (P) which can take up different shapes depending on the voltage supplied to it. (Adapted from Vlasbloem and Kool 1988.)*

2.5. CLASS 5: CREATION OF IMBS BY THE MOVING-BAR TECHNIQUE

A novel way to create 1D IMBs has been proposed by Fiorino *et al* (1995b), Calandrino *et al* (1995) and Cattaneo *et al* (1995). In this technique an

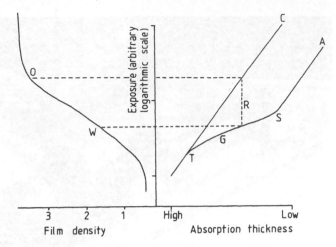

Figure 2.37. *The control curve for AMBER showing how the radiation exposure is controlled to lie on the 'well-exposed' (W) part of the Hurter–Driffield (H–D) curve. On the left is the conventional H–D curve plotted rotated by 90°. On the right is the conventional (C) linear relationship between log exposure and radiological absorption pathlength. AMBER changes this to a curve (A) by reducing the intensity (R) for short radiological pathlengths. The exact shape of the curve can be controlled electronically. It is characterised by a turning point (T) where AMBER begins to function, a region (G) where the gain is reduced and a stopping point (S) specifying the maximum closure of the beam, beyond which the response curve is again linear. It may be seen that this effectively changes the film density which arises from different absorption paths. The dotted lines show how the film density is reduced from (O) to (W) via the exposure reduction (R). (Adapted from Vlasbloem and Kool 1988.)*

absorbing bar is moved across the open field with a time-dependent velocity profile. The bar spends a variable time shielding different parts of the field and thus constructing the intensity-modulation. Since the bar is only a linear block, being 'long' in the direction at right angles to the direction of motion, the modulation is also only 1D. They propose the application to constructing missing-tissue compensators as well as multiple 1D IMBs for conformal radiotherapy. For the latter application the field would be collimated to a narrow beam in the direction orthogonal to the bar movement and 2D conformal distributions would be constructed by repeating the bar translation with different velocity profiles at different gantry angles. 3D conformation would require sequencing the patient through this narrow beam and repeating the translation and rotation movements for each 2D slice.

The proposed advantages of the technique are that it does not require the linac to have a computer-controlled MLC or dynamic collimators; the device attaches to the blocking tray (figures 2.38 and 2.39). In the prototype the block was not focused but a mark 2 device is being constructed with

Figure 2.38.   *The moving-bar technique of Fiorino. The bar is translated across the radiation field spending a variable time shielding each part. In this way some (but not all) IMBs can be constructed. The apparatus is motor controlled and resides in the blocking tray of the linear accelerator. The intensity modulation is only 1D. (From Fiorino et al 1992.)*

a focused absorber (Fiorino *et al* 1995a). In this the 8 cm-thick bar moves on the arc of a circle centred at the source. The absorber is driven by a DC motor with a decoder on its axis which feeds back the location of the bar to the controlling computer. An ultrasound distance-measuring device checks the physical position of the absorber and irradiation stops if the modulator fails.

Fiorino *et al* (1995b) developed the algorithm which starts from the shape of the desired intensity profile and delivers the velocity profile for the bar. The first and simplest solution to the problem considered the movement of the bar in a series of N adjacent static locations x_i (the bar is centred over x_i) but this solution is clearly inefficient because it is both time consuming and also generates penumbra problems.

In this simplest form the algorithm requires the bar to dwell at the ith position for a time t_i where

$$t_i = T \left(1 - \frac{I(x_i)}{I_U} \right) \tag{2.66}$$

where $I(x_i)$ is the required fluence at position x_i, I_U is the unmodulated fluence and T is the total irradiation time. This solution is subject to the condition that

$$\sum_i^N t_i \leqslant T \tag{2.67}$$

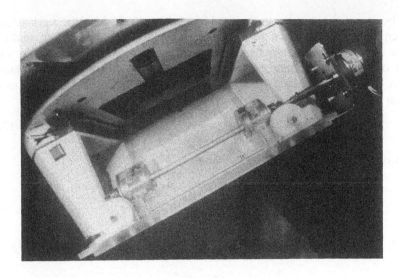

Figure 2.39. *The moving-bar mechanism inserted into the tray holder of a Varian Clinac 6/100 linear accelerator. (From Fiorino et al 1995b.)*

which, substituting from equation (2.66), requires that

$$\sum_{i=1}^{N} I(x_i) > (N-1)I_U.$$ (2.68)

The second solution considered the movement of the bar to be continuous and then approximated this by a stepped velocity profile in which the track was divided into N intervals of width δx wherein the absorber moved at constant speed v_i. In this situation the velocity v_i of the absorber moving between points x_i and x_{i+1}, spaced δx apart, is

$$v_i = \frac{\delta x}{T(1 - (I(x_i)/I_U))}$$ (2.69)

and condition (2.67) becomes

$$\sum_{i=1}^{N} \frac{1}{v_i} \leqslant \frac{T}{\delta x}.$$ (2.70)

The equations were modified for the practical implementation, properly accounting for beam penumbra, scatter, bar transmission and finite maximum speed of absorber.

The major disadvantage of this method is that it cannot create all intensity profiles. There is a physical constraint that the sum of the shielding times has to be less than or equal to the total irradiation time. Also the most accurate beam modulation would be provided by a bar with the smallest width but this would lead to the impossibility of creating the required shielding within the overall time because inequalities (2.67) and (2.70) would be violated.

A given required modulation $I(x_i)$ profile can be expressed in terms of a total modulation

$$M_{tot} = \frac{1}{NT} \sum_{i=1}^{N} t_i \qquad (2.71)$$

which, using equation (2.66), becomes

$$M_{tot} = \left(1 - \frac{\sum_{i=1}^{N} I(x_i)}{N I_U}\right) \qquad (2.72)$$

and, using equation (2.67) or (2.68), the total modulation must satisfy

$$M_{tot} \leqslant \frac{1}{N}. \qquad (2.73)$$

i.e.

$$\left(1 - \frac{\sum_{i=1}^{N} I(x_i)}{N I_U}\right) \leqslant \frac{1}{N}. \qquad (2.74)$$

It can now be seen that, as the width of the bar decreases for a fixed length bar-travel, N increases and it becomes increasingly more difficult to satisfy inequality (2.74). As N increases the right-hand side (rhs) decreases and eventually the total modulation (the left-hand side (lhs)) exceeds the rhs. The same result can be had from inspecting inequality (2.68) where, as N increases, the rhs increases to the point of exceeding the lhs. This is particularly easy to visualise if one considers a constant IMB, say with each $I(x_i) = 0.9 I_U$. Then the total modulation (0.1) is of course independent of N and the inequality (2.74) becomes increasingly more difficult to satisfy as N increases. It would be just satisfied if $N = 10$, easily satisfied if $N < 10$ and not satisfied if $N > 10$. In general the IMB is of course varying spatially, specified by $I(x_i)$.

So the solution has to be a compromise between the degree of accuracy of modulating the profile and the possibility of the modulation itself. From equation (2.68) it is clear that the method works best for shallow modulations. We may recall (section 2.2) that this is precisely the circumstance in which the dynamic-leaf technique has most *difficulties*. Fiorino *et al* (1997) have shown good agreement between measurement and calculation when this technique is used to create a missing-tissue compensator. Finally in the stepped implementation when the profile has multiple maxima and minima these can only be realised if they are spaced apart an integer multiple of the absorber width.

Fiorino *et al* (1995b) compared calculated and experimentally measured IMBs with an agreement of 3%. Fiorino *et al* (1995a) have created seven intensity-modulated fields equispaced in 360° and shown that the dose distribution is much superior to that obtained from six unmodulated fields for the clinical case of a prostate target volume with a concavity containing rectum. They showed an increase in TCP of some 10% for fixed NTCP = 5%.

The above two calculation techniques assumed that the calculation points were spaced apart by the width δx of the moving bar. A difficulty clearly arises if the spatial scale of changes in the IMB profile is smaller than this width (and we must recall that the width itself is set so that the total modulation can be realised (inequality (2.68) or (2.73))). When this is not the case, and even when it is, the velocity profile of the moving bar can be optimised so there is a better fit between the desired and delivered IMB fluence profile. The way this is achieved is to consider points x_i much closer together than the width of the bar. First we can see that this will lead to the same condition for the realisability of the profile.

Suppose there are K points covered at any one time by the bar of width w (so $K\delta x = w$) and as before N points in the whole field of width W (so $N\delta x = W$) then, because of the ability to simultaneously realise in practice groups of M total shielding times t_i, the inequality (2.67) changes to

$$\sum_{i=1}^{N} t_i \leqslant KT \qquad (2.75)$$

and inequality (2.68) becomes

$$\sum_{i=1}^{N} I(x_i) > (N - K)I_U. \qquad (2.76)$$

In turn the inequality for the total modulation becomes

$$M_{tot} = \left(1 - \frac{\sum_{i=1}^{N} I(x_i)}{N I_U}\right) \leqslant \frac{K}{N}. \qquad (2.77)$$

Since

$$\frac{K}{N} = \frac{w}{W} \qquad (2.78)$$

by definition equation (2.77) becomes

$$M_{tot} = \left(1 - \frac{\sum_{i=1}^{N} I(x_i)}{N I_U}\right) \leqslant \frac{w}{W}. \qquad (2.79)$$

So, once again the same physical limitation arises concerning the total modulation and the width of the bar. As the width w of the bar is decreased it becomes progressively more difficult to satisfy the inequality (2.79). Inequality (2.79) is the same as inequality (2.74) when there is just one sampling point corresponding to the width of the bar.

Cattaneo *et al* (1995, 1996, 1997) and Fiorino *et al* (1996) have studied this situation and selected the time profile (the specification of the time spent by the central axis of the bar at each x_i from which the velocity profile follows) by the simulated-annealing method and shown good agreement between desired and delivered profiles. The width of the bar was specified to satisfy inequality (2.79) for each IMB profile studied.

2.6. 1D INTENSITY MODULATION USING CONVENTIONAL COLLIMATORS

Zacarias *et al* (1993) have analysed the precision in setting the Y-jaws of a Philips SL-20 linear accelerator. They used film dosimetry to establish that the Y-jaws could be set to better than 0.3 mm, with (on their machine) the Y-2 jaw less precise than the Y-1 jaw. They investigated the precision as a function of gantry angle and found that field edges moved with gantry rotation for fixed collimator jaw positions. The movement was of the order 0.2 mm over 270° and followed a quadratic curve.

In clinical mode the jaw positions can be set only to the nearest millimetre so they concluded that the accelerator hardware was capable of better performance than the software allowed. The mechanical precision was interpreted as suitable for creating 'segmented conformal radiotherapy' fields, whereby thick transverse segments are separately treated by rectangular transverse abutting fields. Because the independently moving jaws are focused and cross the midline by 125 mm at the isocentre, the matching geometrical divergence can be exploited to create large fields from multiple segments. Since these do not have to be irradiated with the same MUs, intensity-modulated fields could be created.

2.7. SCANNING-BEAM THERAPY

According to Lind and Brahme (1995) the ultimate in delivery of intensity-modulated fields is the scanning-beam technique. Figure 2.40 shows that whilst this could be obtained with a pair of dual dynamic asymmetric jaw pairs the treatment time would be prohibitively long. The viable alternative is the Racetrack Microtron (figure 2.41) in which the electron beam is electronically steered by two orthogonally deflecting bending magnets. Unfortunately the half-width of the photon 'pencil' at the isocentre can be

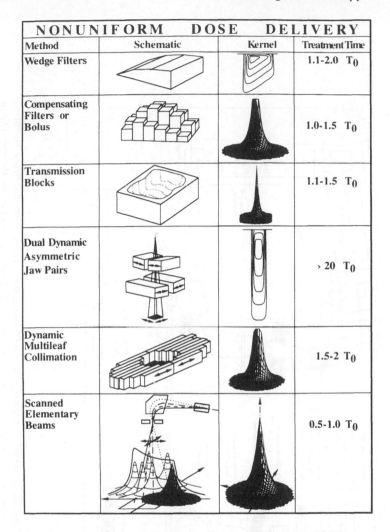

Figure 2.40. *A comparison of six different methods available for delivering non-uniform therapeutic beams. T_0 is the 'standard treatment time' of about 1 minute for uniform dose delivery to the target volume. Only the lower three methods allow dynamic beam shaping but at greatly varying treatment times. Some of the other techniques discussed in this chapter are not illustrated in this figure. (From Lind and Brahme 1995; reprinted by permission of John Wiley and Sons.)*

as large as 4 cm so this is not sufficient collimation alone and the scanning beam is used together with an MLC to define sharper field edges (see section 2.2.7). The Racetrack Microtron is the basis of conformal photon delivery in Stockholm and also at the University of Michigan at Ann Arbor.

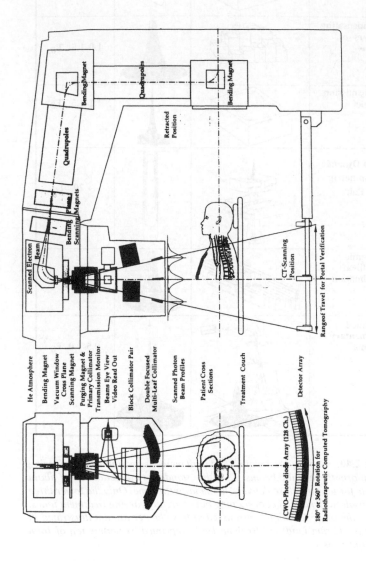

Figure 2.41. *A cross sectional view through the gantry and treatment head of the Racetrack Microtron illustrating the scanning system MLC and the megavoltage CT device used during arc therapy. (From Lind and Brahme 1995; reprinted by permission of John Wiley and Sons.)*

2.8. VERIFICATION OF IMB DELIVERY

It must be acknowledged that one of the major impediments to the development of IMB therapy is the problem of verifying its accuracy. This problem applies as much for 'simple' IMB delivery, such as a dynamic wedge, as for 'full-blown' IMB delivery with a continuously varying beam profile. As is well known, verification is a problem even for 'conventional' therapy but in recent years there has been an escalation of attempts to overcome these difficulties with the development of electronic portal imaging. IMB verification is a different issue. It might even be true to say that there is a real fear of this emergent technology among clinicians and their physics advisers.

A number of possibilities exist and which of them becomes accepted depends to a large extent on one's assessment of the risks. The most obvious 'verification' is to monitor the mechanical movements and radiation intensities at the source side of the patient and accept the therapy if these remain within tolerance. This decouples the issue of the accuracy of beam delivery from any question of patient setup or movement.

A second possible verification is to conduct experiments with phantoms, show that all is well with the delivery and then accept an unverified patient treatment. An issue would be how often the experimental verification would need to be made.

A third possibility would be to make some sort of *in vivo* dosimetry by placing a few detectors on or within the body cavities of the patient. Clearly this is limited by practical difficulties and would only monitor part of the delivered dose distribution.

A fourth possibility would be some analogue of portal imaging with all the known attendant difficulties. The detector would either have to provide a dynamic readout or some sort of integrated signal which would be correlated with the expected integral signal. There is also the issue of the quality assurance of the electronic portal imager itself (Purdy *et al* 1995). Dörner and Neumann (1995) have discussed the possibility of real-time supervision of intensity-modulated fields using a portal fluoroscopic phosphor screen coupled to a video camera. Lutz *et al* (1996) are using the Varian portal imaging system to verify dynamic-leaf therapy in New York. They noted the need to operate the electronic portal imaging device (EPID) in a different mode (Van Herk *et al* 1996) since it requires sampling of the exit distribution at 1 s intervals whereas the leaves move at some 3–4 cm s^{-1} and the imager 'usually' requires 5 s between images.

A fifth and somewhat science-fiction possibility at present is to construct a dose simulator. The actual delivery movements and radiation features would be fed to a computer which would reconstruct the dose delivery as the dose is delivered. The operator would watch the 3D dose distribution build up as time progressed. Such a facility would need computers much faster than any known today but would probably be the ultimate form of verification, limited only by the dose model in use. Certainly, since some accelerators (e.g. Varian) provide a post-treatment record of leaf positions, the actually delivered dose could be reconstructed *after* irradiation and this is the proposed method of IMB verification in New York (Kutcher 1996).

Ling *et al* (1996) have demonstrated extensive QA protocols for dynamic-leaf therapy. The stability of leaf speed was checked by moving jaws at constant speed and creating a uniform profile, 'bumps' on which characterise the leaf-speed stability. The effect of rounded leaf ends was observed by choosing a pattern in which each leaf-pair, otherwise moving at constant velocity, halted for a while at the same *x* location, 'bumps' in the profile characterising the inter-leaf leakage. A special test in which high-dose stripes, normal to the direction of leaf motion, were planted in an otherwise uniform dose characterised the acceleration and deceleration of the leaves.

The clinical work at MSKCC operates with a tolerance of 0.1 mm on leaf position. The actually delivered dose is computed *a posteriori* using data provided by the Varian leaf position file and compared with the planned dose. After initially only using IMBs for a boost dose, the whole treatment to 81 Gy is now delivered this way with dose escalation via the use of IMBs (Ling *et al* 1996). This reduces the treatment-planning effort since only one plan is required.

Purdy *et al* (1995) have alerted the radiation physics community to the issues of quality assurance and safety of the new generation of radiation-therapy devices. They stress three essentials: (i) the existence of rigorous quality assurance (QA) procedures; (ii) continuing education; (iii) the mandatory use of record-and-verify systems. The AAPM Radiation Therapy Task Group 35 has recently reported in detail the necessary QA procedures for computer-controlled therapy (Purdy *et al* 1993). These provide the basis for considering the options for dynamic therapy, which, as we have seen above, are more challenging and potentially more worrying.

2.9. SUMMARY

With the increasing realisation that a combination of IMBs can deliver highly conformal radiotherapy dose distributions even to targets with concave outlines, there has been a rapid development of interest in techniques to deliver such beams. It is generally recognised that the use of milled metal compensators, which have to be specially prepared for each field, is an impediment, especially as the number of fields required may be large. As commercial MLCs have become available, two classes of technique for creating IMBs with them have been proposed: (i) dynamic-leaf-sweep with the beam on; (ii) multiple-static irradiation with the beam off between leaf resettings. Algorithms for optimising the former have been constructed. Some experimental work in a few centres has been done with each technique.

Competing with this is the method of tomotherapy. Independently conceived by two groups, one practical method of delivery already exists and the other is being constructed. Whilst conceptually similar they are not entirely the same. One is analogous to multiple CT slice generation, whilst the other is analogous to spiral CT. Scanning-beam therapy is also under investigation. Contrasting this technical complexity there have been some simpler developments such as the moving-bar technique. There will undoubtedly be much further development in the field of IMB generation well into the next millennium.

REFERENCES

Bortfeld T R 1995 Dynamic and quasi-dynamic multileaf collimation (Proc. ESTRO Conf. (Gardone Riviera, 1995)) *Radiother. Oncol.* **37** Suppl. 1 S16

Bortfeld T R, Boyer A L, Schlegel W, Kahler L and Waldron T J 1994a Experimental verification of multileaf modulated conformal radiotherapy *The Use of Computers in Radiation Therapy: Proc. 11th Conf.* ed A R Hounsell *et al* (Manchester: ICCR) pp 180–1

—— 1994b Realisation and verification of three-dimensional conformal radiotherapy with modulated fields *Int. J. Radiat. Oncol. Biol. Phys.* **30** 899–908

Bortfeld T R, Kahler D L, Waldron T J and Boyer A L 1994c X-ray field compensation with multileaf colimators *Int. J. Radiat. Oncol. Biol. Phys* **28** 723–30

Bortfeld T, Stein J, Preiser K and Hartwig K 1996 Intensity modulation for optimised conformal therapy *Proc. Symp. Principles and Practice of 3-D Radiation Treatment Planning (Munich, 1996)* (Munich: Klinikum rechts der Isar, Technische Universität)

Boyer A L 1994 Use of MLC for intensity modulation *Med. Phys.* **21** 1007

Boyer A L, Bortfeld T R, Kahler L and Waldron T J 1994 MLC modulation of x-ray beams in discrete steps *The Use of Computers in Radiation Therapy: Proc. 11th Conf.* ed A R Hounsell *et al* (Manchester: ICCR) pp 178–9

Boyer A L, Geis P and Bortfeld T 1996 Modulated cone beam conformal therapy *Proc. Intensity-Modulated Radiation Therapy (ICRT) Workshop, sponsored by the NOMOS Corporation, (Durango, CO, 1996)* (Sewickley, PA: NOMOS) pp 11–3

Calandrino R, Cattaneo G M, Fiorino C, Longobardi B and Fusca M 1995 Methods to achieve conformal therapy and therapeutic perspectives (Proc. ESTRO Conf. (Gardone Riviera, 1995)) *Radiother. Oncol.* **37** Suppl. 1 S15

Carol M P 1992a private communication (letter to S Webb, 23 November 1992)

—— 1992b An automatic 3D treatment planning and implementation system for optimised conformal therapy by the NOMOS Corporation *Proc. 34th Ann. Meeting of the American Society for Therapeutic Radiology and Oncol. (San Diego, CA, 1992)*

—— 1993 private communication (letter to S Webb, 4 January 1993)

—— 1994a An automatic 3D conformal treatment planning system for linear accelerator based beam modulation radiotherapy *The Use of Computers in Radiation Therapy: Proc. 11th Conf.* ed A R Hounsell *et al* (Manchester: ICCR) pp 108–9

—— 1994b Integrated 3D conformal multivane intensity-modulation delivery system for radiotherapy *The Use of Computers in Radiation Therapy: Proc. 11th Conf.* ed A R Hounsell *et al* (Manchester: ICCR) pp 172–3

—— 1995 Peacock: a system for planning and rotational delivery of intensity-modulated fields *Int. J. Imaging Syst. Technol.* **6** 56–61 (Special issue on 'Optimisation of the three-dimensional dose delivery and tomotherapy')

—— 1996a Intensity theatre *Proc. Intensity-Modulated Radiation Therapy (ICRT) Workshop, sponsored by the NOMOS Corporation) (Durango, CO, 1996)* (Sewickley, PA: NOMOS) pp 2–3

—— 1996b IMRT in 2003 *Proc. Intensity-Modulated Radiation Therapy (ICRT) Workshop, sponsored by the NOMOS Corporation) (Durango, CO, 1996)* (Sewickley, PA: NOMOS) pp 37–8

Carol M P, Grant W III, Bleier A R, Kania A A, Targovnik H S, Butler E B and Woo S W 1996 The field-matching problem as it applies to the Peacock three-dimensional conformal system for intensity modulation *Int. J. Radiat. Oncol. Biol. Phys.* **34** 183–7

Carol M P and Targovnik H 1994 Importance of the user in creating optimized treatment plans with Peacock *Med. Phys.* **21** 913

Carol M P, Targovnik H, Campbell C, Bleier A, Strait J, Rosen B, Miller P, Scherch D, Huber R, Thibadeau B, Dawson D and Ruff D 1993 An automatic 3D treatment planning and implementation system for optimised conformal therapy *Three-Dimensional Treatment Planning (Proc. EAR Conf. (WHO, Geneva, 1992))* ed P Minet (Geneva: Minet) pp 173–87

Cattaneo G M, Fiorino C, Fusca M and Calandrino R 1995 Dynamic 1D beam modulation by a single absorber: theoretical analysis for optimizing apparatus performances (Proc. ESTRO Conf. (Gardone Riviera, 1995)) *Radiother. Oncol.* **37** Suppl. 1 S17

Cattaneo G M, Fiorino C, Lombardi P and Calandrino R 1997 Optimizing the movement of a single absorber for 1D non uniform dose delivery by fast simulated annealing *Phys. Med. Biol.* **42** 107–21

—— 1996 Application of fast simulated annealing to optimisation of 1D beam modulation by single absorber *Med. Phys.* **23** 1171

Chawla K J, Jaffray D A, Yu C X and Wong J W 1995 Two-dimensional plastic scintillator for dosimetry of dynamic intensity modulation (private communication)

Chui C S, LoSasso T, Ling C and McCormick B 1994 Breast treatment with intensity-modulated fields generated by dynamic multi-leaves *Med. Phys.* **21** 1008

Convery D J and Rosenbloom M E 1992 The generation of intensity-modulated fields for conformal radiotherapy by dynamic collimation *Phys. Med. Biol.* **37** 1359–74

—— 1995 Treatment delivery accuracy in intensity-modulated conformal radiotherapy *Phys. Med. Biol.* **40** 979–99

Curran B 1996 Conformal radiation therapy using a multileaf intensity modulating collimator *Proc. Intensity-Modulated Radiation Therapy (ICRT) Workshop, sponsored by the NOMOS Corporation (Durango, CO, 1996)* (Sewickley, PA: NOMOS) pp 9–10

De Jaeger K, Van Duyse B, De Wagter C, Fortan L and De Neve W 1995 Radiotherapy to targets with concave surfaces: planning and clinical execution by intensity modulation using a static-segmentation technique (Proc. ESTRO Conf. (Gardone Riviera, 1995)) *Radiother. Oncol.* **37** Suppl. 1 S16

De Neve W, De Jaeger K, Van Duyse B, Fortan L and De Wagter C 1995 Virtual simulation of targets with concave surfaces using a static-segmentation technique (Proc. ESTRO Conf. (Gardone Riviera, 1995)) *Radiother. Oncol.* **37** Suppl. 1 S61

Deasy J O 1995 (private communication, letter to S Webb, 22 August 1995)

Djordjevich A, Bonham J, Hussein E M A, Andrew J W and Hale M E 1990 Optimal design of radiation compensators *Med. Phys.* **17** 397–404

Dörner K J and Neumann M 1995 Realtime supervision of dynamic and intensity-modulated radiation fields *Medizinische Physik 95 Röntgen Gedächtnis-Kongress* ed J Richter (Würzburg: Kongress) pp 260–1

Du M N, Yu C X, Taylor R C, Martinez A A and Wong J W 1995 Dynamic wedge: a first step towards clinical application of intensity modulation using MLC (ASTRO Meeting 1995) *Int. J. Radiat. Oncol. Biol. Phys.* **32** Suppl. 1 170

Emami B, Lyman J, Brown A, Coia L, Goitein M, Munzenrider J E, Shank B, Solin L J and Wesson M 1991 Tolerance of normal tissue to therapeutic irradiation *Int. J. Radiat. Oncol. Biol. Phys.* **21** 109–22

Evans P M, Hansen V N and Swindell W 1997 The optimum intensities for multiple static MLC field compensation *Med. Phys.* at press

Fiorino C, Ardesi A, del Vecchio A, Cattaneo G M, Fossati V, Fusca M, Longobardi B, Signorotto P and Calandrino R 1992 Techniche conformazionali mediante il movimento delle barre nei campi pendolari: 1. Realizzazione di prototipi e possibili applicazionila *Radiol. Med.* **84** 310–6

Fiorino C, Cattaneo G M, Corletto D, Fusca M, Mangili P and Calandrino R 1997 Application of a single-absorber dynamic modulation to 1D tissue-defecit compensation *Radiother. Oncol.* at press

Fiorino C, Cattaneo G M, Corletto D, Mangili P and Calandrino R 1996 Dynamic tissue-deficit compensation in mantle field irradiation by a single-absorber modulator *Med. Phys.* **23** 1075

Fiorino C, Cattaneo G M, Lev A, Fusca M, Rudello F and Calandrino R 1995a 1D non-uniform dose delivery by a single dynamic absorber (Proc. ESTRO Conf. (Gardone Riviera, 1995)) *Radiother. Oncol.* **37** Suppl. 1 S59

Fiorino C, Fossati V, Ardesi A, Cattaneo G M, del Vecchio A, Fusca M, Longobardi B, Signorotto P and Calandrino R 1993 Development of a computer-controlled moving bar (CCMB) conformal technique for neck irradiation *Radiother. Oncol.* **27** 167–70

Fiorino C, Lev A, Fusca M, Cattaneo G M, Rudello F and Calandrino R 1995b Dynamic beam modulation by using a single computer controlled absorber *Phys. Med. Biol.* **40** 221–40

Fraass B A, Marsh L, Martel M K, Forster K, McShan D L and Ten Haken R K 1994 Multileaf collimator-based intensity modulation for conformal treatment of prostate cancer *Med. Phys.* **21** 1008

Galvin J M and Han K 1994 A comparison of segmentation techniques for beam modulation *Med. Phys.* **21** 921

Grant W III 1996 Commissioning and quality assurance of an intensity-modulated beam system *Proc. Intensity-Modulated Radiation Therapy (ICRT) Workshop, sponsored by the NOMOS Corporation (Durango, CO, 1996)* (Sewickley, PA: NOMOS) pp 19–20

Grant W III, Bellezza D, Berta C, Bleier A, Campbell C, Carol M and Targovnik H 1994a Leakage considerations with a multi-leaf collimator designed for intensity-modulated conformal radiotherapy *Med. Phys.* **21** 921

Grant W III, Bleier A, Campbell C, Carol M and Targovnik H 1994b Validation of a beam-modelling algorithm for a 3D conformal therapy system using a multi-leaf collimator for intensity modulation *Med. Phys.* **21** 946

Gustafsson A, Lind B K, Svensson R and Brahme A 1995 Simultaneous optimisation of dynamic multileaf collimation and scanning patterns of compensation filters using a generalised pencil beam algorithm *Med. Phys.* **22** 1141–56

Holmes T W 1993 A model for the physical optimisation of external beam radiotherapy *PhD Thesis* University of Wisconsin-Madison

Holmes T, Mackie T R, Swerdloff S, Reckwerdt P and Kinsella T 1993 Tomotherapy: Inverse treatment planning for conformal therapy (Proc. 35th ASTRO Meeting) *Int. J. Radiat. Oncol. Biol. Phys.* **27** Suppl. 1 139

Kalnicki S, Wu A, Berta C, Targovnik H, Chen A and Carol M 1994 Illustrations of Peacock treatment plans for patients with localised diseases (Proc. World Congress on Medical Physics and Biomedical Engineering (Rio de Janeiro, 1994)) *Phys. Med. Biol.* **39A** Part 1 517

Kania A A, Bleier A R and Carol M P 1995 Monitor unit calculation for the multileaf intensity modulating collimator (MIMIC) in the PEACOCKPLAN system (Proc. ESTRO Conf. (Gardone Riviera, 1995)) *Radiother. Oncol.* **37** Suppl. 1 S18

Kool L J S, Busscher D L T, Vlasbloem H, Hermans J, van de Merwe P C, Algra P R and Herstel W 1988 Advanced multiple-beam equalisation radiography in chest radiology: a simulated nodule detection study *Radiology* **169** 35–9

Kutcher G J 1996 private communication (Munich, March)

Leavitt D D, Martin M, Moeller J H and Lee W L 1990 Dynamic wedge field techniques through computer-controlled collimator motion and dose delivery *Med. Phys.* **17** 87–91

Lind B and Brahme A 1995 Development of treatment techniques for radiotherapy optimisation *Int. J. Imaging Syst. Technol.* **6** 33–42 (special issue on 'Optimisation of the three-dimensional dose delivery and tomotherapy')

Ling C C, Burman C, Chui C S, Kutcher G J, Leibel S A, LoSasso T, Mohan R, Spirou S, Stein J, Wang X H, Wu Q, Yang J, Zelefsky M and Fuks Z 1996 Conformal radiation therapy of prostate cancer using inversely-planned intensity modulated photon beams produced with dynamic multileaf collimation *Proc. Intensity-Modulated Radiation Therapy (ICRT) Workshop, sponsored by the NOMOS Corporation (Durango, CO, 1996)* (Sewickley, PA: NOMOS) pp 33–4

Lutz W, Willins J, Hanley J, LoSasso T, Mageras G and Kutcher G J 1996 Portal 'digital dosimetry' for intensity modulated beams *Proc. 4th International Workshop on Electronic Portal Imaging (Amsterdam, 1996)* (Amsterdam: EPID Workshop Organisers) p 52

Lyman J T 1985 Complication probability as assessed from dose–volume histograms *Rad. Res.* **104** S-13–S-19

Mackie T R 1993 private communication (lecture on visit to the Royal Marsden NHS Trust, Sutton; 13 July 1993)

—— 1994 private communication (Rio; 26 August 1994)

—— 1995 private communication (Gardone Riviera; 8 October 1995)

—— 1996 Can conformal radiotherapy simultaneously be better and less expensive? *Med. Phys.* **23** 1050

Mackie T R, Angelos L, Balog J, DeLuca P M, Fang G, Geiser B P, Glass M, McNutt T R, Pearson D, Reckwert P J, Shepard D, Wenman D, Zachman J, Auh Y Y, Cohen G, Jonsson P A and Senzig R F 1996 Design of a tomotherapy unit *Med. Phys.* **23** 1074

Mackie T R, Holmes T W, Reckwerdt P J and Yang J 1995 Tomotherapy: optimized planning and delivery of radiation therapy *Int. J. Imaging Syst. Technol.* **6** 43–55 (Special issue on 'Optimisation of the three-dimensional dose delivery and tomotherapy')

Mackie T R, Holmes T W, Reckwert P J, Yang J, Swerdloff S, Deasy J O, DeLuca P M, Paliwal B R and Kinsella T J 1994a Tomotherapy: A proposal for a dedicated computer-controlled delivery and verification system for conformal radiotherapy *The Use of Computers in Radiation Therapy: Proc. 11th Conf.* ed A R Hounsell *et al* (Manchester: ICCR) pp 176–7

Mackie T R, Holmes T, Swerdloff S, Reckwerdt P, Deasy J O, Yang J, Paliwal B and Kinsella T 1993 Tomotherapy: A new concept for the delivery of conformal radiotherapy using dynamic collimation *Med. Phys.* **20** 1709–19

Mackie T R, Yang J, Reckwerdt P and Deasy J O 1994b Intensity modulation in conformal therapy *Med. Phys.* **21** 1007

Mageras G S, Mohan R, Burman C, Barest G D and Kutcher G J 1991 Compensators for three-dimensional treatment planning *Med. Phys.* **18** 133–40

Mageras G S, Podmaniczky K C and Mohan R 1992 A model for computer-controlled delivery of 3-D conformal treatments *Med. Phys.* **19** 945–53

Mohan R 1995 Field shaping for three-dimensional conformal radiation therapy and multileaf collimation (*Innovations in treatment delivery*) *Semin. Radiat. Oncol.* **5** 86–99

Philips Medical Systems—Radiotherapy 1994 private communication (16 September 1994)

Purdy J A 1996 The development of intensity modulated radiation therapy *Proc. Intensity-Modulated Radiation Therapy (ICRT) Workshop, sponsored by the NOMOS Corporation (Durango, CO, 1996)* (Sewickley, PA: NOMOS) p 1

Purdy J A, Biggs P J, Bowers C, Dally E, Downs W, Fraass B A, Karzmark C J, Khan F, Morgan P, Morton R, Palta J, Rosen I I, Thorson T, Svensson G and Ting J 1993 Medical accelerator safety considerations: report of the AAPM Radiation Therapy Committee Task Group No 35 *Med. Phys.* **20** 1261–75

Purdy J A, Klein E E and Low D A 1995 Quality assurance and safety of new technologies for radiation oncology (*Innovations in treatment delivery*) *Semin. Radiat. Oncol.* **5** 156–65

Spirou S V and Chui C S 1994 Generation of arbitrary intensity profiles by dynamic jaws or multileaf collimators *Med. Phys.* **21** 1031–41

—— 1996 Generation of arbitrary intensity profiles by combining the scanning beam with dynamic multileaf modulation *Med. Phys.* **23** 1–8

Stein J 1993 Techniques to create intensity-modulated radiation beams for conformal therapy. Lecture and discussion at RMNHST; 6 September 1993

Stein J, Bortfeld T, Dörschel B and Schlegel W 1994a Dynamic x-ray compensation for conformal radiotherapy by means of multileaf collimation *Radiother. Oncol.* **32** 163–73

—— 1994b X-ray intensity modulation by dynamic multi leaf collimation *The Use of Computers in Radiation Therapy: Proc. 11th Conf.* ed A R Hounsell *et al* (Manchester: ICCR) pp 174–5

Sternick E S and Sussman M L 1996 Safety management of intensity modulated radiation therapy *Proc. Intensity-Modulated Radiation Therapy (ICRT) Workshop, sponsored by the NOMOS Corporation (Durango, CO, 1996)* (Sewickley, PA: NOMOS) pp 31–2

Svensson R, Källman P and Brahme A 1994a Analytical solution for the dynamic control of multileaf collimators *Phys. Med. Biol.* **39** 37–61

—— 1994b Realisation of physical and biological dose optimisation using dynamic multileaf collimation *The Use of Computers in Radiation Therapy: Proc. 11th Conf.* ed A R Hounsell *et al* (Manchester: ICCR) pp 56–7

Swerdloff S, Mackie T R and Holmes T 1994a Method and apparatus for radiation therapy *US Patent* 5,317,616

—— 1994b Multi-leaf radiation attenuator for radiation therapy *US Patent* 5,351,280

Van Herk M, Boellaard R, Brugmans M and Van Dalen A 1996 Feasibility of the Portalvision system for verification of dynamic therapy *Proc. 4th Int. Workshop on Electronic Portal Imaging (Amsterdam, 1996)* (Amsterdam: EPID Workshop Organisers) p 53

Van Santvoort J, Heijmen B and Dirkx M 1995 Solution for underdosages caused by the tongue and groove effect for multileaf collimators (Proc. ESTRO Conf. (Gardone Riviera, 1995)) *Radiother. Oncol.* **37** Suppl. 1 S38

Vlasbloem H and Kool L J S 1988 AMBER: A scanning multiple-beam equalisation system for chest radiography *Radiology* **169** 29–34

Wang X, Spirou S, LoSasso T, Stein J, Chui C-S and Mohan R 1996 Dosimetric verification of intensity-modulated fields *Med. Phys.* **23** 317–27

Webb S 1989 Optimisation of conformal radiotherapy dose distributions by simulated annealing *Phys. Med. Biol.* **34** 1349–69

—— 1991 Optimisation of conformal radiotherapy dose distributions by simulated annealing: inclusion of scatter in the 2D technique *Phys. Med. Biol.* **36** 1227–37

—— 1994a Tomotherapy and beamweight stratification *The Use of Computers in Radiation Therapy: Proc. 11th Conf.* ed A R Hounsell *et al* (Manchester: ICCR) pp 58–9

—— 1994b Dose optimisation techniques in photon therapy (Proc. World Congress on Medical Physics and Biomedical Engineering (Rio de Janeiro 1994)) *Phys. Med. Biol.* **39A** Part 2 676

—— 1994c Optimising the planning of intensity-modulated radiotherapy *Phys. Med. Biol.* **39** 2229–46

—— 1996 Conformal radiotherapy with intensity modulated radiation; planning issues and slice by slice delivery *Proc. Symp. Principles and Practice of 3-D Radiation Treatment Planning (Munich, 1996)* (Munich: Klinikum rechts der Isar, Technische Universität)

Webb S and Nahum A E 1993 A model for calculating tumour control probability in radiotherapy including the effects of inhomogeneous distributions of dose and clonogenic cell density *Phys. Med. Biol.* **38** 653–66

Webb S and Oldham M 1996 A method to study the characteristics of 3D dose distributions created by superposition of many intensity-modulated beams delivered via a slit aperture *Phys. Med. Biol.* **41** 2135–53

Williams P 1993 Private communication (discussion 27 April 1993)

Woo S Y, Sanders M, Grant W and Butler E B 1994 Does the 'Peacock' have anything to do with radiotherapy *Int. J. Radiat. Oncol. Biol. Phys.* **29** 213–4

Wu A, Berta C, Johnson M, Kalnicki S, Chen A S J and Figura J H 1995 Dose verifications of Peacock system with multileaf intensity modulation collimators *Med. Phys.* **22** 1545–6

Yu C X 1995 Intensity modulated arc therapy with dynamic multileaf collimation: an alternative to tomotherapy *Phys. Med. Biol.* **40** 1435–49

—— 1996 Intensity modulated arc therapy: a new method for delivering conformal radiotherapy *Proc. Intensity-Modulated Radiation Therapy (ICRT) Workshop, sponsored by the NOMOS Corporation (Durango, CO, 1996)* (Sewickley, PA: NOMOS) pp 14–5

Yu C X, Jaffray D, Wong J W and Martinez A A 1995a Intensity modulated arc therapy: dosimetric verification with clinical examples (ASTRO Meeting 1995) *Int. J. Radiat. Oncol. Biol. Phys.* **32** Suppl. 1 186

—— 1995b Intensity modulated arc therapy: a new method for delivering conformal treatments (Proc. ESTRO Meeting (Gardone Riviera, 1995)) *Radiother. Oncol.* **37** Suppl. 1 S16

Yu C X, Symons M J, Du M N, Martinez A A and Wong J W 1995c A method for implementing dynamic photon beam intensity modulation using independent jaws and a multileaf collimator *Phys. Med. Biol.* **40** 769–87

Yu C X, Symons M, Du M N and Wong J W 1994 Photon intensity modulation by dynamic motion of multileaf collimator *Med. Phys.* **21** 1008

Yu C X and Wong J W 1994 Dynamic photon beam intensity modulation

The Use of Computers in Radiation Therapy: Proc. 11th Conf. ed A R Hounsell *et al* (Manchester: ICCR) pp 182–3

—— 1996 Comparison of two dynamic approaches for conformal therapy treatments *Med. Phys.* **23** 1073

Zacarias A S, Lane R G and Rosen I I 1993 Assessment of a linear accelerator for segmented conformal radiation therapy *Med. Phys.* **20** 193–8

CHAPTER 3

THE MULTILEAF COLLIMATOR (MLC) IN CLINICAL PRACTICE

The MLC is having an increasing impact on the way radiotherapy is practised. As we have seen in Chapters 1 and 2 it is central to the planning and delivery of conformal radiotherapy. Early attempts to geometrically shape radiation fields by gravity-assisted shields (Proimos 1960) or dynamic tracking (Davy and Brace 1980) were dogged by complexity and the need for time-consuming individual patient customising. The same problem accompanies custom-made blocks, which, though successful in achieving their purpose, are tedious to fabricate and this problem reduces the likelihood that they will be built for the delivery of multiple fields. The MLC overcomes these difficulties and also, additionally, opens up the option to shape the radiation intensity (see Chapter 2). The MLC has traditionally been the collimator of choice for neutron therapy (Yudelev *et al* 1994). Because the MLC can efficiently tailor the fields to create a treatment volume more like the shape of the target volume than is achievable with a few rectangular fields, it may be that the physical margins added to the clinical target volume to create the PTV could be even larger than with conventional therapy. Jordan and Williams (1994) have highlighted the common misconception that conformal therapy operates with reduced margins and so increased potential for error. This is not necessarily so.

A number of MLCs have been fabricated both by university hospitals and by commercial companies (see tables in the companion Volume, Chapter 5 and in Jordan and Williams (1994) and Stein (1995)). They differ in the number and size of the leaves, on the length of midline over-run, and on the maximum field size. In addition, each MLC has features appropriate to its manufacture which must be investigated by a potential user, e.g. the quality control, the method by which the leaves are set, the technique for linking the leaf placement to the requirements from the planning computer etc.

In the companion Volume, Chapter 5, the evolution of the MLC was described, together with an analysis of some of the commercial pieces of equipment which were becoming available in the early 1990s and the

176

techniques which were being investigated for using them. Some newer MLCs which have become available since that time are reviewed later in this chapter. Since that time also there has been considerable activity in characterising the physical properties of the MLC and in bringing it into clinical service (Uchiyama and Morita 1992, Brahme 1993). These issues are addressed here.

3.1. CHARACTERISING THE PHYSICAL PROPERTIES OF AN MLC

Boyer *et al* (1992) listed the potential advantages of the multileaf collimator:

(i) the time to digitize an MLC field-shaping file is $\frac{1}{3}$ that of the time to mould a Cerrobend block in their centre;

(ii) space requirements are reduced (space for a workstation compared with a Cerrobend block facility);

(iii) a cleaner environment results;

(iv) there is a reduced need for storage space for blocks;

(v) there is no need to change blocks in the blocking tray;

(vi) multiple fields can be set without entering the room.

One could add to this that the use of an MLC is less likely to cause injury to operator or patient since there are no lead blocks to fall, there is no need to lift heavy weights, the MLC is non-toxic and is environmentally friendly. Mohan (1995) has suggested the MLC may even have increased precision in that the designated positions of the leaves can be reproducibly set. Boyer (1995) suggests that there is less likelihood of human error using an MLC compared with the possibility of using or setting the wrong wedge or block. If an MLC were used to create a dynamic wedge, this would eliminate the drawbacks of the physical wedge and indeed, as will be discussed, increase the flexibility of having the wedge at arbitrary orientation to the principal axes of the collimation.

Most published studies have been on characterising either the Varian or the Philips MLC. We begin with the former. Boyer *et al* (1992) made measurements to quantify the penumbra and output factor for a Varian MLC which has 26 leaf-pairs, each leaf projecting to a width of 1 cm at the isocentre and able to over-run by 16 cm. The MLC is on the patient side of secondary conventional collimation. Because the leaf ends are round and the leaf sides are stepped, the penumbra is potentially different from that of Cerrobend blocks. (Shih *et al* (1994) have compared flat-tip and round-tip MLCs.) Field shapes could be created from films by using a digitiser, the leaf settings being transferred to the MLC computer through a terminal server.

Boyer *et al* (1992) showed that at 6 MV the depth–dose curve was identical to within 1% for fields shaped by the regular jaws of an MLC, implying that the 'conventional' depth–dose curves could be used for MLC-collimated fields.

The output factor was defined as the dose per monitor unit for a fixed MLC setting divided by the dose per MU for a 10×10 cm^2 field defined by the jaws alone, both measured at the depth of maximum build-up dose. The main conclusion was that the use of the MLC instead of jaws led to results no different than 1%.

Dose maps were made by scanning a diode over an entire 2D field at a fixed depth and defining penumbra as

$$P_{80/20} = (W_{20} - W_{80})/2 \qquad (3.1)$$

where W_{20} and W_{80} are the mean widths of the field at the 20% and 80% isodose lines. Also

$$P_{90/10} = (W_{10} - W_{90})/2 \qquad (3.2)$$

where W_{10} and W_{90} are the mean widths of the field at the 10% and 90% isodose lines. It was found that in general penumbra increased with depth and energy and was a little larger for the MLC collimated fields than for fields collimated with conventional jaws. This was, however, not always the case and several exceptions to the rule were shown. MLC penumbra was measured for a fixed field size but with the fields at different offsets from the central axis of the beam with notable variations.

Clinical-field dose distributions were measured in planes near d_{max} and at 10 cm depth by films perpendicular to the beam, shaped by either Cerrobend blocks or the MLC. The latter fields were defined so the leaves were inserted into the field by small distances. 'Stair step' irregularities were observed in the 50% isodose contour in the MLC-defined field but were 'washed out' in the 10% and 90% isodose lines.

Galvin *et al* (1993) also evaluated the characteristics of the Varian MLC. The Varian MLC is single-focused, with a double step in the sides of the leaves to prevent leakage radiation. It is a retrofitted option (figure 3.1). The leaves project to a length of 16 cm at the isocentre so all the leaves on one side must be positioned within 16 cm of each other so that gaps do not occur. Interlocks prevent this from happening in practice.

Galvin *et al* (1993) remind us of the possible uses of the MLC which include creating intensity-modulated fields. In this paper the authors studied penumbra width, leaf transmission, inter-leaf leakage and the localisation of leaf ends (figure 3.2). It was claimed that the methods used would be equally appropriate for studying other manufacturers' MLCs.

Firstly the use of silver bromide film for dosimetry was justified. This is potentially a concern because it is well known that silver bromide film has a dramatic change in sensitivity with changing photon energy; the sensitivity increases by a factor of ten below 100 keV. This is *potentially* important because the energy of the beam inside the field is different from that outside.

Figure 3.1. *The Varian MLC is an add-on device which mounts on the existing Clinac accelerator head thereby making it easily retrofitable whilst preserving conventional treatment capabilities. The upper and lower collimator system of the accelerator is unaltered during installation so the need for recommissioning is eliminated. Control and setup of the individual leaf positions are provided from a stand-alone workstation and controller. (From Varian product literature.)*

Inside the field there are primary and scattered photons with the former predominant, whereas, outside, the contribution is predominantly scatter. Silver bromide film, lithium fluoride thermoluminescent dosimetry (LiF TLD) and radiochromic film were compared, the latter two having much less sensitivity to energy. Before comparing the behaviour of the three dosemeters measuring field profiles, the non-linear *dose response* of silver bromide and the linear *dose response* of radiochromic film were established and used to correct all measurements. It was found that for isodose measurements above 20% all three dosemeters more or less agreed and so silver bromide film was used thereafter.

For two beam energies and at two depths (d_{max} and 10 cm) the 20–80% penumbra width in the direction of leaf motion was found to be virtually independent of location of the field relative to the midline because of the shaped MLC leaf ends. The leaf-edge penumbra was found to be the same for both sides (tongue side and groove side) and also the same as for a Cerrobend block and for the leaf end. All four penumbra curves virtually superposed.

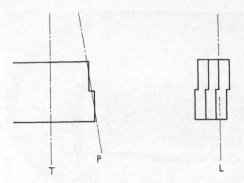

Figure 3.2. *A schematic diagram of an MLC illustrating three of the measurements which must be made to characterise it. On the left is shown a single leaf with a stepped end. Rays such as P contribute to penumbra. (In practice MLCs can have specially shaped ends.) Rays such as T characterise the leaf transmission. On the right is shown three of a set of leaves viewed end-on. The sides are stepped to minimise inter-leaf penetration by rays such as L.*

45° and 15° stepped edges were radiographed and the wavy isodoses plotted. 3D treatment-planning systems must model these sinusoidal isodose lines and to do so will require calculation of dose on 1 mm grids with convolution techniques. Simpler dose calculation models only calculate gross effects and do not model the stepped edges. In practice empirical methods of placing the MLC leaves with respect to the projection of a PTV contour are usually adopted (see also section 3.2).

Leakage was measured. The leaves leaked 1.5% of the open field. The leaf interfaces added between 0.25% and 0.75% more depending on whether this was measured near attachment screws or not. The total leakage was thus never more than 3% which was over 1% less than for standard 3 in ($\sim$ 7.6 cm) thick Cerrobend blocks (figure 3.3).

The leaf ends did not exactly align with the 50% decrement line so hot and cold spots at field joins are possible unless care is taken. This would be a potential problem if creating intensity-modulated fields by field-matching techniques (see Chapter 2) but would be no problem if using field-shrinking methods.

Ihnen *et al* (1995) tabulated the output factors for square fields defined by a Varian multileaf collimator and showed that relative to a factor of unity for a 10×10 cm^2 field the output factor varied from 0.829 for a 2×2 cm^2 field to 1.033 for a 25×25 cm^2 field.

Varian have recently brought out an MLC with 40 pairs of leaves moving parallel to the lower collimator (X) jaws (figure 3.4). The manufacturers specification is <4% leaf transmission, interleaf transmission is <4%. The maximum leaf speed is 1.5 cm s^{-1}. There is redundant readout on all the MLC axes to guarantee positional accuracy of the leaves to ± 1 mm at the

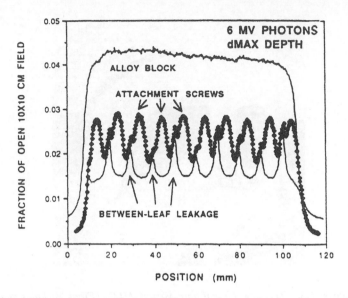

Figure 3.3. *A beam profile for 6 MV x-rays through the closed leaves of a Varian MLC. The lower curve is a profile in the direction normal to the leaf movement and the upper curve is the same but through regions where there are supporting screws. This shows that there is a small increased leakage through the attachment screws (a worse-case situation) and some inter-leaf leakage. However, the total leakage is always less than through a standard 3 in (~ 7.6 cm) Cerrobend block. (Reprinted by permission of the publisher from Galvin et al 1993; © 1993 Elsevier Science Inc.)*

isocentre. With appropriate counterweighting the rotational isocentre is a sphere whose radius is less than 1 mm. The setting of the fields is provided by the 'Shaper' workstation (Varian 1995).

Now we turn our attention to reviewing studies which have characterised the Philips MLC. Jordan and Williams (1994) have provided a fuller account of the evaluation of the prototype Philips MLC in Manchester than in their earlier conference publications, referred to in the companion Volume. Summarising, the Philips MLC has leaves which move in the y direction and the accelerator is equipped wih 7 cm thick x-backup collimators and 3 cm thick y-backup collimators on the patient side of the MLC. The x collimators have to be thicker because they can act as the primary attenuator when part of the field is closed by MLC leaves. The MLC leaves themselves never completely close and the technique is to arrange for the almost-closed leaves to be off-centre and backed up by the x collimator. The y collimators on the other hand are brought up only to the most extended MLC leaves and are there to attenuate leakage radiation through the leaf sides (figure 3.5). Jordan and Williams (1994) provided very useful tables of the transmission through each component and then the transmission through

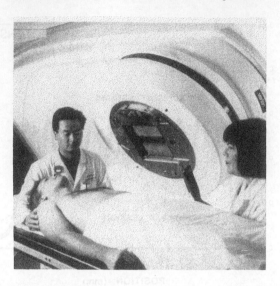

Figure 3.4. *The new 40-leaf-pair Varian MLC. (From Varian product literature 9/95.)*

combined components is easily computable by multiplying the individual transmissions. A table for combinations was also presented. At 6 MV the peak transmission through the components is: leaf-bank maximum (between leaf sides) 4.1%, leaf-bank mean 1.8%, closed leaves 51%, x backup 0.5%, y backup 10.7%. Note the largest penetration is through the nearly closed leaf tips but, multiplied by the x-backup collimation, the total penetration is only 0.3%. Similarly the penetration through leaf sides is $4.1\% \times 10.7\% = 0.4\%$. These figures are for the production model and supercede the figures given in the companion Volume for the pre-production prototype (Jordan 1994). An important element of commissioning an MLC is measurement of the individual transmissions (by shielding all other forms of transmission). A recent software change actually now means the x-backup collimators never align with the edge of the leaves as shown in the lower part of figure 3.5 (Jordan 1996).

Huq *et al* (1995) have made similar measurements for a Philips MLC *retrofitted* to an SL25 linear accelerator. They found that at 6 MV measured individual transmissions were: 2.5% for the mean leakage between leaf sides, 0.9% for the x-backup diaphragms, 8% for the y-backup diaphragms. The transmissions for the combinations were given as: leaves + y diaphragms 0.25%, leaves + x diaphragms 0.10% and leaves + x and y diaphragms 0.05%. These results do not seem to be consistent, given the product rule of transmissions.

The reproducibility of setting the leaf positions must also be measured with the gantry at different orientations to check the effects of gravity on backlash.

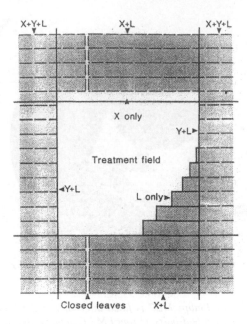

Figure 3.5. *The collimating components involved in defining an irregular field with the Philips MLC. The leaves move in the y direction and are backed up by the y collimation which is thinner than the x collimation in the orthogonal direction since this latter can provide primary collimation. The exact transmission through the collimation will depend on a combination of that through leaves and main jaws. (From Jordan and Williams 1994.)*

The reproducibility can be checked both by increasing and decreasing the size of a field. The Philips MLC, which is internally fitted to the head of the linac (figure 3.6), has leaves which are monitored by a TV system (figure 3.7). Jordan and Williams (1994) found RMS errors less than 0.5 mm and no error greater than 0.75 mm.

The penumbra of an MLC is a complex feature and although it is possible to make simple calculations based on geometry, shape of leaf ends etc, the only true way to assess penumbra is by measurement since this will fold in the effects of finite spot size, scatter and electron transport in tissue. The penumbra is also a function of the angle the leaf tips make to the field edge and when the tips are all in line is also a function of whether the backup collimators are in place. Jordan and Williams (1994) provide very useful figures for the Philips MLC. The almost-closed leaves 'park' off the central axis and when each leaf-pair reaches its extreme position each side of the midline, the theoretical geometric penumbra is the same even though the extreme positions are not symmetrical about the midline (Jordan 1994). Mohan (1995) has pointed out that for commercial MLCs such as both the Philips and the Scanditronix which have single focusing and rounded leaf

Figure 3.6. *The Philips MLC is fitted internally to the head of the linac and has 40 tungsten leaf-pairs. (From PMS Company literature.)*

ends, the motion of the leading edge of the leaf is not linear with the motion of the edge of the radiation field and that small software corrections are called for. Mohan quotes a potential error of about 5 mm for a leaf 20 cm off the central axis.

Huq *et al* (1995) also made measurements of the $P_{80/20}$ and $P_{90/10}$ penumbra under a variety of conditions. First they collimated an 8×8 cm^2 field and positioned it at different y offsets relative to the centre line. Under these circumstances the $\pm x$ border was collimated by the x-backup diaphragm. The $\pm y$ border was however collimated by a combination of the leaves and the y-backup diaphragm so it may be expected that the shape of the leaf ends might influence the penumbra. It was observed that at 6 MV the $P_{80/20}$ penumbra at d_{max} for various field positions varied from 5.5 mm to 6 mm and at 10 cm depth varied from 6.5 mm to 8 mm due to increased scatter. The side of the field furthest away from the central axis had slightly narrower penumbra (by about 1.5 mm at most). This was interpreted as reflecting the effect of the shape of the leaf end since the penumbra for the outer border is determined by the lower portion of the leaf end whereas the penumbra for the inner border is determined by the upper portion of the leaf end. However, Huq *et al* (1995) found that the magnitude of the penumbra increased when the field was collimated by *only the leaves* by 1–2 mm depending on the y location of the field edge. This is an important measurement because it is this penumbra which will govern the setting of an irregularly shaped field when the y-backup diaphragms are set to the

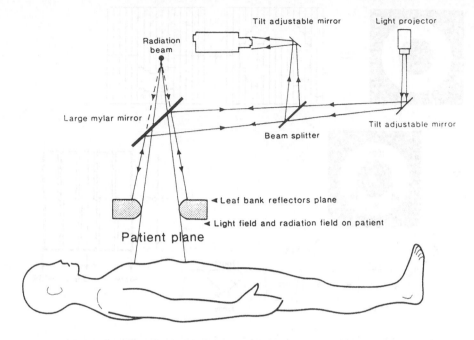

Radiation beam

Tilt adjustable mirror

Light projector

Large mylar mirror

Tilt adjustable mirror

Beam splitter

◄ Leaf bank reflectors plane

◄ Light field and radiation field on patient

Patient plane

Figure 3.7. *The optical arrangement for the video system in the Philips MLC which is used for leaf positioning and verifications. (From Jordan and Williams 1994.)*

position of the most retracted leaf.

Huq *et al* (1995) also compared the penumbras for a 20×20 cm^2 field collimated by either: (i) leaves only (ii) leaves + diaphragms; (iii) a Cerrobend block. They found for both field edges and for both 6 MV and 25 MV beams that the penumbra in conditions (ii) and (iii) were very similar whereas the penumbra for condition (i) was larger. The $P_{90/10}$ in fact increased by nearly 1 cm.

Some further comment on the stepped edge of the sides of each leaf is required. As noted above, manufacturers reduce the transmission between leaves by providing either a step or some kind of tongue-and-groove arrangement. The width of the step is small, usually of the order 1 mm, and as a result can be ignored when planning fixed fields. However, a problem arises when the MLC is used for intensity modulation or to provide internal blocking (Mohan 1995). At the junction of adjacent but opposing leaves there is a region of reduced fluence and therefore reduced dose (figure 3.8). This problem can be particularly troublesome when planning IMBs using dynamic-leaf movement when 'cold-stripes' can develop along the leaf joins. Van Santvoort *et al* (1995) have shown that this problem can be reduced by synchronising the movement of pairs of leaves.

Working with an MLC inevitably then involves paying more attention to

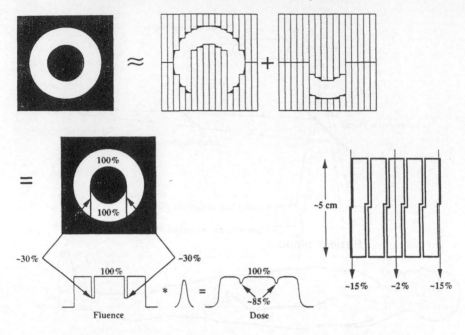

Figure 3.8. *This shows how the tongue-and-groove arrangement may lead to underdosage when an MLC is used to create a centrally blocked field. A circular field with a central block is generated by combining the two fields shown. There are two narrow regions at the boundaries of the constituent fields where either the tongue or the groove is present and reduces the fluence below the expected 100%. In this figurative example the tongue and groove only transmit 15% of the radiation (lower right of figure), giving 30% in total (see lower left of figure). In practice lateral scattering and electron transport (represented by the convolution symbol) reduce the problem but there is still significant underdose to the linear regions shown by the arrows. The same problem would arise if an MLC were used for intensity modulation without synchronisation of leaf movement. (From Mohan 1995.)*

quality control (Hounsell *et al* 1995) and the commissioning process can be lengthy, though hopefully, less so on production models than on prototypes. However, the pay-off will come later when the increased versatility of the MLC begins to be exploited. The vision is that one day the MLC will be the standard method of photon collimation.

3.2. CAN AN MLC REPLACE SHAPED CAST BLOCKS?

LoSasso *et al* (1993) wanted to establish that the use of the MLC is equivalent to the use of shaped Cerrobend blocks for conformal therapy with static fields. They compared plans using visual inspection of 3D dose distributions, the DVH and also predictions of TCP and NTCP, the latter

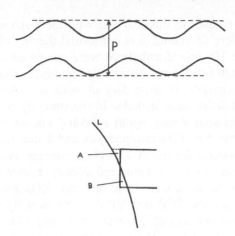

Figure 3.9. *When the MLC leaves make a 'stepped profile' the concept of penumbra is not so simple to define. One proposal has been an 'effective penumbra', the distance P between the line tangent to the crests of one isodose sinusoid and another line tangent to the troughs of a second sinusoid (upper part of figure). Since the stepping of the profile will vary over the set of leaves used, the concept of effective penumbra is a local measurement. The concept of 'transecting' is illustrated in the lower part of the figure. The line L is the edge of the field as collimated by a cast (curved-edge) block. The leaf is brought up to this line such that areas A and B are equal.*

using the Lyman model with data (D_{50}, n, m) from Burman *et al* (1991) (see Chapter 5). TCP was computed by the Goitein model. Both account for inhomogeneous dose distributions.

The MLC investigated was attached to a Scanditronix Racetrack Microtron System MM50. This MLC has 32 pairs of leaves, each projecting to width 1.25 cm at an isocentre 100 cm from the source. The leaves are double-focused. The maximum field size is 30×40 cm^2.

Measurements were made of the isodoses in the penumbral region at d_{max} and at 10 cm depth in tissue, both measurements being at isocentre. Film dosimetry was used. The MLC was set up with the leaves defining a 45° angle to the axes. The *effective penumbra* was defined as the distance between the peaks of the 80% isodose contour and the troughs of the 20% contour (figure 3.9). This penumbra is between 3–5 mm larger than for Cerrobend blocks depending on the depth and on the beam energy. However, Brahme (1993) takes issue with this concept and considers the distance between the 90% and 10% isodoses to be of more significance. This distance does not vary so much between MLC and block. Also this question is influenced by how much daily variations in positioning influence field edges (Brahme 1994).

The MLC was positioned by 'transecting' i.e. the open area treated by each leaf was the *same* as would be treated by the corresponding thin strip of the

Cerrobend block parallel to this leaf and with the same width (figure 3.9). The clinical problem of interest was conformal therapy of the prostate and nasopharynx. The 3D dose distributions were computed using the MSKCC 3DRT system which uses kernel-based calculations for irregular fields. This accounts for the transport of secondary photons and electrons. Treatment-positioning uncertainties were included in the study by convolving the dose kernel with a Gaussian whose width reflected known data about patient movement (this was 3 mm for nasopharynx and 5 mm for pelvic sites).

Scalloping is greatest for the 50% isodose contour. However, in clinical plans, increasing the number of fields and adding in setup uncertainty wipes out the differences between Cerrobend-blocked fields and MLC-collimated fields. The DVH and the TCP and NTCP were virtually identical for both planning sites when comparing Cerrobend blocking with MLC shaping but this is mainly a consequence of using adequate margins ($\simeq 1$ cm) for the PTV.

Frazier *et al* (1993) have compared the use of Cerrobend blocks and the MLC, concluding that the random nature of daily setup variations is large enough to render field-shaping differences between the two methods insignificant.

Webb (1993, 1994a,b) studied the effect on tumour-control probability of varying the margin between the edge of the planning target volume and a 45° edge collimated by a MLC with each leaf set back from its neighbour by a leaf width (figure 3.10). The first step in this task was to measure the sinusoidal penumbra accurately at depth and to interpolate from this to a fine-scale map of the 3D distribution of dose in the penumbra. Then, for a box-shaped PTV variably offset by distance X_c from the line drawn through the projected faces of the leaves (figure 3.11), a model for tumour-control probability when the dose is inhomogeneous (Webb and Nahum 1993) was used (see Chapter 5) to predict the TCP as a function of this distance of offset (figure 3.12). It was determined that a margin of 7 mm around the PTV was adequate to ensure that the MLC behaved as would a Cerrobend block and this is the margin used in clinical practice with the Philips MLC at the RMNHST (Shentall *et al* 1993a). As noted by Mohan (1995), the potential and largely overexaggerated problem of the stair-step penumbra assumes even less importance when fields are combined. Studies of this kind can only be done by simulation since experiments cannot at present establish the variations of dose in 3D, and hence TCP, with sufficient accuracy (see also Mohan 1995).

Burman *et al* (1994) showed that when the MLC was used to collimate fields for tumours very close to OAR such as the cord, brainstem and optical chiasm, that additional Cerrobend block shielding was required. This is consistent with the above studies which only looked at the effect on the tumour.

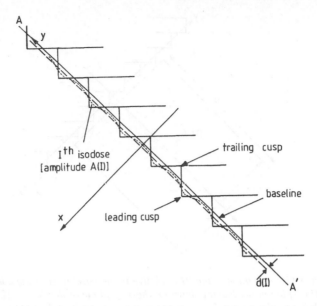

Figure 3.10. *The reference coordinate system and geometry of an MLC set to produce a 45° step pattern. The x, y axes (defining a plane parallel to the MLC) are shown in the (depth) plane $z = 10$ cm. The baseline is the line AA' passing through the projected shadow of the midpoints of the leaf faces. The leaves project to a width $w = 1$ cm at the isocentric distance d_{iso} from the source. The Ith isodose contour is offset from the baseline by $d(I)$ and the Ith isodose contour has amplitude $A(I)$. These displacements and amplitudes are a function of the isodose value and are given in Webb (1993) for the Philips MLC. The area to the bottom left is the irradiated area and the area to the top right is the shielded area. The cusps are labelled as 'leading' or 'trailing' with respect to the position of the leaves relative to the field.*

Frouhar *et al* (1994) also studied the effect of MLC leaf penumbra on the 3D dose distribution and modelled the scalloping effect of the leaves. Welch and Davy (1994) concluded that the MLC is an effective block replacement.

Fernandez *et al* (1993, 1994a) have carried out a detailed study of the block-replacement capabilities of the Philips MLC installed at the RMNHST, Sutton (Shentall *et al* 1993b). A total of 605 simulator films was examined to establish that 37% of fields required blocking. The distribution by site was (first figure per cent of workload, second figure per cent of site requiring blocking): brain (8%, 30%), head and neck (5%, 61%), thorax (21% , 49%), breast (21%, 27%), abdomen (2%, 38%), pelvis (16%, 43%), lymphoma (1%, 100%), spine (20%, 21%), extremities (6%, 46%). It was discovered that only 5.5% of these blocked fields could *not* be achieved satisfactorily with the MLC. This was due to a number of reasons including: (i) for cord shielding the projection of one leaf provided too little shielding but the projection of two leaves provided too much; (ii) when the use of multiple

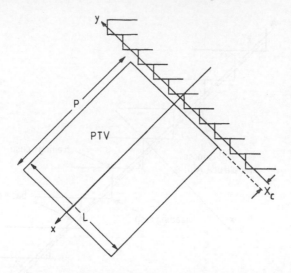

Figure 3.11.   *Location of the PTV relative to the baseline of the shadow of the MLC (plan view; looking down on the x, y plane at depth $z = 10$ cm). The PTV has dimensions $P \times L$ in the x, y plane and is of depth Q in the z direction.*

asymmetric mantle MLC fields introduced potential problems matching the field join; (iii) the limitation that the Philips MLC leaves travel in the y direction which is parallel to the apex of the wedge. This latter feature causes a problem for the following reason. MLC fields are generally longer than they are wide. Brahme's orientation theory (see the companion Volume) has shown that it is best to arrange for the leaves to be parallel to the shortest cross section of the projection of the PTV. Thus the leaves should therefore move in the wedged direction (i.e. perpendicular to the apex of the wedge) for fields wedged in the transverse plane (figure 3.13). (It turns out that Du *et al* (1995) have shown how by dynamic MLC modulation wedges can be created at arbitrary angles to the collimator. However, this is a 'one-off' non-standard development (see Chapter 2)). Whilst the MLC can easily be used to produce internally blocked fields by using two asymmetric fields, this does not produce a matchline problem for fields (e.g. head and neck) where the matchline is under the spine block.

Interestingly 5% of the fields would not have been realisable with the Varian MLC either, because of the necessity for a larger size than 26×40 cm^2. Also because the MLC produces blocking in finite size increments it was found that in 8% of the cases it was necessary to move the central axis in order to achieve satisfactory blocking. Moreover, 52% of cases used the Regular Shape Library, a feature provided by Philips in which several regular shapes can be achieved with very minimal specification (figure 3.14). However, of these only two were found to be frequently called upon (the cross and the rectangle with one straight chamfer).

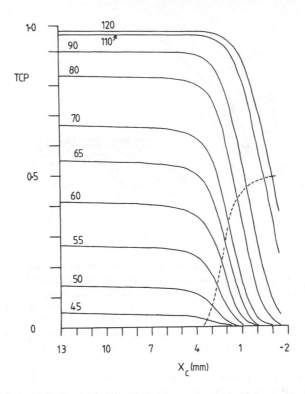

Figure 3.12. *Set of curves of TCP versus X_c for different values of the normalisation dose (in Gy) to the centre of an open field and two-sided collimation. The planning target volume is 64 cm³ and the parameters of the TCP model (see chapter 5) are $\alpha = 0.35$ with $\sigma_\alpha = 0.08$. Note that the TCP is approximately constant for $X_c \geqslant 7$ mm and falls monotonically for smaller X_c. The dotted curve links the point on each curve where the TCP is half the maximum value (at $X_c = 13$ mm) for that curve.*

The MLC in some cases could block a field when cast blocks could not be used because of either: (i) collision between the blocking tray and the couch, or (ii) the required blocks exceeded the weight limitation of the blocking tray. Taken together with the 94% success rate of the MLC at replacing cast blocks and the ergonomic advantages deduced from a work study (Helyer and Heisig 1995, Helyer *et al* 1994) (see section 3.2.2) the MLC has clearly achieved its primary objective of block replacement.

Fernandez *et al* (1994b) have carried out a study of the potential for increased sparing of normal tissue with the Philips MLC. They studied 43 patients who were planned for both conventional and MLC blocking. Traditionally, clinicians have been used to drawing straight-line blocking since this matched the practicalities of delivery. In this study, however, the clinicans drew the ideal PTV on to the simulator films. A 6 mm margin was

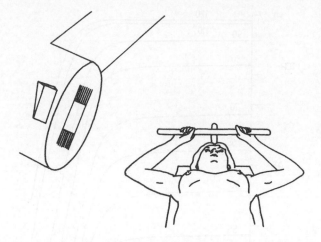

Figure 3.13.   *The delivery of the more anterior of the two tangential fields to treat the left breast. The field is defined by an MLC in which the leaf movement is in the transaxial plane. The field is longer (longitudinally) than it is wide (transaxially). Only a representative number of leaves is shown schematically. In practice more leaves than that shown would be employed. The field is wedged transaxially and this is shown diagrammatically. So the leaf movement must be perpendicular to the apex of the wedge.*

drawn around this contour to represent the 50% isodose line to which the blocking or MLC was matched. The area treated in excess of that enclosed by the 50% contour was then calculated for both the block-method and the MLC method of field creation. Since the study encompassed only parallel-opposed pairs this was taken to represent excess volume treated.

The conclusion from the study was that in every case the MLC treated less than 10% excess tissue and in over 70% of cases the excess was less than 5%. The conventional fields, however, treated over 10% excess tissue in 70% of the cases.

The Varian MLC has also been shown to be an effective block replacement with reduced setup times (Thompson 1995). In a study comparing the differences in area of normal tissue irradiated when using a Varian MLC for conformal therapy and standard block shielding for various head-and-neck tumours it has been shown that considerable normal tissue sparing can be achieved (Bidmead *et al* 1995).

3.2.1.   Setting leaves with respect to the beam's-eye-view of the PTV

Since its inception there has been a good deal of interest in, and activity addressing, the question of how leaves should be placed with respect to a continuous contour representing the PTV or the PTV with some added margin. As reviewed in the companion Volume, Brahme (1988) showed that

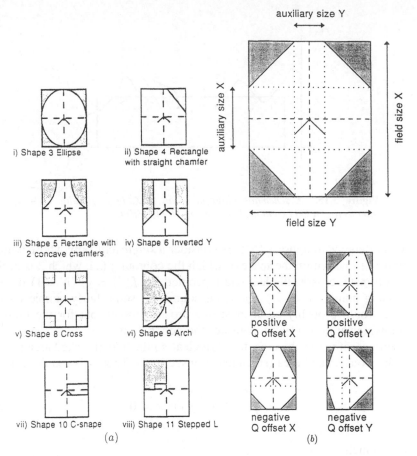

Figure 3.14. *(a) Illustrates some of the shapes in the Philips Shape Library. The shapes are modifiable by specification of a few parameters called offsets and auxiliary sizes as illustrated in (b) for shape number 7: the octagon. (Reprinted from Fernandez et al 1994a, with kind permission from Elsevier Science Ireland Ltd, Bay 15K, Shannon Industrial Estate, Co. Clare, Ireland.)*

when a series of leaves are brought up to a continuous contour, the region overtreated (compared with the use of a block) would be minimised when the direction of motion of the leaf set was normal to that part of the contour with largest radius of curvature. Zhu *et al* (1992) considered the special case of a set of leaves being brought up at 45° to a straight line and identified several options: (i) either the leading leaf corners touch the line or (ii) the trailing leaf corners touch the line or (iii) the edge of the field bisects the leaf ends (so called 'transecting') or (iv) the leaves are set to protrude a fixed distance into the contour.

Yu *et al* (1995) have presented a neat way of characterising the problem.

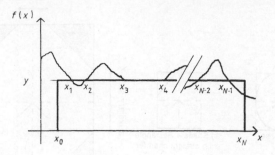

Figure 3.15. *A schematic diagram of an MLC leaf of width $x_N - x_0$ intersecting a field contour $f(x)$. (From Yu et al 1995.)*

With reference to figure 3.15, they consider a *single leaf* only whose width is from x_0 to x_N within the space of which the contour $f(x)$, which it is desired to collimate, crosses the leaf end (specified by $f(x) = y$) $(N - 1)$ times at crossing points $x_1, x_2, \ldots, x_{N-1}$. Of course for some fairly simple contour there would probably be at most one crossing point, but for the moment, following Yu *et al* (1995), consider the general case.

Suppose it is the objective to set the leaf so the underblocked area is equal to the overblocked area (so called 'transecting'). Then it is required that

$$\int_{x_0}^{x_N} (f(x) - y)\, \mathrm{d}x = 0 \qquad (3.3)$$

from which

$$y = \left[\int_{x_0}^{x_N} f(x)\, \mathrm{d}x \right] \Big/ (x_N - x_0). \qquad (3.4)$$

i.e. the leaf end should be set at the *mean* value of $f(x)$ within the width of the leaf. This transection minimises (i.e. sets to zero) the difference between the under- and over-blocked parts of the field.

Yu *et al* (1995), however, suggest a second and alternative strategy. Suppose instead it is required to minimise the total area discrepancy

$$A(y) = \int_{x_0}^{x_N} |f(x) - y|\, \mathrm{d}x. \qquad (3.5)$$

We require that y be set so that

$$\frac{\mathrm{d}A(y)}{\mathrm{d}y} = 0. \qquad (3.6)$$

Expanding equation (3.5)

$$A(y) = \int_{x_0}^{x_1} (f(x) - y) \, dx - \int_{x_1}^{x_2} (f(x) - y) \, dx + \int_{x_2}^{x_3} (f(x) - y) \, dx - \ldots$$

$$+ \int_{x_{N-2}}^{x_{N-1}} (f(x) - y) \, dx - \int_{x_{N-1}}^{x_N} (f(x) - y) \, dx. \tag{3.7}$$

Taking note that the limits of the definite integrals are also functions of y and differentiating the general term in equation (3.7) we find

$$\frac{d}{dy} \left[\int_{x_k}^{x_j} (f(x) - y) \, dx \right] = -(x_j - x_k). \tag{3.8}$$

i.e. the derivative with respect to y of each integral within limits x_k and x_j is simply the negative length between x_k and x_j. Hence when differentiating the whole of equation (3.7) the positive terms (representing the segments where the leaf is outside the field contour) sum to *minus* the total length of the leaf tip *outside* the field contour and the negative terms (representing the segments where the leaf is inside the field contour) sum to *plus* the total length of the leaf tip *inside* the field contour. Setting the whole derivative to zero (equation (3.6)) yields the interesting and rather neat result that to minimise the area discrepancy the total length of the leaf-end segments outside the field contour should be made the same as the total length inside the field contour, i.e. the leaf tip should be set with y equal to the *median* of $f(x)$ within x_0 to x_N, the width of the leaf, so that when the contour $f(x)$ is sampled regularly in x there are as many data points beyond the leaf tip as behind it.

3.2.2. Does the use of the MLC save time?

The promise of the MLC is that with respect to the use of cast blocks it saves time. Is this true? Helyer and Heisig (1995) and Helyer *et al* (1994) have performed a work study using the British Standards Institute (BSI) 'method study' in which the various tasks associated with the use of both blocking and the MLC were broken down into distinct parts which could be individually documented as elements of work and individually quantified. They made a comparative study of the two approaches in two clinical situations: (i) the creation of dog-leg fields for treating germ-cell tumours of the testis, and (ii) conformal pelvic radiotherapy. The study was made by qualified staff at an average rate of work under average conditions over a four month period. Each patient acted as their own control since both methods were worked up for every patient. They included in the study timing of (for the blocking route): (i) mould room casting of blocks; (ii) workshop mounting of blocks; (iii) the physicists' time drawing and checking templates for non-custom

Time (Seconds)

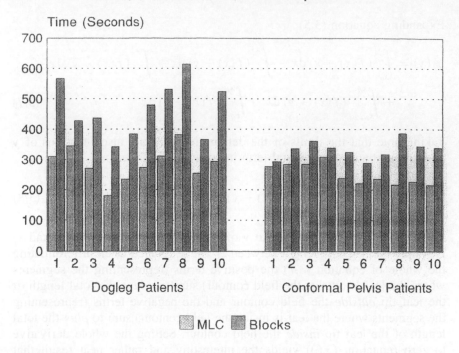

Figure 3.16. *The average treatment times recorded per patient for two types of therapy showing the advantage of using an MLC over conventional lead blocks. (Reprinted from Helyer and Heisig 1995, with kind permission from Elsevier Science Ireland Ltd, Bay 15K, Shannon Industrial Estate, Co. Clare, Ireland.)*

cast blocks; (iv) radiographers' time setting up blocked treatments and (for the MLC route): (a) physicists' time creating beams and associated dose distributions; (b) radiographers' time treating with the MLC.

For dog-leg fields the mean time of the template-drawing process was 27.5 min and the physicists' mean time checking templates was 5 min. However, the mean time to produce MLC dog-leg beams by physics staff was 42.5 min. Hence the time spent prior to treatment was longer for the use of the MLC. However, at the time of treatment, time savings of between 19 and 48% were recorded for the use of the MLC compared with that of blocks (figure 3.16).

For conformal pelvic fields, the mean time for Cerrobend block manufacture was 2 h 7 min. The mean time taken to mount the blocks on templates in the workshop was 38 min. The mean time taken by physics staff to generate conformal block plans was 2 h 30 min. In comparison the mean time taken to produce conformal MLC plans was 1 h 36 min. Thus there was an overall pre-treatment mean saving of 3 h 39 min. At the time of treatment, time savings of between 6 and 44% were recorded for the use of the MLC compared with that of blocks.

In addition to these time savings (which lead to the ability to handle an increased patient workload) it is of course pertinent to note that the use of the MLC: (i) eliminates the need for mould room and workshop activity; (ii) avoids the need for block storage; (iii) avoids lifting difficulties and potential injuries; (iv) avoids the requirement for taking care to avoid the toxic side effects of handling blocks. Similar observations were made by Boyer *et al* (1992).

Whilst the precise times quoted by Helyer and Heisig (1995) obviously relate to the use of a particular MLC at a particular state of development of techniques in a particular centre (RMNHST) there is an inescapable conclusion that the MLC has distinct advantages.

Of course the major time savings using a computer-controlled MLC will eventually come when implementing a treatment with a large number of fields. Kutcher *et al* (1995) have argued that when, for example, non-coplanar fields are used, it will probably be necessary for different fields to have different source-to-tumour distances. This in turn will necessitate changing the height and lateral position of the couch for different gantry orientations to avoid collisions. This, together with the concept of implementing multisegment therapy, would be impossibly time-consuming if performed manually and the automatic MLC, together with remote pedestal and gantry adjustments, is really the only solution.

Whilst the use of an MLC can save time at the treatment stage and even at the planning stage, its use does create more work in some areas. For example, the MLC, being a complex electromechanical device, requires time-consuming quality-assurance. There is also the need for maintenance and interfacing to treatment-planning systems (see section 3.4)

3.3.   THE MLC IN CLINICAL USE

3.3.1.   *The MLC creating multiple fields for conformal therapy*

Powlis *et al* (1993) invoke the evidence that improved local control may reduce distant metastases as an argument for striving for conformal therapy. There is a need for automatic field shaping without entering the room to facilitate the use of multiple MLC non-coplanar fields. They report that this has been standard practice since 1991 at their centre. The MLC in use was the Varian MLC (26 pairs of leaves, each with width 1 cm at isocentre. The over-run was 16 cm and maximum field size was 26×40 cm^2.)

MLC-shaped fields were arranged such that the midpoint of the end of each leaf intersected the block contour (as drawn on simulator films). Portal films were then taken for both Cerrobend-blocked fields and MLC-shaped fields. The 'effective penumbra' as defined by LoSasso *et al* (1993) was found to be some 4.7 mm larger for a 45° edge collimated by an MLC than by a block (at 6 MV).

Doses were calculated using TARGET 2 and also by a kernel-based algorithm. These required computation of dose on a grid with spacing 1–3 mm *and no more*. An inspection of the clinical DVHs demonstrated the improvement with increased number of ports and thus a dependency on the availabilty of the MLC.

3.3.2. Coplanar versus non-coplanar irradiation

Aoki *et al* (1994) compared seven-port coplanar, 17-port coplanar with five-port non-coplanar irradiation of a broncogenic carcinoma. The five non-coplanar ports were chosen to minimise the line-of-sight view of OAR and all fields were BEV-shaped. It was found that the methods gave TCPs of 0.21, 0.64 and 0.67 respectively and NTCPs of 0.02, 0.06 and 0.04 respectively. It was concluded that all techniques were similar from the point of view of NTCP but that the non-coplanar method could give a much higher TCP with fewer fields than when coplanar. Whilst this result is for only a particular case, the message is that use of MLC-shaped non-coplanar ports could lead to the need for fewer fields for the desired clinical outcome. Each particular clinical case should, however, be studied individually. Boyer (1995) suggests that the use of non-coplanar fields is not as great as it might otherwise be because of backlash in the rotations and translations of the couch top, the bearing of which may not be engineered to as high a standard as the bearing determining the gantry isocentre.

3.3.3. The MLC in conformal stereotactic radiotherapy

Schlegel *et al* (1992, 1993) describe the equipment at the German Cancer Research Centre (DKFZ) for fractionated stereotactic radiotherapy of the brain. The tools are: (i) a hinged relocatable face mask in two parts made of self-hardening plastic bandages; the frame has a baseplate which contains marker tubes which show on images and which enable MRI, PET and CT data to be registered and used for treatment planning, the baseplate also serves as the reference for stereotactic coordinates; (ii) the 3D planning system known as VOXELPLAN; (iii) the manual or computer-controlled in-house-constructed MLCs.

3D image datasets of the patient wearing the mask and frame with markers are recorded and registered geometrically. The images are segmented using the TOMAS tool which is part of the Heidelberg treatment-planning system. 3D treatment planning is performed with the VOXELPLAN system and the dose can be calculated for up to 15 irregular non-coplanar fields. The dose algorithm is kernel-based and accounts for scatter.

Figure 3.17. *The microMLC at DKFZ, Heidelberg. The leaves can be retracted by opening the two 'U' shaped levers. A wooden template of the field shape is placed between the leaves and the levers are closed. Finally the other two levers on long tubes are turned to apply pressure to keep the leaves in place once the template is removed. (Courtesy of Professor W Schlegel.)*

The DKFZ have three operational MLCs and others under development. One of these, a so-called microMLC, is a 40 leaf-pair spring-activated MLC with 1 mm leaves; this is used for delivery of very small fields (figures 3.17 and 3.18). The other two have 27 pairs of 3 mm leaves giving a maximum square field of 13.5 cm side at isocentre (the leaves project to 5 mm width at the isocentre); one is a manual MLC and the other is a motorised MLC (figures 3.19 and 3.20). The MLCs are used differently depending on whether a number of static fields are combined or there is dynamic therapy. Schlegel *et al* (1992) describe an interesting technique whereby the envelope of all the possible BEVs of a target volume during arc therapy is established and this is set for that arc with the MLC.

The small-scale micro-MLC works by releasing the spring pressure which keeps the leaves apart and then with the template inserted closing the leaves, tightening clamps and finally removing the wooden template (figures 3.17 and 3.18). The precision of patient positioning is assessed optically. Debus *et al* (1994) have reported that the the use of 11–14 static micro-MLC collimated fields can give a better dose uniformity for irregularly shaped structures than multiple-arc therapy with reduced side effects.

Only the manual MLC with 3 mm leaves, without the envelope technique, is currently in use for large-field irradiations delivering a series of static fields although there are figures in Schlegel *et al* (1992) of the automatic MLC attached to the treatment head and plans to use it. It takes 40 s to reposition

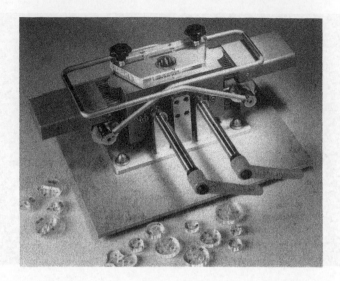

Figure 3.18. *The manual microMLC from Heidelberg is marketed by the Leibinger Company in Germany. Also shown in this picture are the set of standard-shape inserts which can be placed within the leaves to create the irregular field shape. (From Leibinger Company literature.)*

the leaves of the manual large-field MLC. This MLC can be attached to the linac using a portable trolley arrangement (figure 3.21) and fields are set by wooden jigs cut to the BEV of the radiation field. The large-scale MLC works by compressing the leaves on to the jig.

The DKFZ are also constructing an automated MLC with the same geometry as the manual 27 leaf-pair model but with a motor to drive each of the 54 leaves. This will lead to a facility in which leaf setting can be done electronically and also dynamic leaf movement would be possible (Bortfeld 1995) (figure 3.22). The successful development of the computer-controlled modern MLC with small leaf widths has partly been due to the development of powerful miniaturised stepper motors (Kutcher *et al* 1995).

The M D Anderson Cancer Centre also has a special-purpose MLC for stereotactic radiotherapy. This has 15 pairs of tungsten leaves with a projected leaf width of 4 mm at the isocentre, producing a maximum field size of 6×6 cm^2. Each leaf can be driven at 10.5 mm s^{-1} and cross the central axis by 30 mm. The leaf positioning is accurate to 0.02 mm (Shiu *et al* 1994a). Shiu *et al* (1994b) showed that the use of the micro-MLC led to superior dose distributions to those which could be achieved by multiarc therapy with a circular collimator. Forster *et al* (1994) using a Scanditronix MM50 Microtron fitted with an MLC also found advantages for the use of multiple fixed MLC-shaped fields.

Figure 3.19. *The motorised MLC at DKFZ, Heidelberg. Four motors actuate mechanisms to change the positions of four leaves. Two more act as selectors to choose the quadruplet of leaves. Leaf setting is therefore quite time consuming and appropriate to a research setting. (Courtesy of Professor W Schlegel.)*

3.3.4. Sparing dose to normal brain tissue with shaped fields

In the late 1980s and early 1990s several groups developed stereotactic conformal radiotherapy based on the use of a conventional linear accelerator and multiple non-coplanar arc rotations (see the companion Volume, Chapter 3). The technique is becoming widely accepted as an alternative to the gamma knife for treating small spherical tumours.

At much the same time interest also focused on whether *non-spherical* tumours could be adequately treated by a small number of non-coplanar fixed fields geometrically shaped by an MLC or custom block. Laing (1993) assessed the volume of normal brain (non-target) tissue raised to 20%, 50% and 80% of the target dose for targets of varying oblateness and size. For perfectly spherical targets of all sizes, a four-arc rotation technique yielded the lowest doses to non-target tissue. As the degree of oblateness increased and for all sizes, the reverse was observed, that multiple-fixed-field irradiations yielded *lower* volumes of normal tissue irradiated at each dose level; the volume decreased as the number of fixed fields was increased from three to four to six.

Figure 3.20.   *The motorised MLC attached to a Siemens Mevatron 77 linear accelerator (Reprinted by permission of the publisher from Schlegel et al 1992; © 1992 Elsevier Science Inc.)*

The basis for Laing's work was the 3D treatment planning system VOXELPLAN from the German Cancer Research Centre (DKFZ) installed at the RMNHST. This work could have been extended by computing cumulative DVHs for each treatment and converting to a measure of biological response such as NTCP.

3.3.5.   Theoretical studies of normal-tissue sparing in stereotactic radiotherapy using various field-shaping collimators

For several decades stereotactic radiotherapy was only possible using the special-purpose Gamma Knife machine (see the companion Volume, Chapter 3). Then, during the 1980s many groups, worldwide, started to treat brain tumours using multiple arcs of a more or less conventional linac fitted with a circular collimator. The centres differed in the number and disposition of arcs but all such techniques using a single isocentre produced a spherical high-dose volume. Non-spherical high-dose volumes were generated by multiple-isocentric treatments but these do not lead to a very uniform target dose.

Recently Leavitt *et al* (1991) have constructed a new 'dynamic collimator' for stereotactic-multiple-angle radiotherapy (SMART) which, in addition to the circular collimation, has four jaws which are independently translated and rotated to give a variety of special field shapes (see the companion Volume, Chapter 3, figure 3.22). The idea is that as the linac moves in an

Figure 3.21. *The trolley for holding the manual MLC at DKFZ, Heidelberg whilst it is offered up to the inverted head of the linac. The method is akin to that used for changing the collimator on a gamma camera. (Courtesy of Professor W Schlegel.)*

arc, the geometrical shape of the field is repeatedly adjusted to match the projected area of the planning target volume. The jaws, together with the circular collimator are able to define a variety of shapes (figure 3.23).

Nedzi *et al* (1993) have done a detailed analysis of the advantages of this additional field shaping, the conclusions of which we summarise here. In addition to: (i) conventional (if one can use that word of so new a technique) circular collimation; (ii) Leavitt-type dynamic collimation (use of four independently-rotatable jaws), they considered three other field-shaping possibilities (see figure 3.23): (iii) two parallel jaws within the circular field; (iv) four jaws forming a rectangle or square within the circular field; (v) a circular 'ideal' MLC. This is not a physically realisable MLC but an idealisation of the shape of the field matching the projected area of the PTV.

Nedzi *et al* (1993) studied 43 intracranial tumours. For each they created six treatment plans, corresponding to the use of these five techniques plus the so-called 'actual plan', being the plan with one unoptimised circular

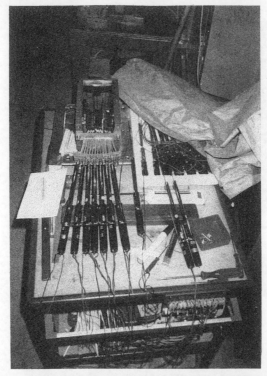

Figure 3.22. *(a) and (b) These pictures show a new motorised MLC under construction in the workshops of the DKFZ, Heidelberg in January 1995. They are interesting as they show the considerable complex engineering which is involved and which is not usually appreciated by users of off-the-shelf commercial MLCs. The motors can be seen being fitted to the leaves. (Courtesy of Professor W Schlegel.)*

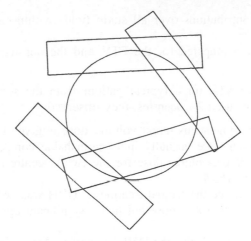

Figure 3.23.   *The method of collimating an irregular field using a circular collimator combined with four independent jaws.*

collimator or with two or more isocentres and circular collimation.

For each plan the procedure was the following:

(i) The SMART radiotherapy was broken down into multiple arcs and each arc was considered as a series of static fields at 10° intervals.

(ii) At each of these pseudostatic orientations the projection of the target volume was formed in collimator coordinates.

(iii) The orientation of the field-shaping collimator was optimised to yield the best placement of the jaws using a merit function which optimised the field boundary relative to the projection of the target volume. This was an important step because it is obvious that there exists a large number of possible choices for setting the collimation. For example, initial parameters were selected based on the direction of maximum displacement from the isocentre of the contour points projected in the plane of the isocentre in the collimator coordinate system. For model (iii) above (two parallel jaws), the initial setting was parallel to the direction of maximum displacement. For model (iv) (four parallel jaws), the initial setting was a square with the diagonal oriented in the direction of maximum displacement. The final jaw settings (for each 10° pseudostatic orientation within each arc) were found by an analytic fitting method known as Levenberg–Marquart (Press *et al* 1986).

(iv) The irregular shape so formed was then divided up into 36 10° angular sectors. The dose to a point in any one sector was approximated by the dose at the same point in a circular field of the same radius as the sector. This is an acceptable approximation given the narrow-beam primary-dose assumption always used in stereotactic radiotherapy (see the companion Volume equation (3.5)) The total dose to any point in the patient is then the sum of these contributions over all static fields within each arc and then over all arcs.

(v) DVHs were prepared for the PTV and the non-target normal brain tissue.

Studying the DVHs for a typical patient from the series of 43 whose 'actual treatment' used 2 isocentres, they observed:

(i) a nearly homogeneous target-volume dose is given by all five model collimations which were virtually indistinguishable compared with a very inhomogeneous target-volume dose for the two-isocentre treatment (whose DVH stretched to 165%);

(ii) in normal tissue the 'actual treatment' DVH was very similar to that using two jaws up to 100% dose but had a significant unwanted high-dose tail up to 140%;

(iii) for all doses $\leqslant 100\%$ the DVH ranked in decreasing volume in the following order: circular, rectangular, four independent jaws, two parallel jaws, ideal MLC.

Nedzi *et al* (1993) defined the Treatment Volume Ratio (TVR) as

$$\text{TVR} = \text{PTV}/\text{TrV} \tag{3.9}$$

where TrV is the treatment volume (i.e. the volume receiving at least the minimum target dose). A truly conformal therapy would have TVR $= 1$. There is a contribution to TrV from both PTV and normal tissue since

$$\text{TrV} = V_{PTV} + V_{OAR} \tag{3.10}$$

where V_{PTV} is the volume of the target receiving the minimum target dose or more and V_{OAR} is the volume of the normal tissue receiving the minimum target dose or more; both can be read off the DVH. Nedzi *et al* (1993) computed the median TVR over the 43 patients for each of the six model collimations. They ranked as follows: (for 'actual' treatments using a single isocentre) actual (0.37), circular collimator (0.37), two parallel jaws (0.44), rectangular jaws (0.46), four rotatable jaws (0.49), ideal MLC (0.55) and (for 'actual' treatments using multiple isocentres) actual (0.36), circular collimator (0.27), two parallel jaws (0.35), rectangular jaws (0.38), four rotatable jaws (0.39), ideal MLC (0.49).

From these figures it can be seen that the use of these special additional collimators can give significant benefits when compared with the use of circular collimators alone. For treating very irregular-shaped tumours, which 'conventionally' would have required the use of two isocentres, the use of the Leavitt-type device, in which all four jaws are independently set, would increase the TVR by some 50% *as well as* decreasing the target dose inhomogeneity. This should lead to reduced morbidity.

However, Nedzi *et al* (1993) did not stratify their data according to the size of the tumour and so an inevitable corollary of their work is that in clinical practice it is necessary to evaluate each case separately. An additional reason to do this is that as well as measuring TVR it is necessary to know the absolute volume of normal tissue spared and even then the same absolute volumes may have different toxicities for different regions in the brain.

Also since the tumours were of different sizes it may be that acceptable conformation could be obtained by a small number of geometrically shaped fixed static fields. They did not include this possibility in their study which was restricted to various methods of *dynamic* therapy. In this respect the work should be cross-compared with that of Laing (1993) and, for example, Graham *et al* (1991).

3.4. COMPUTER CONTROL OF MLC LEAVES AND INTERFACE TO IMAGING

Most manufacturers provide a direct way of controlling the MLC leaves but all do not enable outlines created at the planning stage to automatically set the field. So individual centres have had to develop their own techniques. If the MLC is manual, then the technique of creating a solid template from the planning stage and then arranging for the MLC to 'grip' this is all that is needed (e.g. at DKFZ—see section 3.3.3). Other centres have developed methods to digitise simulator films. Kahler *et al* (1993) create a digital image in a picture archiving and communication system (PACS), then automatically compute the leaf positions and create a file for controlling the leaves. Moore *et al* (1994) describe further developments with the Manchester first production-model of the Philips MLC. Crockford *et al* (1994) describe how MLC leaf prescriptions created by an IGE TARGET planning system are transferred to control the leaves of a Philips MLC at the RMNHST. Bannach *et al* (1995) have transferred the shapes of fields from a TMS-RADIX HELAX AB treatment planning system to a Philips MLC via TCP/IP. The leaf-setting prescription for the Philips MLC can also be obtained from software commercially available from Philips, called 'Multileaf Preparation' (Philips Medical Systems 1995). Ramm *et al* (1996) and Thilmann *et al* (1996) have used the Philips Multileaf Preparation System to create the instructions for placing the leaves with respect to contours. Van Duyse *et al* (1995) create the MLC field shapes via an addition to Sherouse's GRATIS treatment-planning system and transfer the prescription to a Philips MLC via a local-area network. Yu *et al* (1994) describe the chain of events from volume outlining, through BEV computation, optimisation of collimator angle and computer transfer of leaf prescription to a Philips MLC at the William Beaumont Hospital, Royal Oak, USA. Boyer (1995) has commented that it is the non-standard nature of information transfer between equipment from different manufacturers (e.g. planning and treatment unit) which greatly limits progress in communication and control. Fraass *et al* (1995) have made a similar point. They emphasise that the University of Michigan computer-controlled radiotherapy is based on a more sophisticated paradigm that there should be integrated, reliable and verifiable communication between all components of the radiation therapy. They have an MLC attached to a Scanditronix MM50 Racetrack Microtron for which the control is so integrated. Interestingly they also comment that, at least in the USA, the legal and regulatory climate generally limits research in the field of computer-controlled therapy to evaluation of vendor-supplied technology. In this respect the situation is quite unlike the climate in which conformation therapy was first developed as a series of in-house activities.

It is dangerous to reflect too smugly on writing from the past but readers may smile at a quote from the 1970 (3rd) Conference on the Use of Computers in Radiation Therapy. In the paper by Umegaki (1970) we read 'At present three major manufacturers are ready to supply a specially designed multileaf collimator driven by a servo system following a preset pattern. This kind of special collimator is the first step in the realisation of three-dimensional automatic dose-distribution shaping. The next step will be the linkage between the planning system and the treatment machine. The author cannot show actual results at present but this linkage will surely be realised by some additional effort'. A diagram of a six-leaf-pair MLC accompanies the text (figure 3.24). After over two and a half decades this problem is still a hot topic of development.

3.5. VERIFYING MLC LEAF POSITION

As the MLC is used increasingly in clinical practice the need to verify the leaf positions becomes more pressing. At present the leaf positions of commercial MLCs are verified by 'on-board' systems such as a TV, viewing reflected light from the leaf tips, or by potentiometers.

A more independent verification can be obtained by portal imaging. In order to use this in a 'go–no go' manner the delivered field shape must be rapidly compared with the prescription shape. This is especially important when multiple settings of the MLC at the same gantry orientation are being set to create intensity-modulated fields (see Chapter 2).

Zhou and Verhey (1994) have developed such a method. Starting from the portal image, the field-edge lines were extracted by a Hough-type transformation. False-edge lines from noise-corrupted edge points and from the lattice worked on were removed. By taking account of the special knowledge that in MLC fields all edge lines are either parallel or perpendicular to each other it was possible to suppress artefacts. Chamfer matching was then used to find the optimum translation, rotation and scaling to align a delivered portal with the prescription. It was claimed that a $256 \times 256 \times 12$ bit image can be aligned in about 3 s on a Silicon Graphics Indigo workstation.

Purdy *et al* (1995) have emphasised the need for rigorous quality control of the use of MLC systems and set out guidelines from the experience at the Mallinkrodt Institute, St Louis. Electronic portal imaging is an important component. They believe record-and-verify systems are essential but currently do not provide for the MLC. A high level of data security and quality control is needed in electronically passing files to the control of the MLC.

3.6. TWO NEW MLCS

3.6.1. The GEMS MLC

In addition to the commercial MLCs reported in the companion Volume, an MLC has recently become available from General Electric Medical Systems (Briot *et al* 1994). This has two banks of 32 leaves, 10 cm in height with a projected leaf width at the isocentre of 12.5 mm. The leaves can over-run by 10 cm and create a field of maximum size 40×40 cm^2. The leaves travel at up to 2 cm s^{-1} and are double focusing. The zig-zag arrangement of the leaf sides minimises leakage between leaves to less than 1%. The leakage through the leaves is less than 0.5%.

The MLC is a direct replacement for the lower jaws of a Saturne 4 accelerator. The MLC is an option for the Saturne Series linacs and is located inside the radiation head in place of the regular x–y collimation (figures 3.25 and 3.26). The MLC, replacing the lower jaws, provides the 'x' collimation whilst the 'y' collimation is provided by the upper jaws which can be controlled asymmetrically reaching a full over-run of 20 cm. The MLC features programmable irregular fields for block replacements and software and hardware capabilities allow it to perform dynamic treatments (see Chapter 2).

The front face of each leaf and of the jaws is round and the penumbra was about 2–3 mm more than with a rectangular field made by jaws alone. Each leaf is controlled by a separate motor and there are two linear potentiometers per leaf for full redundancy of position checking. Leaves can be set with an accuracy of 1 mm. An interesting feature is the in-room leaf position control from the Saturne hand-control pendant (GE 1994).

MLC planning with TARGET-2 and TARGET-2-PLUS is linked by a network to the MLC control.

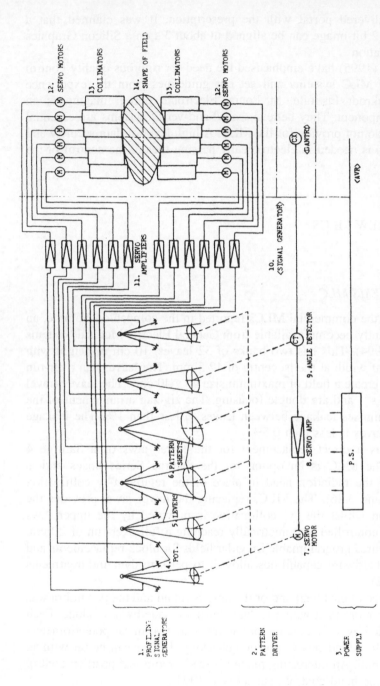

Figure 3.24. *A diagram of some historical significance. This shows an MLC with six leaves collimating an irregular field together with the control apparatus for adjusting the leaf positions. The original caption actually refers to 'conformation radiotherapy' so this is no new terminology. The figures refer to (1) profiling signal generators, (2) pattern driver, (3) power supply, (4) potentiometers, (5) levers for pattern detection, (6) pattern sheets, (7) servo motors to drive patterns, (8) a servo amplifier, (9) angle detector, (10) signal generators, (11) servo amplifiers, (12) servo motors, (13) collimators, (14) shape of field. (From Umegaki 1970.)*

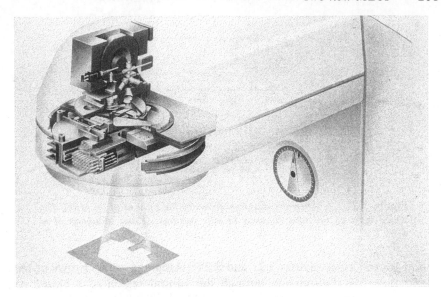

Figure 3.25. *A schematic design of the Saturne Series linac head with the GEMS MLC providing the x-collimation. (Courtesy of General Electric.)*

Figure 3.26. *A view of the leaves of the GEMS MLC showing the distinctive edge pattern of the leaves. (Courtesy of General Electric.)*

3.6.2. *A prototype multi-rod MLC*

Following a design study (Maughan *et al* 1989), a prototype multi-rod collimator (MRC) has been constructed by Maughan *et al* (1995). The MRC comprises 28 rows of tungsten rods of 3.18 mm diameter in a non-

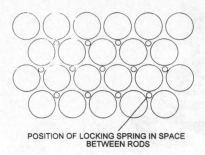

POSITION OF LOCKING SPRING IN SPACE
BETWEEN RODS

Figure 3.27. *A schematic diagram showing the pattern of an MRC. The space between the rods contains locking springs. (From Maughan et al 1995.)*

closed-packed array (figures 3.27 and 3.28). Although only 17 rows of rods would give 5% transmission through the central region of a closed set, equivalent to that through some cast metal blocks, the larger number reduces the transmission to 1.3%. The rods are pneumatically pushed to create the required shape of field by placing two polystyrene templates behind each of the sets of rods. These templates are shaped to the desired field shape, i.e. if they were placed face to face, they themselves would create the required shape. Once the rods are in the correct position locking springs between the individual rods are positioned to prevent the rods moving under gravity. The prototype MRC, weighing some 700 lb (~318 kg) is attached to the blocking tray of a Siemens Mevatron 77 linac. At present the MRC is not computer controlled so can only be used to create static field shapes. However, the shape can be reset in 15–20 s so the MRC could, for example, create IMBs by multisegment therapy with some manual intervention.

Maughan *et al* (1995) measured the 20–80% penumbra both perpendicular to the array of rods and parallel to the array of rods (set to create an almost flat edge). The penumbra in the former arrangement, measured with film dosimetry was 7.6 ± 0.8 mm for 17-rod rows and 5.9 ± 0.2 mm for 28-rod rows. In the latter arrangement the penumbra was smaller, being 6.8 ± 0.5 mm for 17-rod rows and 6.0 ± 0.2 mm for 28-rod rows. This is because the radiation passing tangentially down the edge of the rods does not encounter a flat edge but instead the edge defined by the mesh of the array of circular rods. However, the penumbra for 28-rod rows is almost the same in both directions.

It was shown that irregular field shapes could be easily formed, with a 'spatial resolution' better than an MLC and isodose curves did not resolve the individual rods. However, the MRC was cumbersome to use. The weight restricted the gantry orientations possible and further effort is underway to design a second collimator integral to the treatment head.

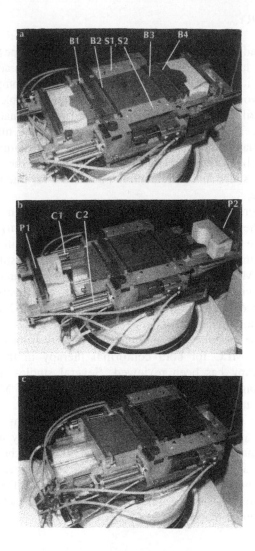

Figure 3.28. *Photographs of the MRC. In (a) the tungsten rods are supported by two brass plates (B) in which an array of circular holes has been drilled. In addition to springs between the individual rods there are springs between plates B1 and B2 and between plates B3 and B4. By moving B2 towards B1 and B3 towards B4 the springs compress and grip the rods. Close-fitting side shielding blocks S1 and S2 provide additional support and shielding. The rods are set to collimate a lateral pelvic field. In (b) the brake is released and the back plates P1 and P2 together with the shaped styrofoam blocks are retracted. The device is driven by pneumatic cylinders C1 and C2. After removal of the templates these drive the rods into the 'restacking' position shown in (c). (From Maughan et al 1995.)*

3.7. SUMMARY

The MLC has had a long history. The first research prototypes were discussed as early as the late 1950s and during the following decade several were built. However, it was not until the mid-1980s that the computer-controlled MLC as we would recognise it today became available. Even at that time and up until the beginning of the 1990s the only machines available were prototypes. This decade, several manufacturers have made commercial collimators available and whilst these are by no means simple off-the-shelf items, the MLC is becoming increasingly considered as the option of choice to geometrically shape fields in the clinic. Their use for intensity modulation is still quite rare.

In this chapter we have reviewed the present position with respect to characterising the physical performance of the MLC and its introduction to clinical use. The characterisation has been concerned with issues of field definition, transmission properties and the ability to substitute for the use of blocks. The MLC offers ergonomic improvements which have received some attention via work studies. There are still controversies concerning how to best place leaves, how to physically control the MLC and how to link it to the planning computer.

Its clinical use is somewhat sporadic and differs from centre to centre. The use of the MLC for both conformal therapy in the body and for stereotactic therapy has been reviewed. Several studies have shown the potential advantages of field shaping. The verification öf MLC leaf shapes continues to be an issue of concern.

At the time of writing it is true to say that the MLC has taken over most roles in field shaping but not all roles. Some fields may be too large for the MLC and mantle fields are still problematic. For other fields the use of the MLC as a block replacement has been conclusively demonstrated. Both financial and cultural limitations may hinder its full introduction but if 3D CFRT is to become a regular fully established practice, its use will undoubtedly become mandatory.

REFERENCES

Aoki Y , Nakagawa K, Onogi Y, Sasaki Y and Akanuma A 1994 Coplanar versus non coplanar dose optimisation with a multileaf collimator *The Use Of Computers in Radiation Therapy: Proc. 11th Conf.* ed A R Hounsell *et al* (Manchester: ICCR) pp 12–3

Bannach B, Doll T, Pape H and Schmitt G 1995 Initiation of conformal radiotherapy with a multileaf collimator—an approach to clinical routine (Proc. ESTRO Conference (Gardone Riviera, 1995)) *Radiother. Oncol.* **37** Suppl. 1 S59

Bidmead A M, Bliss P, Mubata C D and Thompson V 1995 Use of a Varian multileaf collimator in the conformal treatment of head and neck tumours *Proc. Röntgen Centenary Congress 1995 (Birmingham, 1995)* (London: British Institute of Radiology) p 169

Bortfeld T 1995 private communication (visit to DKFZ, 16–20 January 1995)

Boyer A L 1995 Present and future developments in radiotherapy treatment units (*Innovations in treatment delivery*) *Semin. Radiat. Oncol.* **5** 146–55

Boyer A L, Ochran T G, Nyerick C E, Waldron T J and Huntzinger C J 1992 Clinical dosimetry for implementation of a multileaf collimator *Med. Phys.* **19** 1255–61

Brahme A 1988 Optimal setting of multileaf collimators in stationary beam radiation therapy *Strahlenther. Onkol.* **164** 343–50

——— 1993 Optimisation of radiation therapy and the development of multileaf collimation *Int. J. Radiat. Oncol. Biol. Phys.* **25** 373–5

——— 1994 Optimisation of radiation therapy *Int. J. Radiat. Oncol. Biol. Phys.* **28** 785–7

Briot E, Linca S and Chavaudra J 1994 Preliminary study of the physical and dosimetrical characteristics of the GEMS multileaf collimator (Proc. World Congress on Medical Physics and Biomedical Engineering (Rio de Janeiro, 1994)) *Phys. Med. Biol.* **39A** Part 1 527

Burman C, Kutcher G J, Emami B and Goitein M 1991 Fitting of normal tissue tolerance data to an analytic function *Int. J. Radiat. Oncol. Biol. Phys.* **21** 123–35

Burman C, Leibel S A, Mohan R, Kutcher G J and Fuks Z 1994 Comparison of dose distributions for 3D brain plans with Cerrobend and multileaf collimators *Med. Phys.* **21** 885

Crockford D J, Shentall G S and Mayles W P M 1994 The definition and transfer of Philips MLC data from remote computers *The Use of Computers in Radiation Therapy: Proc. 11th Conf.* ed A R Hounsell *et al* (Manchester: ICCR) pp 350–1

Davy T J and Brace J A 1980 Dynamic 3D treatment using a computer controlled cobalt unit *Br. J. Radiol.* **53** 384

Debus J, Engenhart R, Rhein B, Schlegel W, Schad L, Pastyr O and Wannenmacher M 1994 Clinical application of conformal radiosurgery using multileaf-collimators (Proc. 36th ASTRO Meeting) *Int. J. Radiat. Oncol. Biol. Phys.* **30** Suppl. 1 265

Du M N, Yu C X, Taylor R C, Martinez A A and Wong J W 1995 Dynamic wedge: a first step towards clinical application of intensity modulation using MLC (ASTRO Meeting 1995) *Int. J. Radiat. Oncol. Biol. Phys.* **32** Suppl. 1 170

Fernandez E M, Dearnaley D P, Heisig S, Mayles W P M and Shentall G S 1993 The suitability of a multileaf collimator as a standard block replacement for different anatomical sites *Proc. 2nd ESTRO Meeting on Physics in Clinical Radiotherapy (Prague, 1993)* (Prague: ESTRO) p 166

Fernandez E M, Shentall G S, Mayles W P M and Dearnaley D P 1994a The acceptability of a multileaf collimator as a replacement for conventional blocks *Radiother. Oncol.* **36** 65–74

—— 1994b The potential for increased sparing of normal tissue with a multileaf collimator *RMNHST Internal Document*

Forster K M, Martel M K, Ten Haken R K, Sandler H M, Higgins P and Fraass B A 1994 Radiosurgery using segmented conformal therapy with a doubly focused multileaf collimator (Proc. 36th ASTRO Meeting) *Int. J. Radiat. Oncol. Biol. Phys.* **30** Suppl. 1 163

Fraass B A, McShan D L and Kessler M L 1995 Computer-controlled treatment delivery (*Innovations in treatment delivery*) *Semin. Radiat. Oncol.* **5** 77–85

Frazier A, Yan D, Joyce M, Du M, Mazur E, Vicini F, Martinez A and Wong J 1993 The effect of treatment variation on multileaf collimator dosimetry in conventional radiotherapy (Proc. 35th ASTRO Meeting) *Int. J. Radiat. Oncol. Biol. Phys.* **27** Suppl. 1 162

Frouhar V A, Palta J and Yeung D 1994 A methodology for MLC leaf position optimisation for irregular fields *Med. Phys.* **21** 921

Galvin J M, Smith A R and Lally B 1993 Characterisation of a multileaf collimator system *Int. J. Radiat. Oncol. Biol. Phys.* **25** 181–92

GE 1994 Multi-leaf collimator: works in progress Company literature, dated 1994; courtesy of A Mader, GE, Buc, France (letter to S Webb 16 April 1996)

Graham J D, Nahum A E and Brada M 1991 A comparison of techniques for stereotactic radiotherapy by linear accelerator based on 3-dimensional dose distributions *Radiother. Oncol.* **22** 29–35

Helyer S J and Heisig S 1995 Multileaf collimation versus conventional shielding blocks: a time and motion study of beam shaping in radiotherapy *Radiother. Oncol.* **37** 61–4

Helyer S A, Wilcox S, Heisig S and Westbrook K 1994 A study measuring the possible time savings of a multileaf collimator compared to conventional beam shaping in radiotherapy (Proc. 9th Annual Meeting of the British Oncological Association) *Br. J. Cancer* **70** Suppl. 22 32

Hounsell A R, Jordan T J and Williams P C 1995 A proposal for quality assurance of multileaf collimators (Proc. ESTRO Conf. (Gardone Riviera, 1995)) *Radiother. Oncol.* **37** Suppl. 1 S17

Huq M S, Yu Y, Chen Z P and Suntharalingam N 1995 Dosimetric characteristics of a commercial multileaf collimator *Med. Phys.* **22** 241–7

Ihnen E, Melchert C and Richter E 1995 Dosimetry and quality assurance using a multileaf collimator *Medizinische Physik 95 Röntgen Gedächtnis-Kongress* ed J Richter (Würzburg: Kongress) pp 200–1

Jordan T J 1994 private communication

—— 1996 private communication

Jordan T J and Williams P C 1994 The design and performance characteristics of a multileaf collimator *Phys. Med. Biol.* **39** 231–51

Kahler D, Starkschall G, Boyer A L, Li J and Wong J 1993 Extraction of multileaf collimator settings using a radiotherapy picture archival and communication system (Proc. 35th ASTRO Meeting) *Int. J. Radiat. Oncol. Biol. Phys.* **27** Suppl. 1 180

Kutcher G J, Mohan R, Leibel S A, Fuks Z and Ling C C 1995 Computer controlled 3D conformal radiation therapy *Radiation Therapy Physics* ed A Smith (Berlin: Springer) pp 175–91

Laing R 1993 Stereotactic radiotherapy for irregular targets: the advantage of static conformal beams over conventional non-coplanar rotations *Br. J. Radiol. Suppl. Radiother. Oncol.* **66** 92–4

Leavitt D D, Gibbs F A, Heilbrun M P, Moeller J H and Takach G A 1991 Dynamic field shaping to optimise stereotactic radiosurgery *Int. J. Rad. Oncol. Biol. Phys.* **21** 1247–55

LoSasso T, Chui C S, Kutcher G J, Leibel S A, Fuks Z and Ling C C 1993 The use of a multi-leaf collimator for conformal radiotherapy of carcinomas of the prostate and nasopharynx *Int. J. Radiat. Oncol. Biol. Phys.* **25** 161–70

Maughan R L, Blosser G F, Blosser E J, Blosser H G and Powers W E 1989 Transmission measurements in multi-rod arrays: a design study for a multi-rod collimator *Radiother. Oncol.* **15** 125–31

Maughan R L, Powers W E, Blosser G F, Blosser E J and Blosser H G 1995 Radiological properties of a prototype multi-rod collimator for producing irregular fields in photon radiation therapy *Med. Phys.* **22** 31–6

Mohan R 1995 Field shaping for three-dimensional conformal radiation therapy and multileaf collimation (*Innovations in treatment delivery*) *Semin. Radiat. Oncol.* **5** 86–99

Moore C J, Hounsell A R, Sharrock P, Shaw A, Williams P C and Wilkinson J M 1994 A computerised system for conformal therapy using a Philips SL-25 linear accelerator and multi-leaf collimator: first and second generation developments *The Use of Computers in Radiation Therapy: Proc. 11th Conf.* ed A R Hounsell *et al* (Manchester: ICCR) pp 104–5

Nedzi L A, Kooy H M, Alexander E III, Svensson G K and Loeffler J S 1993 Dynamic field shaping for stereotactic radiosurgery: a modelling study *Int. J. Radiat. Oncol. Biol. Phys.* **25** 859–69

Philips Medical Systems 1995 MLP—Multileaf Preparation *Philips Medical Systems Commercial Literature* document 4522 984 41091/ 764 04/95

Powlis W D, Smith A R, Cheng E, Galvin J M, Villari F, Bloch P and Kligerman M M 1993 Initiation of multileaf collimator conformal radiation therapy *Int. J. Radiat. Oncol. Biol. Phys.* **25** 171–9

Press W H, Flannery B P, Teukolsky S A and Vetterling W T 1986 *Numerical Recipes: The Art of Scientific Computing* (Cambridge: Cambridge University Press)

Proimos B S 1960 Synchronous field shaping in rotational megavolt therapy *Radiology* **74** 753–7

Purdy J A, Klein E E and Low D A 1995 Quality assurance and safety of new technologies for radiation oncology (*Innovations in treatment delivery*) *Semin. Radiat. Oncol.* **5** 156–65

Ramm U, Thilmann C, Adamietz I A, Manegold K H, Rahl C G, Mose S, Saran F and Böttcher H D 1996 Clinical implementation of the multileaf preparation system for conventional radiotherapy at a Philips SL15 linear accelerator provided with a multileaf collimator *Proc. Symp. Principles and Practice of 3-D Radiation Treatment Planning (Munich, 1996)* (Munich: Klinikum rechts der Isar, Technische Universität)

Schlegel W, Pastyr O, Bortfeld T, Becker G, Schad L, Gademann G and Lorenz W J 1992 Computer systems and mechanical tools for stereotactically guided conformation therapy with linear accelerators *Int. J. Radiat. Oncol. Biol. Phys.* **24** 781–7

Schlegel W, Pastyr O, Bortfeld T, Gademann G, Gardey K, Menke M and Maier-Borst W 1993 Stereotactically guided fractionated radiotherapy: technical aspects *Radiother. Oncol.* **29** 197–204

Shentall G S, Crockford D J, Fernandez E M and Mayles W P M 1993a The evaluation of a multileaf collimator as a tool for conformation therapy—a progress report *Abstracts of the BIR Meeting (London, 1993)*

Shentall G S, Crockford D J, Fernandez E M and Mayles W P M 1993b The application of a multileaf collimator to conformal therapy *Proc. 2nd ESTRO Meeting on Physics in Clinical Radiotherapy (Prague, 1993)* p 99

Shih J, Chiu-Tsao S and Kim J H 1994 Comparison of flat-tip and round-tip multileaf collimators *Med. Phys.* **21** 885

Shiu A S, Ewton J R, Rittichier H E, Wallace J, Wong J and Tung S S 1994a Evaluation of a dynamic miniature multileaf collimator for stereotactic radiosurgery and radiotherapy *Med. Phys.* **21** 919

Shiu A S, Ewton J R, Tung S S, Wong J and Maor M H 1994b Comparison of miniature multileaf collimation (MMLC) with circular collimation for stereotactic radiation treatment (Proc. 36th ASTRO Meeting) *Int. J. Radiat. Oncol. Biol. Phys.* **30** Suppl. 1 162

Stein J 1995 Einsatz von multileaf-kollimatoren in der konformationstherapie *Dreidimensionale Strahlentherapieplanung* ed W Schlegel, T Bortfeld and J Stein (Heidelberg: DKFZ) pp 43–54

Thilmann C, Ramm U, Adamietz I A, Mose S, Saran F, Manegold K H and Böttcher H D 1996 The use of a multileaf preparation data entry system in palliative treatment *Proc. Symp. Principles and Practice of 3-D Radiation Treatment Planning (Munich, 1996)* (Munich: Klinikum rechts der Isar, Technische Universität)

Thomson V R 1995 Multileaf collimation—our clinical experience *Proc. Röntgen Centenary Congress 1995 (Birmingham, 1995)* (London: British Institute of Radiology) p 111

Uchiyama Y and Morita K 1992 Integrated system of computer-controlled conformation radiotherapy *Japan. J. Radiol. Technol.* **10** 35–46

Umegaki Y 1970 The automation of radiotherapy (Computers in radiotherapy. Proc. 3rd Int. Conf. Computers in Radiotherapy (Glasgow, 1970)) *Br. J. Radiol. Special Report* **5** 147–53

Van Duyse B, Colle C, De Neve W, De Wagter C 1995 Multileaf collimator tool for Sherouse's GRATIS three-dimensional treatment planning system (Proc. ESTRO Conf. (Gardone Riviera, 1995)) *Radiother. Oncol.* **37** Suppl. 1 S60

Van Santvoort J, Heijmen B and Dirkx M 1995 Solution for underdosages caused by the tongue and groove effect for multileaf collimators (Proc. ESTRO Conf. (Gardone Riviera, 1995)) *Radiother. Oncol.* **37** Suppl. 1 S38

Varian 1995 Manufacturers product literature RAD 2260E (September 1995).

Webb S 1993 The effect on tumour control probability of varying the setting of a multileaf collimator with respect to the planning target volume *Phys. Med. Biol.* **38** 1923–36

—— 1994a The effect on tumour control probability of varying the setting of a multileaf collimator with respect to the planning target volume *The Use of Computers in Radiation Therapy: Proc. 11th Conf.* ed A R Hounsell *et al* (Manchester: ICCR) pp 30–1

—— 1994b The effect on tumour control probability of varying the setting of a multileaf collimator with respect to the planning target volume (Proc. World Congress on Medical Physics and Biomedical Engineering (Rio de Janeiro, 1994)) *Phys. Med. Biol.* **39A** Part 1 495

Webb S and Nahum A E 1993 A model for calculating tumour control probability in radiotherapy including the effects of inhomogeneous distributions of dose and clonogenic cell density *Phys. Med. Biol.* **38** 653–66

Welch M E and Davy T J 1994 Is high resolution obtainable on both the blade tip and the edge of the Philips multileaf collimator. What techniques are required to achieve this? (Proc. World Congress on Medical Physics and Biomedical Engineering (Rio de Janeiro, 1994)) *Phys. Med. Biol.* **39A** Part 1 496

Yu C X, Du M N, Wong J W, Symons M, Yan D, Mullins C K, Gustafson G, Matter R C and Martinez A 1994 A new prescription preparation and verification system to drive a multileaf collimator for conventional radiotherapy *The Use of Computers in Radiation Therapy: Proc. 11th Conf.* ed A R Hounsell *et al* (Manchester: ICCR) pp 146–7

Yu C X, Yan D, Du M N, Zhou S and Verhey L J 1995 Optimisation of leaf positions when shaping a radiation field with a multileaf collimator *Phys. Med. Biol.* **40** 305–8

Yudelev M, Maughan R L, Warmelink C, Sharma R and Forman J D 1994 3D shaping of fast neutron therapy by means of the multi-rod collimator

(Proc. World Congress on Medical Physics and Biomedical Engineering (Rio de Janeiro, 1994)) *Phys. Med. Biol.* **39A** Part 1 494

Zhou S and Verhey L J 1994 A robust way for multileaf collimator (MLC) leaf configuration verification *Phys. Med. Biol.* **39** 1929–47

Zhu Y, Boyer A L and Desobry G E 1992 Dose distributions of x-ray fields shaped with multileaf collimators *Phys. Med. Biol.* **37** 163–73

CHAPTER 4

MEGAVOLTAGE PORTAL IMAGING, TRANSIT DOSIMETRY AND TISSUE COMPENSATORS

4.1. THE USE OF PORTAL IMAGES TO ASSESS AND CORRECT PATIENT SETUP ERROR

The primary function of planar portal images is to guide repositioning of the patient so that the treatment position is consistently the same (within an acceptable tolerance) between treatment fractions and with reference to the position of the patient at the planning stage. Traditionally 'by-eye' evaluation was performed using portal films. Recently there have been an enormous number of attempts to use image evaluation tools to assist the process, largely based on assessing field boundaries and identifying internal structures (such as bony landmarks) in relation to these. Electronic portal imaging devices (EPIDs) are vital for these functions to be performed in realistic times (Evans *et al* 1994a, 1995c, Gildersleve *et al* 1994a,b, 1995, Munro 1995, Shalev 1994a,b).

Shalev (1995) has summarised the many potential forms of inaccuracies which can occur in radiotherapy. These may be classified into two categories: (i) systematic field-positioning errors and (ii) random field-positioning errors. The first, systematic errors, may be due to incorrect data transfer between the stage of imaging and planning the patient and the stage of actual treatment setup. Another possible source of error is the incorrect design, marking or positioning of treatment accessories such as compensators, shielding blocks and immobilisation devices. These systematic errors should be correctable once detected. The second category, random field-positioning errors, are due to operator tolerances in setting up each fraction of a treatment together with changes which occur to the patient anatomy during the course of a fractionated treatment, for example due to bladder and rectal filling, tumour growth or shrinkage or respiratory movements (see also Hansen *et al* 1996b).

The simplest setup error to detect is a gross inaccuracy, for example the use of a wedge the wrong way round, or an incorrect shadow tray blocking, an incorrect collimator rotation or a field-size error. These are best determined by on-line EPIDs configured to give an image in the first few MUs. Using automatic methods of field-edge detection the, hopefully rare, occurrence of such gross errors can be linked to instant shutdown of the treatment fraction.

The contrast in electronic portal radiographs is a limiting feature when using them to quantitate and/or correct setup errors. For this reason the use of additional contrast, such as in-tissue markers, is being explored (see section 4.1.1). Without these, the alignment of daily portal images with the reference standard relies on the extraction of field edges in the image and then specific anatomical features. One of two approaches is then used. Either (i) the field edges are aligned, thus determining the parameters of registration, and then the anatomical features of one image are transformed on to the other. The observed misregistration quantifies the error. Alternatively (ii) anatomical features are registered and the transformation parameters are used to superpose the field edges, discrepancies in alignment being interpreted as error.

Field edges are not always easy to detect. A common definition is to identify the 50% intensity level in the image (Evans *et al* 1992). Other methods identify the region of maximum gradient through histogram analysis (Bijhold *et al* 1991) or enhance the edges using the derivative of a Gaussian operator (Leszczynski *et al* 1991).

Correspondingly, identifying internal landmarks in low-contrast images is also tricky and manual placement is often used followed by construction of the transformation operator by singular value decomposition (see the companion Volume, Chapter 1) (Balter *et al* 1992). Gilhuijs and Van Herk (1993) extracted anatomical features automatically and used chamfer matching to construct the transform. Evans *et al* (1992) interactively operated various transforms until the features registered, a so-called visual registration method using a two-image movie loop. Gildersleve *et al* (1995) assessed the reproducibility of patient positioning this way. Portal images have also been correlated by fast-Fourier transform techniques (Scheffler *et al* 1995).

Once the positioning error has been determined for each fraction and specified by a vector indicating the movement which would have to take place to register any particular image with the reference image, it is possible to plot a map of such vectors over a course of treatment. Of course any particular vector does not, and cannot, distinguish random from systematic error, but inspection of the plot *can* make such a distinction. When the centre of gravity of the vectors is not zero there must be systematic error (see section 1.2.3). Whether one or the other error dominates depends on the treatment site and there is disagreement in the literature (Shalev 1995). Some authors have attempted intervention, correcting for the systematic error as the fractionated treatment progressed (Ezz *et al* 1991, De Neve *et al* 1992, Gildersleve *et al* 1994a).

Shalev (1995) has reviewed a large number of measurements made on setup errors in clinical studies. Hunt *et al* (1993) and Kutcher *et al* (1995) have also reported that both systematic and random setup errors were worse for conformal radiotherapy (performed by multisegment techniques) than conventional therapy for nasopharangeal tumours. The DVH for brainstem showed considerable variations between the ideal histogram computed for an ideal static patient and that corresponding to recomputing the plan taking into account measured errors. This emphasises the need to build error analysis into any assessment of the improvements resulting from conformal therapy.

In passing we may note an interesting concept from Moklova where it has been suggested that patients receiving radiotherapy could be treated on a conveyer-belt principle. A patient is set up in one room, positioned and fixated in a second, taken to a third for irradiation and to a fourth for defixation and removal. The concept raises interesting questions concerning the accuracy of positioning, since the main reason for suggesting this would appear to be speed of patient throughput and optimisation of radiation resources (Zakharchenko *et al* 1995).

4.1.1. *The use of radio-opaque markers for patient (re)positioning*

An interesting new development is the use of radio-opaque markers. Two groups have presented similar techniques (with some differences of detail) (Gall *et al* 1993, Lam *et al* 1993). Lam *et al* (1993) positioned 14 tungsten carbide 2.4 mm diameter ball bearings at locations on the surface of a skull (six) and within the brain (eight). Two orthogonal films were taken at the first and subsequent treatments. Techniques for identifying the markers on the films were borrowed from those used to identify microcalcifications in mammograms. The marker images of the two projections at the reference position (first treatment) were used to determine the 3D location of the markers by the same technique as used to localise brachytherapy sources from two film projections. The two reference projection films made a template for searching subsequent pairs. After identifying marker projections on subsequent pairs, the 3D locations were found for these subsequent treatment fractions. By comparing the 3D locations with those in the reference position the appropriate translations and rotations were found to realign the patient. An accuracy of 1 mm translation and 0.3° rotation was reported with an inanimate phantom.

Gall *et al* (1993) presented a similar method using tantalum screws or gold or tantalum seeds. They formed three orthogonal radiographs, any two of which over-determine the 3D position of the marker (four equations from two stereoscopic pairs). The accuracy reported was 1 mm translation and 1° rotation using just three fiducial markers. No special way of locating the markers was reported. The films were simply marked using a back-lit digitising tablet. The information was fed to a computer which, having evaluated the 3D transform, then automatically drove motors to reposition the patient. Gall *et al* (1993) developed their technique for positioning patients receiving proton therapy at the Harvard Cyclotron Centre in Boston. It was suggested that the same method would assist fractionated stereotactic brain radiotherapy without the need for an invasive skull ring.

4.2. TRANSIT DOSIMETRY

Transit dosimetry is a technique whereby the distribution of dose *actually delivered* to the patient is computed from a transmission portal image taken at the time of treatment (Evans *et al* 1993, Hansen and Evans 1994, Hansen *et al* 1993, 1994a, 1995a,b, 1996a, 1997). The purpose is to determine whether this dose distribution matches the dose distribution *computed* at the planning stage from the known position of the radiation source relative to the patient's anatomy as specified by the 3D CT data. The technique is new and only being investigated in a number of research centres. The most extensive such work has been performed at the RMNHST.

There are several reasons why the delivered distribution may even differ from the planned distribution. These include the following.

(i) The patient may be in a different position at treatment from that when radiological data for planning were taken. The position may vary from fraction to fraction.

(ii) Even if the patient's external contour were positioned accurately, internal structures may have moved due to changes in bladder and rectal contents, breathing, growth or shrinkage of the tumour etc. These movements may be rigid-body transformations or 'floppy-body' transformations or elastic transformations.

(iii) The model for planning the dose distribution may not accurately model the actual dose delivery (treatment planning generally involves simple but practically realisable models—see Appendix B).

Transit dosimetry aims to create a model of the treatment which describes the actual treatment as best as possible. To do this it requires several operational stages including: (i) creating a 3D map of the patient at the time of treatment by working out how to transform the planning CT data to a 3D map which is 'locked' to the portal image of fluence exiting the patient and which can be recorded; (ii) reinterpreting the portal image and extracting from it a measure of the transmitted primary fluence; (iii) backprojecting this primary fluence to create a 3D map of terma within the 3D (transformed) CT map of the patient at the time of treatment; (iv) invoking a dose model which accurately transports the terma and results in a best estimate of the

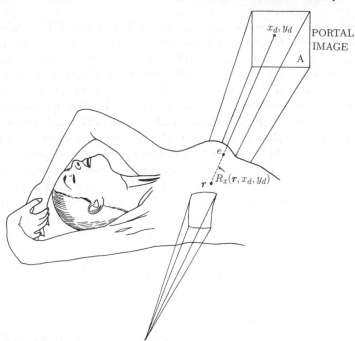

Figure 4.1. *This figure shows how to calculate transit dose. It is a prerequisite of this calculation that the DRR (computed from the 3D CT data) has been registered to the portal image. x_d, y_d labels a general point in the portal image at which the detected signal is $D(x_d, y_d, A)$. r is a general point in the irradiated part of the patient at which the transit dose $D_E(r)$ is computed. $R_X(r, x_d, y_d)$ is the radiological thickness from r to x_d, y_d. e is the point at which the beam from the source S exits the patient. The rectangular field enters on the proximal side of the patient, casting a shadow which will not appear rectangular because of the curved patient contour.*

3D dose distribution at the time of treatment. Thus it may be appreciated that the success of transit dosimetry is dependent on the assumptions in, and accuracy of, the physics going into each of these stages. The separate steps will now be described.

The first step is to create a 2D digital portal image $D(x_d, y_d, A)$ using one of the many types of EPID (see the companion Volume, Chapter 6). The coordinates x_d, y_d describe the detector plane normal to the direction of the axis of the radiation beam and A labels the field size (figure 4.1).

4.2.1. The digital reconstructed radiograph (DRR)

At some time prior to treatment, the patient will have had a series of CT images taken through the region spanning the tumour to be treated with the patient in the treatment position. Let $\rho_{el}(r)$ represent the 3D distribution of electron density determined from these CT data. (The relationship between electron density $\rho_{el,tissue}$ and CT number CT − number$_{tissue}$ is discussed in textbooks of medical imaging; CT − number$_{tissue}$ = $(\rho_{el,tissue} - \rho_{el,water})/\rho_{el,water} \times 1000$). A digital reconstructed radiograph (DRR) DRR(x_d, y_d) is an imaginary image of detected primary photon intensity which can be formed from the 3D CT data by ray-tracing from the source to an (imaginary) detector plane. It is limited in resolution by the finite slice width of the CT scans which is generally much larger than the in-slice spatial resolution. The best DRRs will result from the use of closely spaced CT slices. Spiral CT, which can be reformatted to give coronal and sagittal slices with resolution in the longitudinal direction similar to the in-plane resolution, will lead to excellent DRRs. In the simple mathematics below it has been assumed that the CT data have been reformatted into cubic voxels. Representing the source intensity on the source side of the patient by I_0, the detected primary-photon intensity is

$$I_{x_d, y_d} = I_0 \exp\left(-\sum_i \rho_{el,i} \sigma \Delta\right) \tag{4.1}$$

where σ is the photon attenuation cross section (dimensions L^2), Δ is the CT voxel size (dimensions L), $\rho_{el,i}$ is the electron density (dimensions L^{-3}) in the ith voxel and the sum is over all those voxels which lie on the ray from the source to the element x_d, y_d of the detector. Representing the electron density of water by $\rho_{el,w}$, equation (4.1) may be rewritten

$$I_{x_d, y_d} = I_0 \exp\left(-\sum_i \frac{\rho_{el,i}}{\rho_{el,w}} \rho_{el,w} \sigma \Delta\right). \tag{4.2}$$

Since $\rho_{el,w}\sigma = \mu_w$ where μ_w is the linear-attenuation coefficient of water (dimensions L^{-1}; computed for the effective diagnostic energy of the CT scanner), equation (4.2) becomes

$$I_{x_d, y_d} = I_0 \exp\left(-\mu_w \sum_i \rho_{el,i}^{rel} \Delta\right) \tag{4.3}$$

where $\rho_{el,i}^{rel}$ is the (dimensionless) electron density of the ith voxel *relative* to that of water. By definition, the radiological thickness R_X is

$$R_X = \sum_i \rho_{el,i}^{rel} \Delta \tag{4.4}$$

so a very simple form of equation (4.3) is

$$I_{x_d, y_d} = I_0 \exp(-\mu_w R_X). \tag{4.5}$$

The digital reconstructed radiograph is

$$\mathrm{DRR}(x_d, y_d) = \ln(I_0/I_{x_d,y_d}) = \mu_w R_X. \tag{4.6}$$

Note that the DRR is computed assuming the differential attenuation of primary radiation only. These equations do not model scatter.

Many groups have developed code to create the DRR using this formalism. They differ largely in the way the rays are cast and in spatial resolution. Chaney *et al* (1993) have written code to compute DRRs with 512×512 resolution, portable to a variety of treatment-planning systems and freely available. A nice feature is that the CT number of any structure can be changed artificially. For example, if a bladder were filled with contrast material, it can be 'unfilled' computationally. Conversely the spinal canal can be imaged as a high-contrast structure by artificially increasing the CT numbers in the canal. Other algorithms have been presented (e.g. Dong *et al* 1994, Galvin *et al* 1995, Lovelock *et al* 1994) for the computation of DRRs.

4.2.2. Registration of the DRR with the portal image; creation of treatment-time 3D CT data

The first stage in transit dosimetry is to represent the patient by transformed (planning) CT data with these data registered to the recorded portal image. A rigid-transform operator is obtained by registering the recorded portal image $D(x_d, y_d, A)$ with the computed $\mathrm{DRR}(x_d, y_d)$. This operator is then applied to the planning CT data. When this has been achieved the new 3D CT data, representing the patient at treatment time, can be considered 'locked' to the portal image. The new CT data are only an approximation to the 'true' CT data at the time of treatment because of the rigid-body transform assumption (discussed at length by Hansen *et al* 1996b).

If there were a way of making 3D CT images at the time of treatment with the megavoltage beam, so called megavoltage computed tomography (MVCT), there would be no need for this stage, as the distribution of dose actually delivered could be superposed on the 3D MVCT data, instead of on the registered diagnostic CT data, once again starting with the portal image. Unfortunately although single-slice MVCT has been achieved in a number of centres (Aoki *et al* 1990, Brahme *et al* 1987, Källman *et al* 1989, Lewis and Swindell 1987, Lewis *et al* 1988, 1992, Nakagawa *et al* 1991, 1994, Simpson *et al* 1982, Swindell *et al* 1983) multi-slice CT has not, although plans to do so exist (Rudin *et al* 1994). The MVCT device at the Karolinska Hospital (Rudin *et al* 1994) is capable of executing translation (to form a 2D image at any one gantry angle) and via gantry rotation can generate the data-set needed for cone-beam reconstruction of 3D MVCT images. In principle this is a capability of any portal imaging system of course. As yet this has only been proposed. It would also still be necessary to compare the delivered distribution with the calculated distribution (which is based solely on CT data prior to treatment). A 3D MVCT scanner is under development at the RMNHST (Mosleh-Shirazi *et al* 1994a,b, 1995). Alternatively Jaffray *et al* (1995) have developed cone-beam CT on a radiotherapy accelerator by mounting a diagnostic x-ray tube with its central axis at 90° to that of the megavoltage beam and imaging the beam with a charge-coupled device (CCD)-based fluoroscopic detector. From the projection images obtained at multiple orientations volumetric CT information was obtained with the patient in the treatment position. A CT test phantom was rotated at 4° increments through 360° of rotation. Each exposure to the 100 kVp beam had a total charge of 16 mA s. The detector was a Lanex Fast phosphor screen bonded to a 1 mm thick aluminium plate and a cone-beam CT data-set of a test phantom containing volumes of different attenuation was reconstructed using a cone-beam reconstruction algorithm (Jaffray *et al* 1996). Cone-beam reconstructions were also performed using the *megavoltage* beam. The gantry was rotated through 180° and images were recorded every 2°. Each exposure delivered 2 cGy of dose to the head phantom which was imaged so a total dose of 180 cGy resulted. This experiment demonstrated the feasibility of the technique and it was not proposed that the megavoltage source be used for clinical imaging. Midgley *et al* (1996) have also reconstructed cone-beam MVCT images from projection data recorded on a Varian PortalVision system after suitable calibration of the detector (see section 4.2.3.1).

4.2.3. Portal fluence: primary exit fluence from measured data

4.2.3.1. Detector calibration. Following its use for registration, the portal image $D(x_d, y_d, A)$ is re-used to provide a measure of photon intensity or fluence. This is the second stage of the process of transit dosimetry. It is a complex stage which is discussed here in some detail. For a scanning scintillation detector with separate elements the matrix values may be used directly. For other types of EPID a more complicated relationship exists between the EPID grey-level recorded value and the measurement of total dose at the corresponding point. We first look at studies which have addressed this problem before going on to look at how a measure of primary dose may be extracted from a measure of total dose.

Kirby and Williams (1993a,b) have programmed the Philips SRI-100 EPID to act as an integrating dosemeter by not allowing the system to automatically adjust the gain to obtain the best qualitative image. In this mode it has allowed the verification of segmented modulated treatment fields. The data could form the basis of transit dosimetry.

The use of the Varian PortalVision liquid matrix ion chamber EPID system for portal dose measurements has been investigated by Zhu *et al* (1995).

Firstly they showed the need to correct each pixel value for the variation in response of each detector element to a uniform flood field. This was done in a way which also calibrated out any small non-uniformities in the nominally uniform flood field (about 3%). The overall pixel sensitivity map before this calibration varied by some 30%, indicating the vital necessity for this calibration step.

Then, by comparing with conventional ion-chamber measurements, they showed that the pixel values were directly proportional to the square root of the recorded total dose for two field sizes 10×10 cm^2 and 20×20 cm^2 at 140 cm SSD (figure 4.2). Of course as the field size changes the total dose received to a region-of-interest is itself changing. This is shown in figure 4.3. Figure 4.4 shows the variation in (sensitivity-corrected) pixel value with field size reflecting the same phenomenon. When the data from figure 4.3 and figure 4.4 are combined it was shown that the sensitivity was independent of field size. In figure 4.5 we see that, for each field area, the pixel value has been divided by the square root of the dose. The resulting quantity is independent of field area to within 0.5%. Thus it is possible to use the calibration shown in figure 4.2 *independent of field size* to make a measurement of the total recorded dose from a measurement of the (calibrated) pixel value recorded at the detector. It was also shown that the system was stable with time to within a couple of per cent over a two month period. Boellaard *et al* (1995) showed a similar (almost) square-root response between recorded pixel value and dose-rate for a liquid-filled EPID for low and medium dose-rates but the response became more linear with higher dose-rates. Stucchi *et al* (1995) also found a square-root dependence with a liquid-filled EPID. Herman *et al* (1996) found a power law in 0.526 dependence and long-term stability to ±1.8% over a 3.5 month period. In contrast, a diode detector showed a linear response (Fiorino *et al* 1995). Hensler and Auer (1996) have also developed a 2D portal imaging system based on scanning a 1D array of 87 semiconductor diodes and have used the system for portal dosimetry.

Roback and Gerbi (1995) have also investigated the suitability of the Varian PortalVision EPID as a dosemeter. They performed a number of experiments concluding from each of them that the variation in response with different irradiation conditions was essentially negligible. In the first experiment the variation in response with changing SSD between 96 and 124 cm was determined to be within 4.6% whatever the energy and with measurements made with the detector both horizontal and vertical. A second experiment determined the variability of response over the detector area and found that provided the detector had been calibrated for the relevant gantry orientation the distribution of pixel values was within 1% at 6 MV. A third experiment showed that the response to the position of a slab inhomogeneity in a slab phantom (and using slabs representing air, bone and lung), did not vary by more than 2.2%. A fourth experiment imaged a tungsten ball at

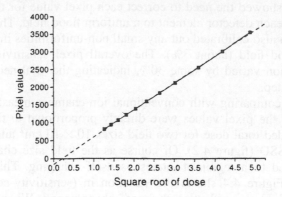

Figure 4.2. *Pixel values versus the square root of the radiation dose given in 24 MUs for the calibrated and corrected Varian PortalVision detector. The line shows a linear fit to the data. (From Zhu et al 1995.)*

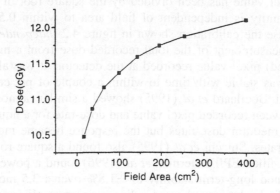

Figure 4.3. *The measured dose in solid water at the depth of maximum dose for different field sizes at 140 cm SSD showing how the total recorded dose will increase with increasing field size. (From Zhu et al 1995.)*

isocentre and showed that the image of the ball wandered by up to 0.5 cm projected back to the isocentre (figure 4.6).

Whilst these variations are not zero, the authors concluded they could be neglected and that the measurement of image pixel value could be converted into a measurement of missing tissue thickness for the design of a compensator. They imaged two simple configurations of a step wedge and a wedge, then used the EPID data to compute the geometry of a compensator, built the compensator and with compensator and phantom in place re-irradiated, showing almost uniform exit profiles at the EPID (figure 4.7). The subject of the use of EPIDs for compensator construction is further considered in section 4.3.

Heijmen *et al* (1995a) have quantified the performance of a Philips' SRI-100 EPID for portal dosimetry. The portal imager consists of a fluorescent

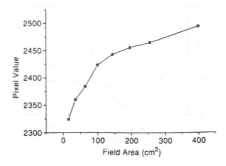

Figure 4.4. *Pixel values versus field area for different collimator field sizes from 4×4 cm² to 20×20 cm² at 140 cm SSD for the Varian PortalVision detector. (From Zhu et al 1995.)*

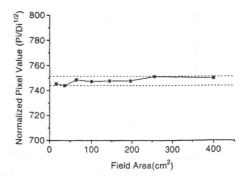

Figure 4.5. *Normalized pixel value (pixel value per unit square root of dose) at d_{max} as related to field area. The two horizontal lines indicate a ±0.5% variation from the mean and from the data it is clear that the pixel sensitivity is independent of field area. (From Zhu et al 1995.)*

screen viewed by a CCD camera via two mirrors. The screen is a 1.65 mm-thick stainless-steel plate coated with a fluorescent layer. Optical images are digitised using a CCD camera and charge integration is performed on the CCD chip and multiple frames are added together in a frame processor to improve image quality. Dark current was measured and subtracted from the images used for calibration. Thus high-quality relatively noise-free images were obtained for calibration purposes.

Heijmen *et al* (1995a) characterised the performance of the EPID via the ratio (which they called G/D_p) between the grey level EPID value G and the actual dose D_p at a point established by an ionisation chamber. To form the ratio, grey-level values were grouped for those EPID image pixels corresponding in area to the size of the chamber. G/D_p was measured under

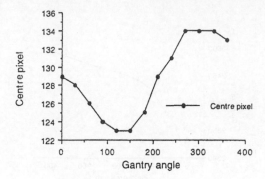

Figure 4.6. *A plot of the imaged location of a tungsten ball at the isocentre relative to the beam central axis as a function of the gantry angle for the Varian PortalVision system. (From Roback and Gerbi 1995.)*

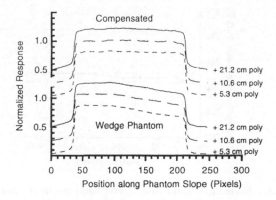

Figure 4.7. *In-plane profiles for a wedge phantom. The lower three curves represent central-axis profiles for a wedge phantom placed on top of different thicknesses of polystyrene, 5.3, 10.6 and 21.2 cm. The upper three curves represent central-axis profiles for a wedge phantom together with the compensator. The compensation is better than 3%. (From Roback and Gerbi 1995.)*

a variety of circumstances leading to the following useful results.

A prerequisite of the use of any EPID for portal dosimetry is an adequate time independence of response. Measurements of G/D_p were made on 14 occasions in a 38 day period for 28 locations (14 along each principal axis) in a 30×30 cm^2 beam. It was found that the variation in the ratio had a standard deviation of 0.4% once due account had been taken of the unavoidable sequential nature of the separate measurements of G and D_p and uncertainty in the measurement of D_p. The day-to-day variation in local

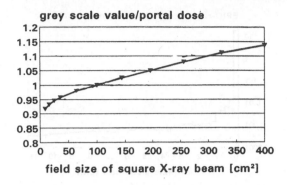

grey scale value/portal dose

field size of square X-ray beam [cm²]

Figure 4.8. *The solid line plus markers shows the EPID response (G/D_p ratio) for a Philips SRI-100, measured at the beam axis, as a function of the applied field size for square irradiation fields. The presented data have been normalised to the G/D_p ratio for a 10×10 cm² field. There was no other absorber in the beam. The figure indicates the considerable crosstalk present due to light scattering (see text). (Adapted from Heijmen et al 1995a.)*

response between different parts of the detector was 0.2%.

The linearity of response with respect to changing incident dose was established by measuring the ratio G/D_p, on the beam axis, for a 10×10 cm² field with radiation passing through 20 cm of flat water-equivalent absorber. The ratio G/D_p was constant to better than 0.5% over a range of MUs between 80 and 220. No saturation effects were observed.

Measurements of the ratio G/D_p were made, for the beam axis, for a number of different square field sizes (6×6, 10×10, 15×15 and 20×20 cm²) for 100 MU of radiation passing through water-equivalent slabs of thickness varying between 10 and 29 cm. The EPID response was almost independent (better than 1%) of absorber thickness, for a fixed field size, indicating that no spectral effects were contributing to the measurements.

However, the ratio G/D_p showed a strong dependence on field size with no absorber in the beam and 100 MUs of radiation. Compared with a value normalised to unity for a 10×10 cm² field, the ratio, measured at the beam axis, varied from 0.92 at 3×3 cm² to 1.14 at 20×20 cm² (figure 4.8). Experiments established that this was due to increases in the EPID signal from scattered light. Visible photons produced by the x-ray beam at a point of the fluorescent screen contribute not only directly to the corresponding point in the image but also to other points in the image.

Heijmen *et al* (1995a) derived a method to take care of this crosstalk. They computed a factor $k(r)$ to describe the increase in EPID response at the beam axis per square centimetre of irradiated surface of the fluorescent screen at a distance r from the beam axis. Essentially this is a point spread function for the system (figure 4.9). This was measured by differentiating the graph of the variation of the ratio G/D_p normalised to unity for a 10×10 cm² field. It then followed that the function so generated could be used to *predict* the

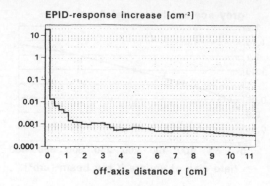

Figure 4.9. *The function $k(r)$ describing the increase in EPID response at the beam axis per square centimetre irradiated fluorescent screen at a distance r from the beam axis. This function was derived by differentiating the data in figure 4.7. (From Heijmen et al 1995a.)*

greyscale response at any point in the EPID image via the equation

$$G(x, y) = \left. \frac{G}{D_p} \right|_{10 \times 10}$$
$$\times \int_{(x', y') \in f} D_p(x', y') S(x', y') k(\sqrt{(x' - x)^2 + (y' - y)^2}) \, dx' \, dy'$$
(4.7)

where $G(x, y)$ is the predicted greyscale value at (x, y) of the EPID image, $D_p(x', y')$ is the absolute portal dose at point (x', y') of the portal-dose image, $G/D_p|_{10 \times 10}$ is the absolute on-axis EPID response for an open 10×10 cm^2 field and $S(x', y')$ is the local relative EPID response at (x', y') (i.e. response relative to a value of unity at the beam axis $(x = y = 0)$: this function was measured by imaging small 3×3 cm^2 fields centred at each of a rectangular grid of points, with no absorber in the field of view and the response averaged over 1 cm^2). The integral is made over all the (x', y') in the field 'f'. Heijmen *et al* (1995a) used this equation to predict the portal dose for an anthropomorphic phantom and showed good agreement with measurements.

Equation (4.7) is a 'forward' equation predicting the greyscale response from the known portal dose. Of course in practice the reverse operation is required to determine portal dose from the greyscale information. This requires a deconvolution that in principle could be done but was not the subject of the paper by Heijmen *et al* (1995a). Other applications of EPID imaging for portal dosimetry have been presented by Heijmen *et al* (1992, 1993, 1994, 1995b).

4.2.3.2. Extraction of primary exit fluence. At this stage we pause to note that each of these studies gives the information to extract from electronic portal images a measure of dose to each point, independent of field size, attenuating material etc. The analysis is quite different for the different types of EPIDs. Of course these data alone do not give a measure of *primary* dose which must be determined by knowledge of the scatter-to-primary ratio and its field-size and depth-of-phantom dependence.

Unfortunately the detected signal will be a mix of response to primary and scattered photons. For transit dosimetry the primary-only fluence $P(x_d, y_d, A)$ is required. The two are related through the scatter-to-primary ratio SPR, i.e.

$$P(x_d, y_d, A) = \frac{D(x_d, y_d, A)}{(1 + \text{SPR}(A))} \tag{4.8}$$

for measured field size A. Swindell and Evans (1993), using Monte Carlo calculations in water slabs for photon beams of different areas A and for patients of different thickness T, showed that:

(i) when the detector was a large distance ($\geqslant 50$ cm) from the exit surface of the patient, as is the case in practice, the SPR was virtually independent of position x_d, y_d in the detector plane;

(ii) for a given and wide range of patient thickness, the SPR was approximately linear with beam area for all practical detector-to-isocentre distances;

(iii) for a given and wide range of beam area, the SPR was approximately linear with patient thickness for all practical detector-to-isocentre distances;

(iv) for realistic beam areas and patient thicknesses the SPR is small (of the order of a few per cent) when the detector is a long way (of the order 100 cm) from the isocentre.

From these observations it was possible to say that

$$\text{SPR} \simeq kAT \tag{4.9}$$

with the constant k depending on the beam spectrum and on the distance between the detector and the isocentre and on the source-to-isocentre distance, e.g. for a 6 MV spectrum, an isocentric distance of 100 cm, a Compton detector, a detector-to-isocentre distance of 100 cm and measurement on the central axis, $k = 1.064 \times 10^{-5}$. Equation (4.9) is a first-order expression.

Swindell and Evans (1996) have extended this simple theory to show that one may write the SPR in more general terms with parameters for the geometry. Strictly, after considering a number of different definitions of SPR, they defined a quantity SPR* as the SPR for a *Compton* detector responding to the *mean* values of primary and scattered radiation and with the measurements made on the *central axis* of the radiation beam. They considered a beam emanating from a point source defining a circular field of area A at an isocentric distance of L_1 with the detector at a distance L_2

from the isocentre. A water phantom of thickness T was arranged with its mid-depth at the isocentre.

The general expression for SPR* is then

$$SPR^* = k_0 A T (1 + k_1 T)(1 + k_2 A) \tag{4.10}$$

where the constant k_0 is dependent on the geometry via

$$k_0 = 0.0266 \frac{(L_1 + L_2)^2}{L_1^2 L_2^2}. \tag{4.11}$$

The parameter k_1 is a factor which weakly depends on L_2 and is approximated well by

$$k_1 = 0.002 \tag{4.12}$$

and the parameter k_2 is given by

$$k_2 = \left(-\frac{1}{2\pi}\right)\left[\frac{1}{L_1^2} + \frac{1}{L_2^2} + \left(\frac{1}{L_1} + \frac{1}{L_2}\right)^2 \left(\frac{2}{3} + \frac{3\alpha}{2}\right)\right]. \tag{4.13}$$

The parameter α is the mean spectral energy in units of electron rest mass.

The expression for k_0 is exact for vanishingly small scattering volume, i.e. as both A and T tend to zero. It was derived by taking the ratio of the forward-scattered photon fluence from a point scatterer at the isocentre to the primary fluence arriving at the detector. The constant 0.0266 (dimensions $[L]^{-1}$) is the product of the water electron density $n_e = 3.346 \times 10^{23}$ and the zero-deflection-angle differential cross section $(d\sigma/d\Omega)_{\theta=0} = 79.4 \times 10^{-27}$ cm^2. This constant is independent of beam energy.

The expression for k_2 was derived considering an infinitely thin but large-area scatterer and integrating the ratio of scattered-to-primary fluence. Swindell and Evans (1996) showed very elegantly that it is possible to derive the functional form $(1 + k_1 T)$ from first principles by considering the ratio of first- and second-scattered photons to primary photons for a beam of infinitely small area but passing through a thickness T of material.

However, the thrust of their work was to show that, with the parameters k_0, k_2 fixed by the above geometrical relationships, Monte Carlo generated data for SPR* under a variety of scattering conditions (different values of A and T) could be fitted to the above empirical formula with k_1 used as a free fitting parameter. k_1 turned out to be only loosely tied to L_2 and, rather remarkably, the contributions to SPR* from the 'area effect' and the 'thickness effect' could be decoupled multiplicatively. The fit to the Monte Carlo data was better than 0.5%. They also separately validated both the Monte Carlo model and the empirical model by making measurements related to SPR* with the RMNHST portal imaging system. It turns out that $k_1 > 0$ and $k_2 < 0$

reflecting the shapes of the Monte Carlo and experimental curves for SPR*
versus T or versus A.

Using the above information it is therefore possible to deduce, at least
to first order, a measure of the primary exit fluence from the recorded
total signal including scatter since we have the basis in equation (4.10)
for calculating the SPR to insert into equation (4.8). The beam area A is
known. An estimate of the patient thickness T is determined as follows:

Evans *et al* (1994b,c, 1995a) showed that for a field with reference area
A_{ref} (= 25×25 cm^2)

$$\log \left[\frac{\text{OFS}(A_{ref})}{D(x_d, y_d, A_{ref})} \right] = \alpha T(x_d, y_d) + \beta T^2(x_d, y_d) \tag{4.14}$$

where α and β are constant parameters (which were determined for the
RMNHST EPID), $T(x_d, y_d)$ is the thickness of tissue for the ray detected at
(x_d, y_d), OFS(A_{ref}) is the open-field detector signal and $D(x_d, y_d, A_{ref})$ is
the recorded signal after radiation passes through the material. This equation
was established by imaging a series of plastic blocks of known radiological
thickness with a reference field of area A_{ref}.

When a patient is imaged, the measured data, the total dose at a point
on the detector, $D(x_d, y_d, A)$ with field-size A can be related (to first order;
see also equation (4.25)) to the corresponding data $D(x_d, y_d, A_{ref})$, as if the
patient had been imaged with the reference field, via

$$D(x_d, y_d, A_{ref}) = D(x_d, y_d, A) \times O \tag{4.15}$$

where a field factor is defined as

$$O = \frac{P(x_d, y_d, A_{ref})}{P(x_d, y_d, A)} \tag{4.16}$$

and was determined by measurements with no object (patient) in the fields.
Using equations (4.14), (4.15) and (4.16) the value of T can be established
for each point in the image. A refinement of this procedure to obtain T is
given in section 4.3.

Thus knowing A and estimating T, SPR follows from equation (4.9)
or (4.10) and the primary fluence follows from equation (4.8). Strictly
equations (4.15) and (4.16) as written are very slightly inconsistent. If
equation (4.15) is replaced by the strictly correct equation (4.25), the process
to obtain $T(x_d, y_d)$ and SPR(A) becomes iterative as described in section 4.3.

Since $D(x_d, y_d, A)$ depends on position x_d, y_d, different values of T will
apply for each element of the detector, reflecting the different radiological
thicknesses. There is a second-order error because making the correction
this way would imply that the scatter arriving at this element arises from a
patient whose radiological thickness is everywhere the same as it is for the

ray connecting that element to the source. These small difficulties are noted and then ignored in practice.

The theory summarised here is a model of the axial scatter which is interpreted as applying throughout the image. Later work by this group has solved the problem of deconvolving the image scatter-response function from the portal image to yield a measurement of primary dose. It was shown by using Monte Carlo-generated data as a gold-standard benchmark that the scatter can be predicted rather well as the convolution of the primary fluence with a kernel. By solving this problem of extracting the primary signal by repetitive forward convolutions and subtractions (a well-known technique in medical imaging for approximating deconvolution) the space-variant nature of the scatter is accounted for (Hansen 1996, Hansen *et al* 1996c). Boellaard *et al* (1996) have also developed a convolution–superposition method of extracting the primary signal from the total signal for a liquid-filled EPID. It was claimed that the accuracy was about 2% for homogeneous phantoms and was only a little worse for inhomogeneous phantoms.

In passing we may note that the presence of scatter also degrades broad-area contrast in portal imaging from actual contrast C to measured contrast C' where

$$C' = \frac{C}{(1 + \text{SPR})} \tag{4.17}$$

and also affects the visibility of small detail.

4.2.4. *Backprojection to obtain a 3D map of the primary fluence in the patient*

The next (third) stage in transit dosimetry is to backproject the primary fluence $P(x_d, y_d, A, E)$ along the ray path to the source, computing the fluence $\Gamma(r)$ at each position in the irradiated part of the patient. Making use of the CT data (and remembering this is now 'locked' to the portal image) the radiological thickness $R_X(r, x_d, y_d)$ from r to x_d, y_d is computed (figure 4.1). Hence the fluence $\Gamma(E, r)$ is

$$\Gamma(E, r) = P(x_d, y_d, E)(r_0/r)^2 \exp(\mu_{w_E} R_X(r, x_d, y_d)). \tag{4.18}$$

where r_0 is the source-to-detector distance and r is the magnitude of the vector position r at which the fluence is computed. In equation (4.18) the specific energy dependence is shown and μ_{w_E} is the linear-attenuation coefficient for *megavoltage* photons of energy E (see also section 4.2.6). The area label can now be dropped since this information was only required in the determination of $P(x_d, y_d, E)$.

Hansen *et al* (1993, 1994a) used equation (4.18) (with the approximation $P = D$) to compute the fluence $\Gamma(E, r)$ and interestingly noted that, if the ray tracing is performed right back to the entrance port of the patient, a constant entrance fluence should be the result. The small departure from constancy which they observed was a measure of the inaccuracy of the technique.

Returning to the theme at the end of section 4.2.2 we may note that a similar map of fluence throughout a 2D section could be made by backprojecting the portal image primary exit fluence $P(x_d, y_d, E)$ through *MVCT data* using equation (4.18) and with $\mu_{w_E} R_X$ replaced by the line integral of the *measured* MVCT data. This would avoid the stage of registration with a DRR computed from diagnostic CT data, and indeed would avoid the use of diagnostic CT data altogether. It would also avoid the need to elucidate μ_{w_E} from the photon spectrum and would lead directly to a map of fluence superposed on the patient CT data at the time of treatment. An example of such a map of fluence (which they called relative collision kinetic energy distribution) has been presented by Aoki *et al* (1990).

4.2.5. Determination of terma

$\Gamma(E, r)$ determines the terma $T_E(r)$ via

$$T_E(r) = \mu/\rho(E, r)E\Gamma(E, r) \tag{4.19}$$

where $\mu/\rho(E, r)$ is the mass-attenuation coefficient of the primary photons of energy E at point r. Terma is the total energy released per unit mass by primary photon interactions. Note since P can be scaled to primary photons per unit detector area, P has dimensions L^{-2}. So fluence Γ has dimensions L^{-2} and, since mass-attenuation coefficient μ/ρ has dimensions $L^2 M^{-1}$, terma T_E has dimensions energy per unit mass, i.e. the same as dose. To obtain the dose from the terma, the effect of secondary photon and electron transport must be introduced.

4.2.6. Determination of the 3D dose distribution

Finally the fourth stage in transit dosimetry is to obtain a 3D dose map $D_E(r)$. The terma must be transported via the use of an energy-deposition kernel $h(E, r)$

$$D_E(r) = \iiint T_E(s)h(E, r - s)\,\mathrm{d}^3 s. \tag{4.20}$$

This convolution equation assumes dose deposition is shift invariant. Convolution dosimetry has been discussed in detail in the companion Volume, Chapter 2. Evans *et al* (1993) and Hansen *et al* (1994a) have implemented the technique for transit dosimetry using kernels provided by Andreo. The earliest calculations they performed assumed the beam was monoenergetic with mean energy 2 MeV for the 6 MV spectrum and they also ignored tissue inhomogeneities. More accurate calculations take account of the beam spectrum. A phantom study (Hansen *et al* 1995a) has determined the degree of complexity required in the physical model for dose calculation. Changing the resolution from 1 to 5 mm changes depth–dose curves by less than 2%. Similar work has been reported by Papanikolaou and Mackie (1994).

Equations (4.18) to (4.20) strictly only apply to irradiation with monoenergetic photons and the total dose distribution results from integrating equation (4.20) over the energy variable, weighted by the spectrum. Of course, in practice the calculated primary fluence at each exit point $P(x_d, y_d)$ is a superposition of fluence from photons with an energy spectrum. Since the detector has no energy discrimination $P(x_d, y_d)$ cannot be decomposed into separate components labelled by energy. Hence in practice (Hansen *et al* 1996b), $P(x_d, y_d, E)$ in equation (4.18) was replaced by the composite measurement $P(x_d, y_d)$ and E in equations (4.18), (4.19), (4.20) was taken as an effective single energy determined from the spectrum. Specifically, in equation (4.20) the convolution kernel is a weighted-over-spectrum kernel so that the dose distribution, labelled by effective energy, is determined without the need for a fourth integral over energy.

Hansen *et al* (1996b) tested the model for transit dosimetry with a clinical pelvic case irradiated by a four-field technique and obtained very good correlation between the planned dose distribution and that determined by transit dosimetry. The largest deviations were at the 50% isodose level (figure 4.10). They point out, however, that the beam model used in the planning computer (Bentley–Milan measurement-based algorithm) is not the same as that in the transit dosimetry technique (convolution of a kernel). Using humanoid phantoms they also showed good agreement between transit dose calculations and measurements with thermoluminescent dosimetry (TLD). This latter measurement bears on the question of how accurately the convolution model represents the actual treatment beam.

Alternatively to this methodology it is possible to combine the knowledge of CT data with a convolution method of dose calculation to predict the dose profile at the plane of a portal imager. Papanikolaou *et al* (1995) showed a good correlation between the prediction of this method and the measurement of dose by a film. McNutt *et al* (1996) use an iterative process in which repeated iterations of the predicted portal dose are compared with the measured portal dose until convergence is reached, at which stage the dose in the patient is then determined.

It is important in transit dosimetry to keep a close track on high-quality benchmarks. The methodology described makes assumptions at each stage and necessarily some approximations. The ultimate verification of methodology is to compute and measure a fully 3D dose distribution (see next section 4.2.7). Then the question arises of the accuracy of measurements. Against all these difficulties authors tend to demonstrate

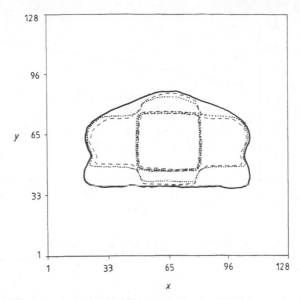

Figure 4.10. *Dose distributions for a pelvic patient compared with the TARGET planning system (dotted line) and with transit dosimetry (broken line), 90%, 80% and 50% isodose curves are shown. (From Hansen et al 1996b.)*

consistency between results and within certain limits, which is probably the best one can do given the fairly unquantifiable uncertainties in a multistage calculation.

4.2.7. Verification of verification, MRI of dose-sensitive gels

It is important to be able to verify that the method of treatment verification is accurate. To do this it is necessary to irradiate some phantom, compute the dose distribution from the portal radiograph by the methods described and then *make measurements* within the same phantom at the time the portal image is taken to correlate with these predictions. Whilst this could be done with ion-chamber, TLD, diode or film dosimetry, a better way would be to use gels which respond to radiation, 'store in memory' the delivered dose, and which can then be magnetic resonance (MR) imaged to give a 3D map of delivered dose.

There have been many studies showing that ferrous sulphate ions held in a gel can be used as a radiation dosemeter with MR readout (Appleby *et al* 1987, Gore *et al* 1984, Hazle *et al* 1991, Johansson *et al* 1995, Olsson *et al* 1990, Prasad *et al* 1991, Podgorsak and Schreiner 1992, Schulz *et al* 1990). This is because under irradiation Fe^{2+} ions are oxidised to Fe^{3+} ions with the amount of oxidation proportional to the dose. For example, recently Chan and Ayyangar (1995) made a gel containing 7.5% gelatin by weight

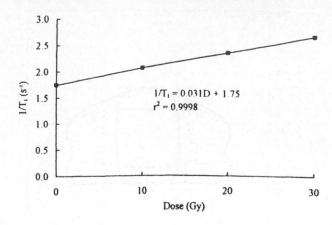

Figure 4.11. *The linearity of a dosimetry gel irradiated by 6 MV x-rays. Dose is plotted against the reciprocal of T_1. (From Chan and Ayyangar 1995.)*

with about three quarters of the total desired volume of double deionised and distilled water. A solution of ionic compound was prepared by adding to the remaining quarter sodium chloride (1 mM), ferrous ammonium sulphate (1 mM) and sulphuric acid (0.1 N). Once the gel, heated to 62 °C had completely melted it was cooled to 35 °C and the ionic solution was added. The density of the resulting FeMRI gel was found to be 1.005 g cm^{-1} and the solution was shown to be water-equivalent from a radiation viewpoint. When samples of the solution were irradiated, a straight-line relationship was determined between $1/T_1$ and dose

$$1/T_1 = aD + b \qquad (4.21)$$

where T_1 is the longitudinal MR relaxation time, D is the dose and $a = 0.031$ s^{-1} Gy^{-1} and $b = 1.75$ s^{-1} (figure 4.11).

The remaining gel was used to fill a head phantom. Five non-coplanar conformal treatment beams were planned and delivered to this phantom which was then MR scanned using the same inversion-recovery sequence. Using the above equation the 3D map of T_1 was converted to a map of dose and compared with the planned 3D dose distribution on a pixel-by-pixel basis. The correspondence was accurate to 5% at worst, errors being attributed to: (i) image distortion; (ii) chemical-shift and magnetic susceptibility artefacts; (iii) environmental temperature variations.

Maryanski *et al* (1993, 1994) discuss a new tissue-equivalent polymer-gel which is made of aqueous gelatin infused with acrylamide and N, N'-methylene-bisacrylamide monomers, and made hypoxic by nitrogen saturation. This gel has a specific gravity of 1.03 g cm^{-1} and an electron density of 3.32×10^{26} kg^{-1}. Irradiation of the ('BANG') gel causes localized

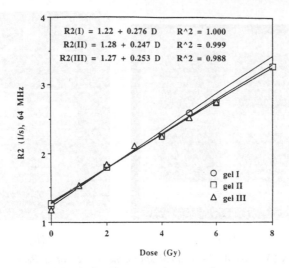

Figure 4.12. *Shows the linearity of relaxivity $1/T_2$ versus dose for three separately prepared BANG gels, over the range of dose from 0 to 8 Gy. The separate linear fits are consistent with the average behaviour expressed in equation (4.22). (From Maryanski et al 1994.)*

polymerization of the monomers, which, in turn, reduces the transverse nuclear magnetic resonance (NMR) relaxation times T_2 of water protons. The dose dependence of the NMR transverse relaxation rate is reproducible to better than 2% and its reciprocal, relaxivity, is linear up to about 8 Gy,

$$1/T_2 = cD + d \qquad (4.22)$$

where the slope $c = 0.25$ s^{-1} Gy^{-1} at 1.5 T (figure 4.12). The measurement is independent of dose-rate between 0.05 and 16 Gy min^{-1}. The linear relationship is almost independent of the temperature at the time of irradiation but changes significantly with the temperature at the time of MR imaging which must therefore be strictly controlled. MRI may be used to obtain accurate 3D dose distributions with high spatial resolution (Baldock *et al* 1995) (figure 4.13). The technique is patented. It has been shown that the sensitivity of the technique can be varied by varying the composition of the BANG gel (Baldock *et al* 1996). This can overcome problems of limited spatial resolution, sensitivity and time needed to use other methods of dosimetry. Indeed this experimental method could be applied to verifying the delivery of 3D dose distributions with 'unconventional' shapes from superposed IMBs (Chapter 2). This requires experimental verification before such techniques can be clinically implemented. This work has already started. De Wagter *et al* (1995) have shown that MR imaging of gel dosemeters can be used to measure the dose distributions developed by superposing conformal therapy using multiple segmented fields. Bogner

(a)   *(b)*

Figure 4.13. *(a) A T_2-weighted spin-echo image of a BANG gel in a phantom irradiated to four dose levels 5, 10, 15 and 20 Gy; (b) The visual image of the irradiated gel corresponding to (a). (From Maryanski et al 1994.)*

(1996) has used BANG-gel dosimetry to measure the distribution of dose for ^{90}Sr and ^{192}Ir applicators for head-and-neck treatments as well as for recording proton dose distributions at the Swiss PSI Facility. He demonstrated no dependence on linear energy transfer, no diffusion (unlike the Fricke gel which diffused at 2 mm h^{-1}) and linearity of $1/T_2$ up to 10 Gy. Alternatively the BANG-gel dose distributions can be obtained by optical tomography (Maryanski *et al* 1996).

4.2.7.1. *PET imaging for verifying proton irradiations.*

When high-energy *protons* interact with tissue they produce a number of positron emitters by nuclear reactions with oxygen, nitrogen and carbon in the body. The positron activity characterises the dose distribution (Del Guerra and Di Domenico 1993, Vynckier *et al* 1993). The shortest living positron emitters decay too fast for any subsequent imaging of the irradiated region but longer-living isotopes could be imaged to verify the delivered dose distribution although the sensitivity would be poor. This technique has been tried but the depth–dose profile did not correlate as well with the depth–activity profile as would be desired (Oelfke *et al* 1996). An alternative scenario would be to perform on-line imaging of the short-lived isotopes during the proton irradiation in which case the count-rate would be several orders of magnitude higher. The schematic illustration in figure 4.14 shows the concept.

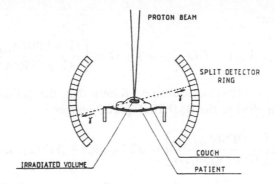

Figure 4.14. *A schematic diagram showing the concept of on-line imaging of the short-lived positron emitters created by proton irradiation. The image of the activity distribution would characterise the 3D dose distribution. (From Ion Beam Applications Company literature.)*

4.3. USE OF PORTAL IMAGING FOR COMPENSATOR CONSTRUCTION

The measurements from a portal imager can be used to compute radiological thickness as shown in section 4.2.3. From these measurements an appropriate compensator can be constructed, for example, for the breast, so that dose homogeneity in the breast might be improved. This is in addition to the use of EPIDs for verification of the position of the field with respect to the breast (Lirette *et al* 1995).

Many authors have noted that for large-breasted women in whom dose inhomogeneity can be large, the results of simple tangential radiation therapy with wedges alone is suboptimal (Yarnold and Neal 1995) although others disagree (Rodger 1995). The method of determining a compensator has been derived by Evans *et al* (1994b,c, 1995a,d). From equation (4.8) we have for a beam of area A:

$$D(x_d, y_d, A) = P(x_d, y_d, A) \times (1 + \mathrm{SPR}(A)) \qquad (4.23)$$

where we recall $D(x_d, y_d, A)$ is the measured intensity at (x_d, y_d), $P(x_d, y_d, A)$ is the primary intensity at (x_d, y_d) and $\mathrm{SPR}(A)$ is the scatter-to-primary ratio for a beam with area A. For some reference area A_{ref} equation (4.23) becomes correspondingly

$$D(x_d, y_d, A_{ref}) = P(x_d, y_d, A_{ref}) \times (1 + \mathrm{SPR}(A_{ref})). \qquad (4.24)$$

Defining a field factor as

$$O = \frac{P(x_d, y_d, A_{ref})}{P(x_d, y_d, A)} \qquad (4.16)$$

it follows that

$$D(x_d, y_d, A_{ref}) = D(x_d, y_d, A) \times O \times \left[\frac{(1 + \text{SPR}(A_{ref}))}{(1 + \text{SPR}(A))} \right]. \qquad (4.25)$$

Evans *et al* (1994b,c, 1995a,d) found that the following calibration expression applied to the RMNHST portal imager

$$\log \left[\frac{\text{OFS}(A_{ref})}{D(x_d, y_d, A_{ref})} \right] = \alpha T(x_d, y_d) + \beta T^2(x_d, y_d) \qquad (4.14)$$

where α and β are parameters and $T(x_d, y_d)$ is the breast thickness. Combining equations (4.25) and (4.14) we have

$$\log \left[\frac{\text{OFS}(A_{ref})}{D(x_d, y_d, A) \times O \times [(1 + \text{SPR}(A_{ref}))/(1 + \text{SPR}(A))]} \right]$$

$$= \alpha T(x_d, y_d) + \beta T^2(x_d, y_d). \qquad (4.26)$$

Equation (4.26) taken with equation (4.9) or (4.10) yields the following iterative algorithm to obtain $T(x_d, y_d)$:

(i) set all SPR $= 0$ initially and use equation (4.26) to obtain a first estimate of $T(x_d, y_d)$;
(ii) use equation (4.9) or (4.10) to obtain both SPR(A) and SPR(A_{ref});
(iii) put these into equation (4.26) to obtain a refined estimate of $T(x_d, y_d)$. Repeat stages (ii) and (iii) to convergence.

In practice, because SPRs are small, the process converges in one or two quick iterations. Note the one-step procedure to obtain T given in section 4.2.3.2 was a (fairly good) approximation to this iterative procedure, consequent on ignoring $[(1 + \text{SPR}(A_{ref}))/(1 + \text{SPR}(A))]$ in equation (4.15) which is almost unity.

Evans *et al* (1994b,c, 1995a,d) and Hansen *et al* (1994b, 1997) used the map of $T(x_d, y_d)$ to design a 'missing-tissue compensator'. By relating the measurement of radiological thickness to the true physical thickness of the breast on a point-by-point basis and creating a 'pseudo-CT' slice in which the lung tissue in the field was set symmetrically about the midline of the breast, Evans *et al* (1995a,d) were able to show that the use of compensators created DVHs in the breast with greater homogeneity (figure 4.15). Evans *et al* (1995b) have shown how suitable intensity-modulated 2D beams may be achieved by dynamic MLC field shaping with the dose inhomogeneity in the breast being reduced from greater than $\pm 8\%$ to less than $\pm 4\%$. This technique does not take into account the normal variation of breast volume through the menstrual cycle which in one study (Hussain *et al* 1995) has been measured, using serial MR imaging, to be as great as 41%, the variation

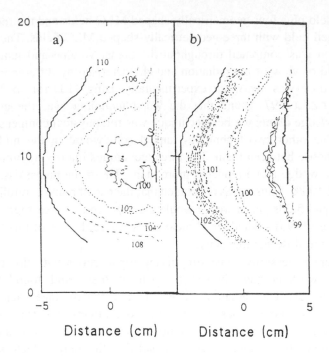

Figure 4.15. *A comparison of calculated dose distributions for the IMB and for a wedge whose angle has been chosen to give maximal dose uniformity within the transaxial slice passing through the isocentre. (a) The sagittal slice passing through the isocentre for the wedge with dose variation 10%; (b) the sagittal slice passing through the isocentre for the IMB with dose variation reduced to 3%. (Reprinted from Evans et al 1995a, with kind permission from Elsevier Science Ireland Ltd, Bay 15K, Shannon Industrial Estate, Co. Clare, Ireland.)*

of adipose tissue exhibiting two peaks, one at ovulation and the other at menses. However, this translates into not large linear changes. Evans *et al* (1997) showed that in practice such compensators could be formed by multiple-static-MLC-shaped fields and gave the algorithm for computing the optimum intensity levels. This was based on minimising the least-square quantisation error when a fixed number N of quantisation levels is used instead of the continuous histogram of levels representing the true IMB.

Hansen *et al* (1996b, 1997) verified the methodology of Evans *et al* (1995a,d) to study the inhomogeneity in dose to the breast when ideal 2D intensity-modulations, created by the above method, were realised in practice. They studied four specific implementations. The first shaped the intensity of each of two tangential fields by a simple wedge as is common clinical practice. The second used a wedge to create 'most of' the intensity modulation and then tuned this with a physical compensator. The third used an open field plus four geometrically-shaped MLC fields and the fourth used one wedged field with three geometrically-shaped MLC fields. The 3D dose distribution was computed throughout the breast volume and lung volume in the field of view of the radiation and characterised by differential DVHs. Also the dose was *measured* experimentally using TLD for six locations. Hansen *et al* (1997) found that the worst technique was the use of the physical wedge alone; the best technique was the physical compensator plus wedge. The other two techniques ranked somewhere in the middle with little to choose between them. The three techniques improving on the use of the simple wedge alone rendered more than 97% of the breast tissue within ±1.5% of 100%. Thus it was concluded that the use of a few carefully chosen MLC-shaped fields plus a wedge would greatly improve breast radiotherapy. Evans *et al* (1996) studied ten clinical cases where the use of this technique improved the dose homogeneity.

There are necessarily some simplifying assumptions with this technique but these workers have looked at each of these and found they are small problems. The radiation 'tubes' do not strictly face each other at 180°; the dose model does not include axial effects; the true outline of the breast is not available but could be with use of a CT attachment to a radiotherapy simulator (see Chapter 7); the output factors for the geometrically shaped subfields may differ slightly from the largest field and there are 'resolution effects' due to the finite sizes of detector element and leaf compensation. However, the overall objective of improving the homogeneity of radiotherapy of the breast is being achieved.

In their experimental measurements Hansen *et al* (1997) constructed a physical compensator using shaped 0.5 mm thin lead sheet bonded to make the appropriate thickness map. Alternatively the concept of using multiple-static fields was discussed. Another possibility is to construct a 'multileaf compensator' (Shentall 1995) in which the leaves comprise aluminium rods of rectangular cross section. There would be several planes of such rods with the planes normal to the axis of radiation. Each plane would collimate a different geometrical fieldsize determined as described above and the planes would be mounted one above the other creating a 'reusable' compensator. This could provide a cheap, and in principle simple, alternative to the concept of using multiple MLC-shaped fields, at present a rather time-consuming activity. However, it transfers the time to a different area, the requirement to construct such a device and possibly the need for several if a few patients are on treatment at any one time. There are also questions of the accuracy of resetting the device.

Yin *et al* (1994) and Pouliot *et al* (1994) have reported similar developments using data from portal imaging to design a compensator.

4.4. SUMMARY

Megavoltage portal imaging has come a long way since its first use as a substitute for film imaging for positional verification. A prime use of EPID images is still the recording of data on patient positioning and the guiding of possible repositioning to reduce treatment errors. Automatic methods of image analysis are now possible given the digital nature of the data but it is still quite hard to use the images in real time.

An exciting new development is the use of EPID-generated images for transit dosimetry which has been discussed in some detail here because it is an emerging discipline and the problems associated with each stage of the process can be separated out for study. It is important to recognise that at each stage there are approximations made and so there is still considerable scope for improving transit dosimetry. Also it is still some way from being a tool with immediate clinical use but one must anticipate this in due course.

The data from EPIDs have been used to design missing-tissue compensators for improving the homogeneity of the dose distribution in the breast. This is an important area of application since there is still much concern over the effects of inhomogeneous distributions of dose in the breast.

REFERENCES

Aoki Y, Akanuma A, Evans P M, Lewis D G, Morton E J and Swindell W 1990 A dose distribution evaluation utilizing megavoltage CT imaging system *Radiat. Med.* **8** 107–10

Appleby A, Christman E A and Leghrouz A 1987 Imaging of spatial radiation dose distribution in agarose gels using magnetic resonance *Med. Phys.* **14** 382–4

Baldock C, Burford R P, Billingham N C, Cohen D and Keevil S F 1996 Polymer gel composition in magnetic resonance imaging dosimetry *Med. Phys.* **23** 1070

Baldock C, Patval S, Keevil S, Summers P, Piercy A, Graddon J, Prior D, Batey S and Lutkin J. 1995 Investigation of field strength dependencies in MRI dosimetry *Proc. Röntgen Centenary Congress 1995 (Birmingham, 1995)* (London: British Institute of Radiology) p 469

Balter J, Pelizzari C and Chen T 1992 Correlation of projection radiographs in radiation therapy using open curve segments and points *Med. Phys.* **19** 329–34

Bijhold J, Gilhuijs K G A, van Herk M and Meertens H 1991 Radiation field edge detection in portal images *Phys. Med. Biol.* **36** 1705–10

Boellaard R, Van Herk M and Mijnheer B J 1995 Investigation of the dosimetric characteristics of a liquid-filled electronic portal imaging device (Proc. ESTRO Conf. (Gardone Riviera, 1995)) *Radiother. Oncol.* **37** Suppl. 1 S21

—— 1996 Exit dosimetry using a liquid-filled electronic portal imaging device *Proc. 4th Int. Workshop on Electronic Portal Imaging (Amsterdam, 1996)* (Amsterdam: EPID Workshop Organisers) p 47

Bogner L 1996 Measurement of 3D-dose distributions; Dose-volume effects *Proc. Symp. Principles and Practice of 3-D Radiation Treatment Planning (Munich, 1996)* (Munich: Klinikum rechts der Isar, Technische Universität)

Brahme A, Lind B and Näfstadius P 1987 Radiotherapeutic computed tomography with scanned photon beams *Int. J. Radiat. Oncol. Biol. Phys.* **13** 95–101

Chan M F and Ayyangar M 1995 Confirmation of target localization and dosimetry for 3D conformal radiotherapy treatment planning by MR imaging of a ferrous sulphate gel head phantom *Med. Phys.* **22** 1171–5

Chaney E L, Thorn J S, Tracton G, Cullip T, Rosenman J G and Tepper J E 1993 A portable software tool for computing digitally reconstructed radiographs (Proc. 35th ASTRO Meeting) *Int. J. Radiat. Oncol. Biol. Phys.* **27** Suppl. 1 180

De Neve W, van der Heuvel F, Coghe M, Verellen D, de Beukeleer M, Roelstraete A, de Roover P, Thon L and Storme G 1992 Interactive use of online portal imaging in pelvic radiation *Int. J. Radiat. Oncol. Biol. Phys.* **25** 517–24

De Wagter C, De Deene Y, De Jaeger K, Van Duyse B, De Neve W and Achten E 1995 Verification of conformal radiotherapy using a polymer-gel dosimeter and MR-imaging (Proc. ESTRO Conf. (Gardone Riviera, 1995)) *Radiother. Oncol.* **37** Suppl. 1 S40

Del Guerra A and Di Domenico G 1993 Positron emission tomography as an aid to *in vivo* dosimetry for proton radiotherapy: a Monte Carlo study *TERA Report* 93/10 TRA 9

Dong L, Kachilla D and Boyer A 1994 A ray tracing program based on Bresenhem algorithm for generating digitally reconstructed radiographs *Med. Phys.* **21** 886

Evans P M, Gildersleve J Q, Morton E J, Swindell W, Coles R, Ferraro M, Rawlings C, Xiao Z R and Dyer J 1992 Image comparison techniques for use with megavoltage imaging systems *Br. J. Radiol.* **65** 701–9

Evans P M, Gildersleve J Q, Rawlings C and Swindell W 1994a The implementation of patient position correction using a megavoltage imaging device on a linear accelerator *Br. J. Radiol.* **66** 883–8

Evans P M, Hansen V N, Mayles W P M, Swindell W, Torr M and Yarnold J R 1995a Design of compensators for breast radiotherapy using electronic portal imaging *Radiother. Oncol.* **37** 43–54

Evans P M, Hansen V N, Mayles W P M, Torr M, Swindell W and Yarnold J R 1995b Applications of electronic portal imaging in breast dosimetry

Proc. Röntgen Centenary Congress 1995 (Birmingham, 1995) (London: British Institute of Radiology) p 229

Evans P M, Hansen V N and Swindell W 1993 The inclusion of scatter into transit dosimetry measurements using a portal imaging system *Proc. BIR Conf. on Megavoltage Portal Imaging (London, 1993)* (London: British Institute of Radiology); *Br. J. Radiol.* **66** 1083

—— 1994b The use of a portal imaging system to design multiple MLC fields for compensation in conformal radiotherapy (Proc. World Congress on Medical Physics and Biomedical Engineering (Rio de Janeiro, 1994)) *Phys. Med. Biol.* **39A** Part 1 488

—— 1995c Megavoltage imaging *Proc. Röntgen Centenary Congress 1995 (Birmingham, 1995)* (London: British Institute of Radiology) p 242

—— 1995d Some aspects of the design of intensity modulated beams for breast radiotherapy (Proc. ESTRO Conf. (Gardone Riviera, 1995)) *Radiother. Oncol.* **37** Suppl. 1 S46

—— 1997 The optimum intensities for multiple static MLC field compensation *Med. Phys.* at press

Evans P M, Hansen V N, Swindell W, Torr M, Mayles W P M, Neal A J, Brown S and Yarnold J R 1994c The use of a portal imaging system to design tissue compensators for radiotherapy of the breast *The Use of Computers in Radiation Therapy: Proc. 11th Conf.* ed A R Hounsell *et al* (Manchester: ICCR) pp 118–9

Evans P M, Johnson U, Shentall G, Hansen V N, Helyer S J, Yarnold J R and Swindell W 1996 Evaluation of a technique for designing compensators using electronic portal imaging *Proc. 4th Int. Workshop on Electronic Portal Imaging (Amsterdam, 1996)* (Amsterdam: EPID Workshop Organisers) p 48

Ezz A, Munro P, Porter A T, Battista J, Jaffray D A, Fenster A and Osborne S 1991 Daily monitoring and correction of radiation field placement using a video-based portal imaging system: a pilot study *Int. J. Radiat. Oncol. Biol. Phys.* **22** 159–65

Fiorino C, Uleri C, Cattaneo G M and Calandrino R 1995 On-line portal dosimetry by a diode linear array (Proc. ESTRO Conf. (Gardone Riviera, 1995)) *Radiother. Oncol.* **37** Suppl. 1 S20

Gall K P, Verhey L J and Wagner M 1993 Computer-assisted positioning of radiotherapy patients using implanted radiopaque fiducials *Med. Phys.* **20** 1153–9

Galvin J M, Sims C, Dominiak G and Cooper J S 1995 The use of digitally reconstructed radiographs for three-dimensional treatment planning and CT simulation *Int. J. Radiat. Oncol. Biol. Phys.* **31** 935–42

Gildersleve J, Dearnaley D P, Evans P M, Law M, Rawlings C and Swindell W 1994a A randomised trial of patient repositioning during radiotherapy using a megavoltage imaging system *Radiother. Oncol.* **31** 161–8

Gildersleve J, Dearnaley D P, Evans P M, Morton E J and Swindell W 1994b Preliminary clinical performance of a scanning detector for rapid portal imaging *J. Clin. Oncol.* **6** 245–50

Gildersleve J, Dearnaley D P, Evans P M and Swindell W 1995 Reproducibility of patient positioning during routine radiotherapy as assessed by an integrated megavoltage imaging system *Radiother. Oncol.* **35** 151–60

Gilhuijs K and van Herk M 1993 Automatic on-line inspection of patient setup in radiation therapy using digital portal images *Med. Phys.* **20** 667–77

Gore J, Kang Y and Schulz R J 1984 Measurement of radiation dose distribution by nuclear magnetic resonance (NMR) imaging *Phys. Med. Biol.* **29** 1189–97

Hansen V N 1996 private communication

Hansen V N and Evans P M 1994 Transit dosimetry *Rad Mag.* **20**(232) 33–4

Hansen V N, Evans P M, Helyer S J, Yarnold J R and Swindell W 1997 The implementation of compensation in radiotherapy of the breast: MLC intensity modulation and physical compensators *Radiother. Oncol.* at press

Hansen V N, Evans P M, Knight R and Swindell W 1995a Evaluation of transit dosimetry by phantom study *Proc. Röntgen Centenary Congress 1995 (Birmingham, 1995)* (London: British Institute of Radiology) p 476

Hansen V N, Evans P M and Swindell W 1993 Transit dosimetry using an on-line imaging system *Proc. BIR Conf. Megavoltage Portal Imaging (London 1993)* (London: BIR) *Br. J. Radiol.* **66** 1083

—— 1994a Transit dosimetry—computer generated dose images for verification *The Use of Computers in Radiation Therapy: Proc. 11th Conf.* ed A R Hounsell *et al* (Manchester: ICCR) pp 116–7

—— 1994b The role of EPI in the improvement of breast radiotherapy *Proc. 3rd Int. Workshop on Electronic Portal Imaging (San Francisco, 1994)*

—— 1995b Transit dosimetry (Proc. ESTRO Conf. (Gardone Riviera, 1995)) *Radiother. Oncol.* **37** Suppl. 1 S45

—— 1996a The application of transit dosimetry to precision radiotherapy *Med. Phys.* **23** 713–21

—— 1996b The implementation of EPID-designed compensators in radiotherapy of the breast *Proc. 4th Int. Workshop on Electronic Portal Imaging (Amsterdam, 1996)* (Amsterdam: EPID Workshop Organisers) p 12

Hansen V N, Swindell W and Evans P M 1996c Primary signal extraction from EPIDs for homogeneous objects *Proc. 4th Int. Workshop on Electronic Portal Imaging, Amsterdam (June, 1996)* (Amsterdam: EPID Workshop Organisers) p 44

Hazle J D, Hefner L, Nyerick C E, Wilson L and Boyer A L 1991 Dose response characteristics of a ferrous-sulfate-doped gelatin system for determining radiation absorbed dose distributions by magnetic resonance imaging (FeMRI) *Phys. Med. Biol.* **36** 1117–25

Heijmen B J M, Pasma K L, Kroonwijk M, Althof V G M, de Boer J C J, Visser A G and Huizenga H 1995a Portal dose measurement in radiotherapy using an electronic portal imaging device (EPID) *Phys. Med. Biol.* **40** 1943–55

Heijmen B J M, Pasma K L, Kroonwijk M, Visser A G and Huizenga H 1995b Prediction of portal images for *in vivo* dosimetry in radiotherapy *Med. Phys.* **22** 992

Heijmen B J M, Storchi P R M and Van Der Kamer J B 1994 A method for prediction of portal dose images *The Use of Computers in Radiation Therapy: Proc. 11th Conf.* ed A R Hounsell *et al* (Manchester: ICCR) pp 112–3

Heijmen B J M, Stroom J C, Huizenga H and Visser A G 1993 Application of a fluoroscopic portal imaging system with a CCD camera for accurate *in vivo* dosimetry *Med. Phys.* **20** 870

Heijmen B J M, Visser A G and Huizenga H 1992 *In vivo* dose measurements using an electronic portal imaging device: a feasibility study *Radiother. Oncol.* **24** Suppl. S25

Hensler E and Auer T 1996 A silicon diode array for portal verification *Proc. 4th Int. Workshop on Electronic Portal Imaging (Amsterdam, 1996)* (Amsterdam: EPID Workshop Organisers) p 51

Herman M G, Nicholson M and Cong Y S 1996 Exit dose verification using an electronic portal imaging device *Proc. 4th Int. Workshop on Electronic Portal Imaging (Amsterdam, 1996)* (Amsterdam: EPID Workshop Organisers) p 50

Hunt M A, Kutcher G J, Burman C, Fass D, Harrison L, Leibel S and Fuks Z 1993 The effect of positional uncertainties on the treatment of nasopharynx cancer *Int. J. Radiat. Oncol. Biol. Phys.* **27** 437–47

Hussain Z, Noble N, Betal D, Walton J, Roberts N and Whitehouse G 1995 Magnetic resonance image analysis of the variation in breast volume during the menstrual cycle *Proc. (Work In Progress) Röntgen Centenary Congress 1995 (Birmingham, 1995)* (London: British Institute of Radiology) p 19

Jaffray D A, Drake D G, Pan C and Wong J W (1996) Conebeam CT using a clinical fluoroscopic portal imager *Proc. 4th Int. Workshop on Electronic Portal Imaging (Amsterdam, 1996)* (Amsterdam: EPID Workshop Organisers) p 58

Jaffray D A, Yu C X, Yan D and Wong J W 1995 SL/CT: volumetric imaging on a medical linear accelerator for accurate radiotherapy (Proc. ESTRO Conf. (Gardone Riviera, 1995)) *Radiother. Oncol.* **37** Suppl. 1 S62

Johansson S A, Magnusson P, Olsson L E, Montelius A, Fransson A and Holmberg O 1995 Verification of dose calculations in external beam treatment planning using a gel dosimetry system (Proc. ESTRO Conf. (Gardone Riviera, 1995)) *Radiother. Oncol.* **37** Suppl. 1 S31

Källman P, Lind B, Iacobeus C and Brahme A 1989 A new detector for radiotherapy computed tomography verification and transit dosimetry *Proc. 17th Int. Congress Radiology (Paris, 1989)* (Paris: ICR) p 83

Kirby M C and Williams P C 1993a The use of an electronic portal imaging device for dosimetry and quality control measurements (Proc. 35th ASTRO Meeting) *Int. J. Radiat. Oncol. Biol. Phys.* **27** Suppl. 1 163

—— 1993b Measurement possibilities using an electronic portal imaging device *Radiother. Oncol.* **29** 237–43

Kutcher G J, Mohan R, Leibel S A, Fuks Z and Ling C C 1995 Computer controlled 3D conformal radiation therapy *Radiation Therapy Physics* ed A Smith (Berlin: Springer) pp 175–91

Lam K L, Ten-Haken R K, McShan D L and Thornton A F 1993 Automatic determination of patient setup errors in radiation therapy using spherical radio-opaque markers *Med. Phys.* **20** 1145–52

Leszczynski K W, Shalev S and Cosby N S 1991 The enhancement of radiotherapy verification images by an automated edge detection technique *Med. Phys.* **19** 611–21

Lewis D G, Morton E J and Swindell W 1988 Precision in radiotherapy: a linear accelerator based CT system in Megavoltage Radiotherapy 1937– 1987 *Br. J. Radiol. Suppl.* **22** 24

Lewis D G and Swindell W 1987 A megavoltage CT scanner for radiotherapy verification *The Use of Computers in Radiation Therapy* ed I A D Bruinvis *et al* (Amsterdam: Elsevier Science North-Holland) pp 339–40

Lewis D G, Swindell W, Morton E, Evans P and Xiao Z R 1992 A megavoltage CT scanner for radiotherapy verification *Phys. Med. Biol.* **37** 1985–99

Lirette A, Pouliot J, Aubin M and Larochelle M 1995 The role of electronic portal imaging in tangential breast irradiation: a prospective study *Radiother. Oncol.* **37** 241–5

Lovelock D M, Mohan R and Obrien J 1994 The interactive generation of digitally reconstructed radiograms for use in a 3D planning system *Med. Phys.* **21** 886

Maryanski M J, Gore J C, Kennan R P and Schulz R J 1993 NMR relaxation enhancement in gels polymerised and cross-linked by ionising radiation: a new approach to 3D dosimetry by MRI *Magn. Res. Imag.* **11** 253–8

Maryanski M J, Gore J C, Schulz J and Ranade M 1996 Three-dimensional dose distributions from optical scanning of polymer gels *Med. Phys.* **23** 1069

Maryanski M J, Schulz R J, Ibbott G S, Gatenby J C, Xie J, Horton D and Gore J C 1994 Magnetic resonance imaging of radiation dose distributions using a polymer-gel dosimeter *Phys. Med. Biol.* **39** 1437–55

McNutt T R, Mackie T R, Reckwerdt P J and Paliwal B R 1996 Modeling and reconstruction of dose distributions in an extended phantom from exit dose measurements *Proc. 4th Int. Workshop on Electronic Portal Imaging (Amsterdam, 1996)* (Amsterdam: EPID Workshop Organisers) p 46

Midgley S M, Dudson J F and Millar R M 1996 A feasibility study for the use of megavoltage photons and a commercial electronic portal imaging area detector for beam geometry CT scanning to obtain 3D tomographic data sets of radiotherapy patients in the treatment position *Proc. 4th Int. Workshop on Electronic Portal Imaging (Amsterdam, 1996)* (Amsterdam: EPID Workshop Organisers) p 60

Mosleh-Shirazi M A, Swindell W and Evans P M 1994a A combined electronic portal imaging device and large-volume megavoltage CT scanner *Proc. 3rd Int. Workshop on Electronic Portal Imaging (San Francisco, 1994)* (San Francisco: EPID Workshop Organisers)

—— 1994b Development of a large-volume, megavoltage computed tomography scanner for radiotherapy verification *Br. J. Radiol.* **67** WIP 2

—— 1995 A combined 3D megavoltage CT scanner and portal imager for treatment verification in radiotherapy *Proc. Röntgen Centenary Congress (Birmingham, 1995)* (London: British Institute of Radiology) p 475 (Winner of the IPSM Röntgen Poster Prize)

Munro P 1995 Portal imaging technology: past, present and future (*Innovations in treatment delivery*) *Semin. Radiat. Oncol.* **5** 115–33

Nakagawa K, Aoki Y, Akanuma A, Onogi Y, Karasawa K, Terahara A, Hasezawa K and Sasaki Y 1991 Development of a megavoltage CT scanner using linear accelerator treatment beam *J. Japan. Soc. Therap. Radiol. Oncol.* **3** 265–76

Nakagawa K, Aoki Y, Onogi Y, Terahara A, Sakata A, Muta N, Sasaki Y and Akanuma A 1994 Real time beam monitoring in dynamic conformation radiation *The Use of Computers in Radiation Therapy: Proc. 11th Conf.* ed A R Hounsell *et al* (Manchester: ICCR) pp 18–9

Oelfke U, Lam G K Y and Atkins M S 1996 Proton dose monitoring with PET: quantitative studies in Lucite *Phys. Med. Biol.* **41** 177–96

Olsson L E, Fansson A, Ericsson A and Mattsson S 1990 MR imaging of absorbed dose distributions for radiotherapy using ferrous sulphate gels *Phys. Med. Biol.* **35** 1623–32

Papanikolaou N and Mackie T R 1994 The extended phantom concept: modelling the treatment machine modifiers and portal imaging dosimetry system using a convolution code *Med. Phys.* **21** 877

Papanikolaou N, Mackie T R and Gehring M 1995 A convolution based algorithm for dose computation in radiation therapy *Medizinische Physik 95 Röntgen Gedächtnis-Kongress* ed J Richter (Würzburg: Kongress) pp 254–5

Podgorsak M B and Schreiner L J 1992 Nuclear magnetic relaxation characterization of irradiation Fricke solution *Med. Phys.* **19** 87–95

Pouliot J, Beaudoin L, Blais R, Letourneau D and Tremblay D 1994 Application of digital portal imaging systems for the evaluation of missing tissue compensators *Med. Phys.* **21** 964

Prasad P V, Nalcioglu O and Rabbani B 1991 Measurement of three-dimensional radiation dose distribution using MRI *Radiat. Res.* **28** 1–13

Roback D M and Gerbi B J 1995 Evaluation of an electronic portal imaging device for missing tissue compensator design and verification. *Med. Phys.* **22** 2029–34

Rodger A 1995 Tangential breast irradiation *Br. J. Radiol.* **68** 936

Rudin B-I, Lind B K, Iacobaeus C, Lundgren A, Kihlen B and Brahme A 1994 A detector array for radiotherapeutic computed tomography, transit dosimetry and 3D treatment verification (Proc. World Congress on Medical Physics and Biomedical Engineering (Rio de Janeiro, 1994)) *Phys. Med. Biol.* **39A** Part 1 488

Scheffler A, Aletti P, Wolf D, Pennequin J C and Diller M L 1995 Correlation of portal images using fast-Fourier transform (Proc. ESTRO Conf. (Gardone Riviera, 1995)) *Radiother. Oncol.* 37 Suppl. 1 S34

Schulz R J, de Guzman A F, Nguyen D B and Gore J C 1990 Dose response curves for Fricke-infused gels as obtained by nuclear magnetic resonance *Phys. Med. Biol.* **35** 1611–22

Shalev S 1994a Progress in the evaluation of electronic portal imaging— taking one step at a time *Int. J. Radiat. Oncol. Biol. Phys.* **28** 1043–5

—— 1994b The design and clinical application of digital portal imaging systems (Proc. 36th ASTRO Meeting) *Int. J. Radiat. Oncol. Biol. Phys.* **30** Suppl. 1 138

—— 1995 Treatment verification using digital imaging *Radiation Therapy Physics* ed A Smith (Berlin: Springer) pp 155–73

Shentall G 1995 private communication (14 November 1995)

Simpson R G, Chen C T, Grubbs E A and Swindell W 1982 A 4 MV CT scanner for radiation therapy: the prototype system *Med. Phys.* **9** 574–9

Stucchi P, Conte L, Mordacchini C, Bianchi C, Monciardini M, Novario R and Cassani E 1995 The use of an electronic portal imaging device for exit dosimetry (Proc. ESTRO Conf. (Gardone Riviera, 1995)) *Radiother. Oncol.* **37** Suppl. 1 S21

Swindell W and Evans P M 1993 Scattered radiation in portal images *Proc. BIR Conf. Megavoltage Portal Imaging (London 1993)* (London: British Institute of Radiology); *Br. J. Radiol.* **66** 1082

—— 1996 Scattered radiation in portal images; a Monte-Carlo simulation and an empirical model *Med. Phys.* **23** 63–74

Swindell W, Simpson R G and Oleson J R 1983 Computed tomography with a linear accelerator with radiotherapy applications *Med. Phys.* **10** 416–20

Vynckier S, Derreumaux S, Richard F, Bol A, Michel C and Wambersie A 1993 Is it possible to verify directly a proton-treatment plan using positron emission tomography? *Radiother. Oncol.* **26** 275–7

Yarnold J R and Neal A J 1995 Authors reply *Br. J. Radiol.* **68** 936

Yin F F, Schell M C and Rubin P 1994 Input/output characteristics of a matrix ion chamber electronic portal imaging device *Med. Phys.* **21** 1447–54

Zakharchenko G S, Novicova N V and Suichmezov V S 1995 New succession of technological actions in external beam radiotherapy and process of densitive structure of material medium changing under it *Medizinische Physik 95 Röntgen Gedächtnis-Kongress* ed J Richter (Würzburg: Kongress) pp 162–3

Zhu Y, Jiang X-G and Van Dyk J 1995 Portal dosimetry using a liquid ion chamber matrix: dose response studies *Med. Phys.* **22** 1101–6

CHAPTER 5

CONVERTING 3D DOSE TO BIOLOGICAL OUTCOMES: TUMOUR CONTROL PROBABILITY (TCP) AND NORMAL-TISSUE-COMPLICATION PROBABILITY (NTCP)

5.1. TUMOUR CONTROL PROBABILITY: GENERAL CONSIDERATIONS

The basis of converting from a 3D distribution of dose into a tumour control probability (TCP) is a knowledge of the shape of the dose–response function, i.e. the predicted TCP for a known number of clonogenic cells irradiated to a given uniform dose. Including the effects of inhomogeneous dose is straightforward since the volume can be divided up into subvolumes small enough so that within these subvolumes the dose can be considered locally constant. When the TCP for these subvolumes TCP_i is computed, the overall TCP is simply

$$TCP = \prod_i TCP_i. \tag{5.1}$$

Sometimes TCP_i is called the voxel control probability (VCP_i) since the subdivision can be conveniently done using the voxels of a 3D dose distribution. Whilst a somewhat artificial concept, this is as if there were a tumour of the size of the voxel.

There is a remarkable paucity in reliable dose–response data for human tumours *in vivo*, mainly because not many trials in which the dose levels have been varied have been carried out and properly monitored. The reasons are fairly obvious. Clinical practice has established a fairly narrow dose range in which specific tumours can be treated for ethical reasons—it would be unethical to treat some patients with a low probability of cure—and limited by normal-tissue tolerances. Deacon reviewed the available literature in 1987 and concluded that 12 tumour types showed a varying dose-response.

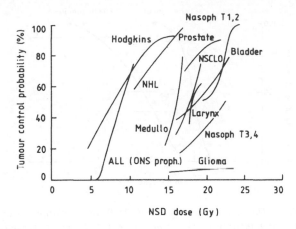

Figure 5.1. *A collection of curves which are approximately sigmoidal in shape showing the tumour control probability as a function of nominal standard dose (NSD) in Gy. It can be appreciated that some tumours are highly curable whereas others are not. The doses required for 50% TCP vary from around 6–8 Gy for some lymphomas to over 20 Gy for some epithelial tumours. Gliomas are so radioresistant that, not only is the 50% TCP not reached within this dose range, it is hard even to see whether there is a dose–response relationship. (From Steel 1993 after J M Deacon unpublished.)*

The basic dose–response curve is sigmoidal in shape (figure 5.1). TCP is not only a function of dose but also of tumour size. For example, figure 5.2 shows both the size and dose dependence of local TCP for skin cancers. There are several mathematical parametrisations. In order to be realistic and match the observed shapes, the effects of inter- and intra-tumour heterogeneity must be included.

5.2. MODELS FOR TCP WITH INTER- AND INTRA-TUMOUR HETEROGENEITY

Suit *et al* (1992) observed that heterogeneity of response to radiation of human tumours of the same type but across a patient population clearly exists. This heterogeneity of response determines the slope of the dose–response curve. The TCP versus dose curve also depends on tumour size (i.e. number of 'tissue rescue units' (TRU) or clonogenic cells) and the cellular radiation sensitivity. Suit *et al* (1992) list other factors as well, such as gender, haemoglobin concentration, cellular repair capacity etc. For example, normal tissues of patients with ataxia telangiectasia respond excessively to radiation.

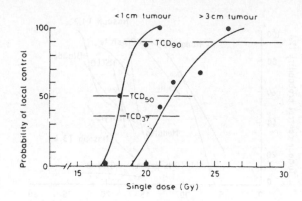

Figure 5.2. *The dependence of local TCP on radiation dose after single-dose radiotherapy of skin cancers of 0.5–1 cm in diameter and 3–4 cm in diameter. (Reprinted from Trott 1989, with kind permission from Elsevier Science-NL, Sara Burgerhartstraat 25, 1055 KV Amsterdam, The Netherlands.)*

Let us first consider the simplest situation in which all the tumour cells of a given type respond uniformly to dose both within any one tumour and across a patient population. Let us also consider uniform-dose irradiation of such a tumour. In the absence of heterogeneity between tumours, the slope of the dose–response curve is maximally steep and simply dependent on the Poisson statistics of killing cells. Any heterogeneity amongst tumours causes a flattening of the dose–response curve as we shall see later.

The parameter SF_2, for each tumour type, is the surviving fraction of TRUs after a uniform dose of 2 Gy. So, by definition, a formula for the TCP, after dose D, based on Poisson statistics of the survival of TRUs, is

$$TCP = \exp(-TRU\ SF_2^{(D/2)}) \qquad (5.2)$$

or, alternatively, using the notation of Nahum and Tait (1992)

$$TCP = \exp(-TRU \exp(-\alpha D)) \qquad (5.3)$$

where α is a radiosensitivity parameter, characteristic of a particular tumour type. The β term in the α–β model of cell kill (Wheldon 1988) can be ignored for fractionated radiotherapy. From equations (5.2) and (5.3) it follows that

$$SF_2 = \exp(-2\alpha). \qquad (5.4)$$

In this model the number of tissue rescue units $TRU = \rho V$ where V is the tumour volume and ρ is the density of clonogenic cells. Johnson *et al* (1995) confirm the proportionality of number of clonogenic cells to tumour volume.

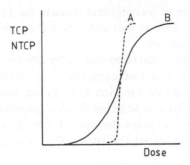

TCP
NTCP

Dose

Figure 5.3. *Curves of TCP and NTCP versus dose are observed to be sigmoidal in shape. So empirical equations model this shape. When a prediction of TCP or NTCP is made using a physical model it is found initially that the predicted shape of the curves is too steep (like curve A). Invoking a distribution of patient radiosensitivities across a population can yield instead more realistic shallower curves such as B.*

Another widely used characteristic of dose–response curves is the γ factor. The γ factor is related to the slope of the dose–response curve, i.e. $\gamma = D \times (\partial\mathrm{TCP}/\partial D)$. γ_{50} is the particular value where the TCP $= 0.5$ (50%) and $D = D_{50}$. It is easily shown, by differentiating equation (5.2) that, in the absence of any inter- or intra-tumour variability of radiation sensitivity,

$$\gamma_{50} = (\ln 2/2) \times (\ln \mathrm{TRU} - \ln\ln 2) \tag{5.5}$$

so when there are 10^7 TRU, $\gamma_{50} \simeq 6$. Suit *et al* (1992) showed a straight-line plot of γ_{50} versus $\ln(\mathrm{TRU})$.

In the language of Nahum and Tait (1992), using a TCP model (equation (5.3)) specified by TRU, α and D, it follows, by differentiating equation (5.3), that

$$\gamma_{50} = \alpha \ln 2 D_{50}/2. \tag{5.6}$$

Equations (5.5) and (5.6) are equivalent to each other through the use of equation (5.3).

However, for *in vivo* local control of real tumours it is observed that $\gamma_{50} \simeq 2$ and Suit *et al* (1992) interpreted this (i.e. the difference from $\simeq 6$) as a reflection of the heterogeneity of radiosensitivity response (figure 5.3). Nahum and Tait (1992) have made the same observation for their model (equation (5.3)). There is evidence for the heterogeneity of response.

The variability of radiosensitivity, SF_2, shows wide ranges for tumours of the same type and for normal tissues across a patient population. For normal tissue cells positive correlation has been found between skin fibroblast radiosensitivity and the sensitivity of late normal-tissue reactions (Peters *et al* 1995, Steel 1993). A study at the Christie Hospital in Manchester (West and Hendry 1995) showed the mean SF_2 determined by culture growth correlated with the success or failure of locoregional disease for 118 cases of stages 1 to 3 cervical carcinoma. This is indicative that dose should be tailored to the individual's radiosensitivity.

Inter-tumour heterogeneity can be modelled by allowing a spread of values of radiosensitivity. In the model of equation (5.2), heterogeneity of SF_2 is described by the coefficient of variation CV. In the model of Nahum and Tait (1992) (equation (5.3)) α is selected from a Gaussian distribution with mean α_0 and variation σ_α. σ_α plays the role of CV. To obtain TCP, a large number of values are considered for either SF_2 or α and a mean over a population is computed.

When this over-population averaging is added to equations (5.2) and (5.3) the predicted slopes become more like observed slopes. Suit *et al* (1992) distinguish between *inter-* and *intra-*tumour variability and show that *inter-*tumour variation in the CV for SF_2 affects the *slope* of the dose–response curve but *intra-*tumour variation primarily affects D_{50}.

Suit *et al* (1992) give the equations allowing calculation of TCP including variability in SF_2. Specifically the equation

$$TCP = \int_0^1 \exp(-TRU\ SF_2^{D/2}) f(s)\,ds \qquad (5.7)$$

gives TCP when SF_2 is distributed *inter-*tumour according to a function $f(s)$. Other equations are given for *intra-*tumour heterogeneity. Correspondingly the formulism of Nahum and Tait (1992) incorporates inter-patient heterogeneity by averaging over a Gaussian population of α values in equation (5.3) (see also equation (5.10)). When this is included in equations (5.2) and (5.3) the curves have a gentler slope (see also the companion Volume, Chapter 1).

Deasy (1995a) presents a model in which (among other features modelled) the TCP includes the effects of both inter-patient and intra-tumour variations in both the mean sensitivity and the coefficient of variation and shows that the flattening of the dose–response curve can be derived from either or both of these phenomena. Inclusion of inter-patient variability in the coefficient of variation of α differs from the approach of Suit *et al* (1992). Deasy (1995a) also includes the effects of distributions in tumour-doubling time. Unfortunately there are few data to tie down many of the parameters involved and this modelling can go little further at present than demonstrate potential influences on TCP. Simpler models with fewer variables have been matched to observed data (see section 5.3).

Zagars *et al* (1987) studied the change in slope of the dose–response curve when two factors are distributed in a population of tumours. The first factor is the number of tumour cells. The second factor is the surviving fraction. They considered three situations:

(i) number of tumour cells fixed; surviving fraction normally distributed;
(ii) vice versa;
(iii) both distributed normally.

The concept of stochastic fraction naturally arises. The equations are exactly the same as presented by Suit *et al* (1992) for inter-tumour variability when SF_2 is varied. Zagars *et al* (1987) did *not* consider intra-tumour variability.

Equations (5.2) and (5.3) have been discussed, considering the dose D and the density of TRUs to be uniform. When either or both the dose and clonogenic cell densities are non-uniform, these equations are used instead for small subvolumes in which they are locally constant and the overall TCP is then given by equation (5.1).

5.3. MECHANISTIC PREDICTION OF TCP

As interest continues to grow in predicting TCP and basing optimisation of the physical basis of radiotherapy on biological outcome, it is increasingly important to have a reliable way to predict TCP. To date the methods to do so have been either (i) empirical, for example the Logit equation for the sigmoidal shape, or (ii) mechanistic, based on biological considerations. The latter are inherently more satisfying.

A mechanistic equation for predicting TCP when a volume containing N clonogenic cells is uniformly irradiated to dose D is

$$TCP = \exp[-N \exp(-\alpha D - G\beta D^2)] \tag{5.8}$$

where α and β are the usual parameters in the linear-quadratic model and G is a quantity ($\leqslant 1$) depending on the fractionation scheme (Brenner 1993). If the radiotherapy is given in a large number of 2 Gy fractions, then a model with just the α term is usually used because the effective slope of the cell-survival curve is very nearly given by the α term alone. In this case equation (5.8) reduces to equation (5.3). To yield a realistic slope TCP must be computed from equation (5.8) by averaging over a population of α values (see also equation (5.10)). For the moment let us continue to consider uniform irradiation of uniform-density clonogenic cells. At the end of this section is the most general formalism for non-uniform irradiation of non-uniformly irradiated cells with inter-patient averaging of radiosensitivities.

Nahum and Tait (1992), Webb and Nahum (1993) and Webb (1994a,b) have used a generalisation of equation (5.8), without the term in β (representing fractionated radiotherapy), and with averaging over a population of α values, in forms which include either or both an inhomogeneous dose distribution and an inhomogeneous distribution of clonogenic cells, but with the additional assumption

$$N = \left(\frac{\pi \rho}{6}\right) d^3 \tag{5.9}$$

that the number of clonogenic cells is proportional to the volume (here written for a spherical tumour of diameter d). ρ is the density of clonogenic cells. The main advantage of this equation is that the volume effect on TCP is both understandable and mechanistically predictable.

Whilst the assumption in equation (5.9) that the number of clonogenic cells to kill is proportional to the volume of the tumour is intuitively acceptable it has only recently been verified. Brenner (1993) has analysed a large number of TCP data and has provided a validation. Brenner (1993) studied four tumour groups for which dose-volume-TCP data were available: (i) breast tumours; (ii) malignant melanoma; (iii) tumour control of squamous cell carcinoma of the digestive tract; (iv) nodal control of squamous cell carcinoma of the digestive tract. 3D plots were formed of TCP (vertical axis) against dose and tumour diameter (two horizontal axes) (figure 5.4). These data were fitted to a TCP surface using the method of simulated annealing and equations (5.8) and (5.9) and allowing both α and ρ to be free parameters. *No* averaging over a population was done. The fits yielded the following values for α: (i) $\alpha = 0.061$ Gy^{-1}, (ii) $\alpha = 0.070$ Gy^{-1}, (iii) $\alpha = 0.118$ Gy^{-1}, (iv) $\alpha = 0.039$ Gy^{-1}. The values for $(\pi\rho/6)$ were: (i) 0.346, (ii) 69.2, (iii) 51.2, (iv) 0.060. Brenner (1993) presented the arguments which show that these results were statistically significant and the hypothesis that the number of clonogenic cells is proportional to volume could not be rejected. The conclusion then was that equations (5.8) and (5.9) are valid.

The implication is that with this knowledge radiobiologically based dose corrections for tumour volume effects can be made both for the treatment as a whole and for variations in tumour size from fraction to fraction. Brenner (1993) concluded that whereas other cellular effects such as oxygenation status and inter-cell communication may be important in other contexts, they were of lesser importance in determining the volume effect.

There was an outstanding difficulty, namely that the values for the two free parameters resulting from the best fits would appear to be quite small. It was concluded that this is probably due to inter-patient heterogeneity ignored in equation (5.8). The hypothesis was rejected that the low values are a consequence of using equation (5.9) because even lower values were obtained when the data were fitted with a volume-dependent 'constant'. In that values of α are generally lower for clinical data than for the corresponding *in vitro* experiments, small observed values for α might have been expected. However, since the number of tumour cells per unit volume is of the order $10^{8}-10^{9}$ cm^{-3} the observed values of $(\pi\rho/6)$ imply that the clonogenic (stem) cell density is only some $10^{-6}-10^{-10}$ of the tumour cell density. This clonogenic fraction is very small. Others (see Steel 1993, Steel *et al* 1987, Steel and Stephens 1983) have used much higher fractions. For example, Nahum and Tait (1992) assume there are 10^{7} clonogenic cells per cubic centimetre which implies a fraction 10^{-2} if the total number of tumour

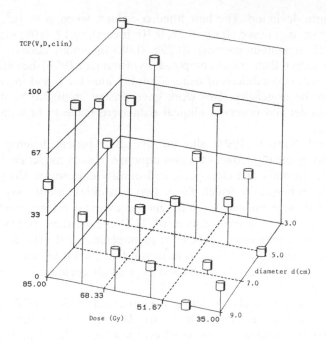

Figure 5.4. *Plot of the TCP (V, D; clin) data as a function of dose D in Gy and diameter d of breast tumour. Note that the data do not always exhibit the expected behaviour that TCP (V, D; clin) decreases with decreasing dose and increasing tumour diameter. For example, for tumours with diameter 7 cm, the TCP (V, D = 55 Gy; clin) > TCP (V, D = 65 Gy; clin). (Data replotted from Brenner 1993.)*

cells is 10^9 per cubic centimetre. Brenner argues that a fivefold increase in α would lead to a several orders of magnitude increase in the clonogenic cell fraction.

Webb (1994, 1995) has reanalysed all the data provided by Brenner (1993), fitting the data for each of the four tissue types to the Nahum and Tait (1992) model and using the same G factors and β/α ratio as Brenner (1993). This improved upon the work of Brenner by: (i) using equation (5.9) for the initial number of clonogenic cells and *not* allowing this to be a free parameter; (ii) including in equation (5.8) averaging α over a Gaussian distribution centred on a mean α_0 and randomised via a standard deviation σ_α. This incorporates the inter-patient heterogeneity of radiosensitivity.

A sequential fitting routine was used whereby α_0 and σ_α were varied in steps of 0.01 with α_0 in the range 0–0.8 and σ_α in the range 0–0.6. For each of the 4800 (α_0, σ_α) pairs the TCPs were computed for each of the data points given by Brenner (variable volume and dose for each tissue type). The RMS deviation between these fitted values and the clinical Brenner data was computed and that (α_0, σ_α) pair was selected which had the minimum deviation. The best fitted α values, when $\rho = 10^7$, became, labelled by ($\alpha, \sigma_{\alpha_0}$; tissue type): (0.27, 0.10; breast), (0.17, 0.08; melanoma), (0.305, 0.07; squamous tumour), (0.295, 0.08; squamous nodes). These α_0 values are larger than those computed by Brenner (1993) because of the new optimisation conditions of using a larger value for ρ and incorporating inter-patient heterogeneity. This work gave greater credibility to the model since the model and observed clinical data were made to fit with realistic parameters.

Webb and Nahum (1993) showed that inter-patient heterogeneity of radiosensitivity greatly reduces the consequence of even moderate deviations from dose uniformity and clonogenic cell-density uniformity. Using models of spherical tumours in which these two quantities could vary radially, plots of TCP were created for different geometrical arrangements. Large values of σ_α lead to shallow curves of TCP against dose and because of the logarithmic dependence of TCP on clonogenic cell density and double-logarithmic dependence on dose, it turns out that even if the clonogenic cell density falls at the edges of tumours only a small decrease in dose can be tolerated for iso-TCP.

The work described above, fitting the observed clinical TCP data using a physically sound, mechanistic model for TCP, thus determining optimum radiosensitivity parameters, was based on the clinical data which described the irradiations as being of uniform stated dose to stated volumes. In other circumstances where the dose distribution is inhomogeneous and the DVH is known, the model, extended via the use of equation (5.1) and also including the possibility of inhomogeneous clonogenic-cell density, will give a quantitative measure of TCP. Including the averaging over inter-patient population of sensitivity α, we have explicitly:

$$\text{TCP} = (1/K) \sum_{i=1}^{K} \prod_{j=1}^{M} \exp[-\rho_j V_t f_j \exp(-\alpha_i D_j)]. \tag{5.10}$$

where the total volume V_t is subdivided into M subvolumes with clonogenic cell density ρ_j, dose D_j and fractional volume f_j. K values of α_i are sampled to model the inter-patient heterogeneity. Equation (5.10) averages over the set of values of TCP computed for each α_i selected from the Gaussian with mean α_0 and standard deviation σ_α. K should be large, of the order 10^4. Deasy (1995b,c) argues for the robustness of this model even at the voxel level.

5.4.   DVH REDUCTION TECHNIQUES AND FORMULAE FOR COMPUTATION OF NTCP

There is continual effort put into simplifying the vast number of data produced by 3D treatment planning. We have seen in sections 5.2 and 5.3

how dose in the tumour is converted to TCP. A 3D dose distribution in a normal organ (OAR) is often simplified into a DVH. There are two forms; the integral DVH gives the fractional volume V_i raised to dose D_i *or greater*; the differential DVH gives the fractional volume v_i raised to dose D_i. Thus a complicated 3D distribution is simplified to one of two types of 2D graph (see the companion Volume, Chapter 1).

There have been several methods suggested to further reduce such 2D graphs into a single number which can be further used to predict the normal-tissue-complication probability (NTCP) (Kutcher *et al* 1991). These are based on the concept of the volume effect of normal-tissue damage. Trott (1996) has, however, strongly criticised the concept of the volume effect. He argues that there are no data supporting the postulates of Emami *et al* (1991). Specifically his argument is that whereas the tolerance of the *patient* may depend on the irradiated volume, this does not imply a volume dependence of the *tissues* involved. He further argues that since an OAR may contain a variety of structures with different purposes (e.g. the heart) the simple specification of the volume of such an organ receiving a specific dose is oversimplistic because this ignores the *spatial location* of such dose. A specific dose may be harmless for one part of an organ but harmful for another part. Put another way, the same DVH may give quite different complications if distributed spatially in different ways. Trott (1996) argues that the main advantage of 3D treatment planning is the ability to firm up some of the presently shaky biological data entering the models. However, despite such criticisms, modellers press on with their attempts to understand biological outcome and to link it to dose.

Methods suggested to further reduce such 2D DVH graphs into a simple form are summarised here following Hamilton *et al* (1992). Let the integral DVH be described by the histogram steps (V_i, D_i), $i = 1, 2, \ldots, N$ where the lowest index represents the highest dose. Each method starts with the histogram step furthest to the right on the graph (V_1, D_1; corresponding to the highest dose) and proceeds one step at a time until the histogram is reduced to a single step (figure 5.5).

The methods reduce the integral DVH to an effective dose D_{eff} to the full volume $V_N = 1$ (Lyman and Wolbarst 1987, 1989).

(i) The 'volume-weighted-dose' (VWD) algorithm is

$$D_i' = \frac{V_{i-1}}{V_i} D_{i-1}' + \left(1 - \frac{V_{i-1}}{V_i}\right) D_i \text{ for } i = 2, 3, \ldots, N. \quad (5.11)$$

In this and subsequent equations $D_1' = D_1$. After $(N - 1)$ sequential operations the resulting D_N' is the effective dose D_{eff} at full volume $V_N = 1$. The NTCP is then given by any functional form of NTCP($V_N = 1, D_{eff}$), for example the Lyman equations or the Logistic model.

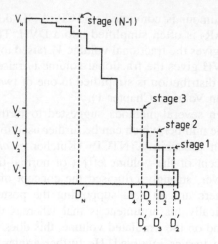

Figure 5.5. *The reduction of an N-step integral DVH to a single-step DVH with an effective dose D'_N at the full volume $V_N = 1$. The reduction is performed by $(N - 1)$ intermediate steps which at each step generate an effective dose D'_i. Four different algorithms are discussed in section 5.4. Note that in algorithms (iii) and (iv) the computation of D'_i involves the computation of an intermediate quantity D^*_i. This is to ensure that D'_i is never smaller than D_i which it could otherwise be mathematically.*

The Lyman equations (Lyman 1985; see the companion Volume) are:

$$\text{NTCP}(v, D) = 1/\sqrt{2\pi} \int_{-\infty}^{t} \exp(-x^2/2)\, dx \qquad (5.12)$$

with

$$t = [D - D_{50}(v)]/[m/D_{50}(v)] \qquad (5.13)$$

and

$$D_{50}(v) = D_{50}(v = 1)v^{-n}. \qquad (5.14)$$

Here D_{50} is the dose which results in a 50% complication probability for some specified complication or end-point; $D_{50}(v = 1)$ is the value appropriate to irradiating *all* the volume and $D_{50}(v)$ is the value for partial volume v irradiation. m is a parameter governing the slope of the dose–response curve and n describes the 'volume effect' (figure 5.6). The alternative Logistic formula (see also section 5.4.1) is

$$\text{NTCP}(v = 1, D) = \frac{1}{1 + (D_{50}/D)^k} \qquad (5.15)$$

with

$$k = 1.6/m \qquad (5.16)$$

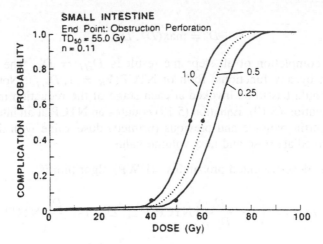

Figure 5.6. *A typical complication probability plot versus dose for necrosis of the small bowel for whole, two-thirds and one-third uniform irradiation. The dots represent clinical tolerance data and the curves represent fits with equation (5.12). We see the typical pattern that as the volume of tissue irradiated decreases, the NTCP curve shifts to the right on the dose axis. (Reprinted by permission of the publisher from Kutcher and Burman 1989; © 1989 Elsevier Science Inc.)*

giving the relation between the parameter k and the shape parameter m in the Lyman model (see proof in section 5.4.1). Note the Logistic formula has one fewer fitting parameter than the Lyman model.

(ii) The 'volume-weighted-probability' (VWP) algorithm is

$$\text{NTCP}(V_i, D_i') = \frac{V_{i-1}}{V_i}\text{NTCP}(V_i, D_{i-1}') + \left(1 - \frac{V_{i-1}}{V_i}\right)\text{NTCP}(V_i, D_i)$$

$$\text{for } i = 2, 3, \ldots, N. \tag{5.17}$$

Again after all steps are completed, $D_{eff} = D_N'$ and the NTCP is then given by any functional form of $\text{NTCP}(V_N = 1, D_{eff})$. As clarified by Hamilton *et al* (1992), the first algorithm computes D_{eff} directly and finally invokes a formula for NTCP, whilst the second operates inversely, the NTCP is given by equation (5.17) at each step and the effective dose is computed inversely from the formula for $\text{NTCP}(V_i, D_i)$.

(iii) The 'dose-weighted-volume' (DWV) algorithm is

$$V_i' = \left(1 - \frac{D_i}{D_{i-1}'}\right)V_{i-1} + \frac{D_i}{D_{i-1}'}V_i \qquad \text{for } i = 2, 3, \ldots, N \tag{5.18}$$

D_i^* is defined by

$$\text{NTCP}(V_i, D_i^*) = \text{NTCP}(V_i', D_{i-1}') \tag{5.19}$$

and

$$D_i' = \max(D_i^*, D_i).\tag{5.20}$$

Again on completion of all steps the result is $D_{eff} = D_N'$. The NTCP is then given by any functional form of $\text{NTCP}(V_N = 1, D_{eff})$. Note that an NTCP formula has to be invoked at each stage of the reduction in order to operate equation (5.19). Equation (5.19) equates an NTCP at an intermediate (primed) partial volume and previous (primed) dose value with the NTCP at an intermediate dose and next volume value.

(iv) The 'dose-weighted-probability' (DWP) algorithm is

$$\text{NTCP}(V_i', D_{i-1}') = \left(1 - \frac{D_i}{D_{i-1}'}\right)\text{NTCP}(V_{i-1}, D_{i-1}') + \frac{D_i}{D_{i-1}'}\text{NTCP}(V_i, D_{i-1}')$$

$$\text{for } i = 2, 3, \ldots, N\tag{5.21}$$

where (as with algorithm (iii)) D_i^* is defined by

$$\text{NTCP}(V_i, D_i^*) = \text{NTCP}(V_i', D_{i-1}')\tag{5.19}$$

and

$$D_i' = \max(D_i^*, D_i).\tag{5.20}$$

Again on completion of all steps the result is $\text{NTCP}(V_N = 1, D_{eff})$ where $D_{eff} = D_N'$.

(v) The 'integral probability model' (IPM), based on Schultheiss *et al* (1983), assumes that the NTCP is formed by considering that the probability of damage in each part of the histogram is independent. Hence the algorithm becomes

$$\text{NTCP}(V_N, D_{eff}) = \text{NTCP}(V_N, D_N)\tag{5.22}$$

hence

$$\begin{aligned}\text{NTCP}(V_N, D_{eff}) =\ &1 - [(1 - \text{NTCP}(V_1, D_1)) \times (1 - \text{NTCP}(V_2 - V_1, D_2))\\ &\times (1 - \text{NTCP}(V_3 - V_2, D_3)) \times \ldots\\ &\times (1 - \text{NTCP}(V_N - V_{N-1}, D_N))].\end{aligned}\tag{5.23}$$

Note that since

$$v_i = V_i - V_{i-1}\tag{5.24}$$

this algorithm essentially follows from the *differential* rather than the *integral* DVH.

Since the five algorithms are all based on slightly different assumptions, one would not expect them to give exactly the same predictions. However, as Niemierko and Goitein (1991) have shown, (the proof is reproduced in the companion Volume, Chapter 1), the IPM and the VWP model approximate to each other to first-order expansion. Hamilton *et al* (1992) analysed a series of DVHs showing that, with the exception of the VWD model all five algorithms gave similar results and specifically the IPM and the VWP model gave almost the same numerical results.

Kutcher and Burman (1989) have presented a sixth technique (KB) for computing NTCP. The differential DVH is reduced according to

$$v_{eff} = \sum_i v_i (D_i/D_1)^{1/n} \tag{5.25}$$

Niemierko and Goitein (1991) showed this is also equivalent to the IPM (provided one sets the value of k in the equation (5.15) to $k = 1/n$) and so this also establishes that the KB technique is equivalent to the VWP model for small NTCP.

The proof of equivalence of the IPM and the KB algorithm is as follows. The proof makes use of the Logistic form of the NTCP formula (equation (5.15)). Using the IPM model, the NTCP for an inhomogeneously irradiated OAR can be expressed as

$$\text{NTCP}_{inhom} = 1 - \prod_{i=1}^{N} [1 - \text{NTCP}(v_i, D_i)] \tag{5.26}$$

which becomes (see proof in the companion Volume)

$$\text{NTCP}_{inhom} = 1 - \prod_{i=1}^{N} [1 - \text{NTCP}(1, D_i)]^{v_i}. \tag{5.27}$$

With the DVH reduced to a single effective volume v_{eff} at maximum dose $D_{max} = D_1$ the NTCP could alternatively be written

$$\text{NTCP}_{inhom} = 1 - [1 - \text{NTCP}(v_{eff}, D_1)] \tag{5.28}$$

which applying the same rule as above becomes

$$\text{NTCP}_{inhom} = 1 - [1 - \text{NTCP}(1, D_1)]^{v_{eff}}. \tag{5.29}$$

Equating the right-hand sides of equations (5.27) and (5.29) we have

$$\prod_{i=1}^{N} [1 - \text{NTCP}(1, D_i)]^{v_i} = [1 - \text{NTCP}(1, D_1)]^{v_{eff}}. \tag{5.30}$$

Taking logs of each side

$$\sum_{i=1}^{N} v_i \ln[1 - \text{NTCP}(1, D_i)] = v_{eff} \ln[1 - \text{NTCP}(1, D_1)]. \tag{5.31}$$

Hence

$$v_{eff} = \sum_{i=1}^{N} v_i \frac{\ln[1 - \text{NTCP}(1, D_i)]}{\ln[1 - \text{NTCP}(1, D_1)]}. \tag{5.32}$$

For small ($\ll 1$) NTCP (and since for small x, $\ln(1 - x) \simeq -x$), equation (5.32) becomes

$$v_{eff} = \sum_{i=1}^{N} v_i \frac{\text{NTCP}(1, D_i)}{\text{NTCP}(1, D_1)}. \tag{5.33}$$

Using the Logistic formula (equation (5.15)) when the doses D_i and D_1 are small with respect to dose D_{50}

$$v_{eff} = \sum_{i=1}^{N} v_i \frac{(D_i/D_{50})^k}{(D_1/D_{50})^k} \tag{5.34}$$

and so

$$v_{eff} = \sum_{i=1}^{N} v_i \left(\frac{D_i}{D_1} \right)^k \tag{5.35}$$

which is the KB algorithm (equation (5.25)) when $k = 1/n$.

The analyses which show the equivalence of certain algorithms when the NTCP is small (as it should be in clinical practice) provide very useful insight into the physical meaning of the algorithms. For example, the IPM algorithm has a precise biological basis (independent voxel probabilities) and this basis can then be translated onto some of the algorithms which came into being more as mathematical concepts.

5.4.1. *Equivalence of the Lyman expression and the Logistic formula for NTCP*

Both formulae generate sigmoidal curves. Their equivalence is demonstrated by matching the gradient at dose $D = D_{50}$. From equation (5.12), and differentiating, noting that the variable D appears in both the integrand and in the upper limit,

$$\left. \frac{d\,\text{NTCP}}{dD} \right|_{D=D_{50}} = \left[1/\sqrt{2\pi} \int_{-\infty}^{t} \frac{d}{dD} \exp(-x^2/2)\,dx \right] \Bigg|_{x=t(D_{50})}$$
$$+ \left[1/\sqrt{2\pi} \frac{dt(D)}{dD} \exp(-x^2/2) \right] \Bigg|_{x=t(D_{50})} \tag{5.36}$$

The first term is identically zero and the second yields

$$\frac{d\,NTCP}{dD}\bigg|_{D=D_{50}} = \frac{1}{\sqrt{2\pi}\,m\,D_{50}}. \tag{5.37}$$

From the Logistic algorithm (equation (5.15)),

$$\frac{d\,NTCP}{dD}\bigg|_{D=D_{50}} = \left[(-1)\left(\frac{1}{1+(D_{50}/D)^k}\right)^2 \times (-k)D_{50}^k D^{-k-1}\right]\bigg|_{D=D_{50}} \tag{5.38}$$

so

$$\frac{d\,NTCP}{dD}\bigg|_{D=D_{50}} = \frac{k}{4D_{50}}. \tag{5.39}$$

Equating equations (5.37) and (5.38) we have

$$k = \frac{4}{\sqrt{2\pi}\,m} \simeq 1.6/m. \tag{5.40}$$

5.5.   A MODEL FOR NTCP BASED ON PARALLEL TISSUE ARCHITECTURE

5.5.1.   Homogeneous irradiation

The popular way to compute NTCP is to use the Lyman equation (integrated normal or probit model) which gives NTCP (v_{eff}, D_{max}) in terms of the four parameters $D_{50}(v = 1)$, n, m and v_{eff} (at maximum dose D_{max} in the DVH.) Or, alternatively, the Logistic model may be used. This is especially useful on occasions as it is in closed form. These models are based on fitting a curve through experimental or clinical-judgmental data. They are not based on a biological mechanistic response model. In this section we review some attempts to create a mechanistic model of NTCP which would yield the observed dose–volume dependence.

When the OAR can be considered built of functional subunits (FSUs) arranged in *series*, the so-called 'critical element architecture', the NTCP for the whole organ NTCP(1, D) comprising N FSUs can be simply given in terms of the probability p of damage of any one FSU:

$$NTCP(1, D) = 1 - (1 - p)^N \tag{5.41}$$

the assumption being that damage to each of the FSUs is statistically independent. This model was extensively analysed by Niemierko and Goitein

(1991) and is discussed in detail along with the Lyman four-parameter model in the companion Volume, Chapter 1. It is applicable to structures such as the spinal cord, the bowel or indeed any chain-like structure whose performance is impaired if any link in the chain is weakened or broken. However, not all normal organs function this way. Many are 'parallel' structures meaning that, despite some damage to part of the organ, it remains viable.

Yorke *et al* (1993) have provided a model for computing the NTCP for whole and partial organ irradiation when the organ is modelled as a set of *parallel* FSUs. The NTCP is computed modelling radiation response and the mechanistic expression was derived from first principles as follows.

Let there be N FSUs in the whole organ. Let p be the probability of completely eradicating one FSU with dose D. Let there be a uniform dose D to fraction v of the whole organ. Then the probability of destroying m FSUs (out of N) is given by the binomial distribution

$$\text{prob}_{m,N} = b(vN, m)p^m(1 - p)^{vN-m} \tag{5.42}$$

where $b(vN, m)$ is the binomial coefficient. Now assume that if fewer than L FSUs are destroyed the complication is not manifested but that it only arises if more than L FSUs are eradicated. It follows that the NTCP for partial volume v irradiation at dose D is

$$\text{NTCP}(v, D) = 1 - \sum_{m=0}^{L-1} \text{prob}_{m,N}. \tag{5.43}$$

Since N and L will be large the binomial distribution can be replaced by a Gaussian distribution with mean

$$\langle M \rangle = vNp \tag{5.44}$$

and standard deviation

$$\sigma = \sqrt{vNp(1 - p)} \tag{5.45}$$

and so equation (5.43) becomes

$$\text{NTCP}(v, D) = 1 - \frac{1}{\sqrt{2\pi}\sigma} \int_0^L \exp\left(-\frac{(m - \langle M \rangle)^2}{2\sigma^2}\right) dm. \tag{5.46}$$

By changing the variables to

$$t = (m - \langle M \rangle)/\sigma \tag{5.47}$$

we have

$$\text{NTCP}(v, D) = 1 - \frac{1}{\sqrt{2\pi}} \int_{-\langle M \rangle/\sigma}^{U} \exp(-t^2/2) dt \tag{5.48}$$

where the upper limit of integration is

$$U = (L - \langle M \rangle)/\sigma. \tag{5.49}$$

Provided p is large compared with $1/vN$, the lower limit can be set to $-\infty$ so that altogether

$$\text{NTCP}(v, D) = 1 - \frac{1}{\sqrt{2\pi}} \int_{-\infty}^{U} \exp(-t^2/2)\, dt. \tag{5.50}$$

Equation (5.50) holds for $vN > L$. If $vN < L$, $\text{NTCP}(v, D) = 0$. Supposing there are N_c clonogenic cells per FSU, then the probability of inactivating an individual FSU with dose D is

$$p = \exp(-N_c \exp(-\alpha R D)) \tag{5.51}$$

where

$$R = \left(1 + \frac{\beta}{\alpha} d\right) \tag{5.52}$$

is the 'relative effectiveness factor' and d is the dose per fraction. α and β are the usual coefficients in the linear-quadratic (L-Q) model.

It may now be observed that this chain of equations enables NTCP to be calculated from basic biological data. If N_c, α, β, d and D are known, p follows from equations (5.51) and (5.52). Knowing N and v, $\langle M \rangle$ and σ follow from equations (5.44) and (5.45). Introducing the fraction

$$f = L/N, \tag{5.53}$$

(called the 'functional reserve' since no damage manifests itself if less than this fraction of FSUs is eradicated), U follows from f and equation (5.49). Hence $\text{NTCP}(v, D)$ follows from equation (5.50). It may be observed that once the upper limit is calculated, NTCP can be looked up in tables of the error function.

Specifically when $L = \langle M \rangle$ (i.e. at least the mean number of functional subunits must be killed to obtain the complication), we have $U = 0$ and $\text{NTCP}(v, D) = 0.5$. When $L = \langle M \rangle$, using equation (5.44),

$$p = f/v \tag{5.54}$$

so from equation (5.51)

$$-N_c \exp(-\alpha R D_{50,v}) = \ln(f/v) \tag{5.55}$$

where $D_{50,v}$ is the dose in Gy to partial volume v which gives $\text{NTCP}(v, D) = 0.5$.

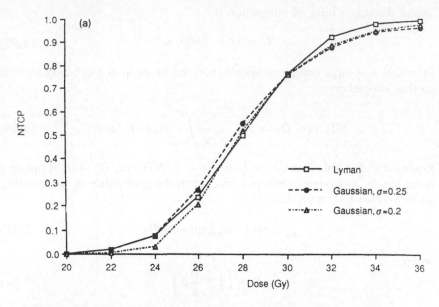

Figure 5.7. *The NTCP versus dose curve for whole-kidney irradiation. The open squares show the predictions of the Lyman function and the open triangles and solid circles show the predictions of the parallel model with two different distributions of functional reserve given by truncated Gaussian distributions. In this example no radiobiological broadening was assumed and $\alpha = 0.19 \text{ Gy}^{-1}$. (Reprinted from Yorke et al 1993, with kind permission from Elsevier Science Ireland Ltd, Bay 15K, Shannon Industrial Estate, Co. Clare, Ireland.)*

From the equation for U for full-volume irradiation

$$U = \frac{L - Np}{\sqrt{Np(1 - p)}} \qquad (5.56)$$

p may be had as a function of U by inversion of the quadratic in p. We obtain

$$p = \frac{f + U^2/2N \pm [(U^2/N)(f - f^2 + (U^2/4N))]^{0.5}}{1 + (U^2/N)}. \qquad (5.57)$$

Because p is a quadratic function of U and dose D is a double-logarithmic function of p it turns out that, for reasonable biological data, U (and hence $\text{NTCP}(v, D)$) is a very steep function of dose D. In fact the NTCP for lung changes from 5% to 95% over an interval of less than 1 Gy! The parallel architecture model predicts a greatly oversteep dose–response curve compared with clinical observation. The four-parameter fit of the Lyman model has (being a fit) no such difficulty. To overcome this problem Yorke *et al* (1993) have to invoke the familiar concept of population averaging to give a less-steep dose–response curve (figure 5.7).

Yorke *et al* (1993) point out that the collated NTCP data of Emami *et al* (1991) include many anecdotal data concerning partial volume irradiation; they are not based on the use of 3D dose distributions correlated to CT-defined anatomy. Hence Yorke *et al* (1993) choose to adjust the population distribution parameters to get a best fit to the low-NTCP full-volume-irradiated dose–response curve. Only when this fit is achieved do they go on to examine the implications for partial-volume irradiation.

It follows from equation (5.55) that the parameters α, N_c and R can be eliminated to give

$$\frac{D_{50,2/3} - D_{50,1}}{D_{50,1/3} - D_{50,1}} = \frac{\ln(\ln f/(\ln f + \ln 1.5))}{\ln(\ln f/(\ln f + \ln 3))}. \tag{5.58}$$

Since there are published values of $D_{50,1}$, $D_{50,2/3}$ and $D_{50,1/3}$ (Emami *et al* 1991), f can be deduced from equation (5.58). Yorke *et al* (1993) found the predicted values of the functional reserve for lung were much too small.

They found that by averaging over a population distribution for the radiosensitivity parameter α and for the functional reserve f, the parallel model could be brought into line with the observed dose-response of NTCP for full-volume irradiation but that the model did not always agree with the Lyman prediction of NTCP for partial-volume irradiation. For large partial volumes v and low doses the parallel model could be seen to almost agree with the Lyman power-law prediction but at high doses and small fractional volumes v the NTCP saturated at a low and sometimes zero value whilst the Lyman power-law equation predicted a much larger NTCP. This is described as a 'volume threshold effect'. The behaviour is in stark contrast to the predictions of a power-law model. The effect arises because of the population distribution of functional reserve.

The identity of an FSU is uncertain; the model assumes that they are small relative to the organ volume (typically there may be 10^4–10^6 FSUs per organ and an FSU might contain of the order 10^3 clonogenic cells). There have been several suggestions for the identity of the FSUs for the parallel organs lung and kidney.

5.5.2. Inhomogeneous irradiation

Jackson *et al* (1993a) considered how to calculate the NTCP when the irradiation is inhomogeneous and the FSUs are in parallel architecture. Once again it was assumed that, in order for a complication to be manifest, L FSUs from a total population in the organ of N FSUs must be destroyed. The NTCP was then defined as the probability that L or more FSUs are destroyed. Let prob_m be the probability that exactly m FSUs are destroyed. Then

$$\mathrm{NTCP} = 1 - \sum_{m=0}^{L-1} \mathrm{prob}_m = \sum_{m=L}^{N} \mathrm{prob}_m. \tag{5.59}$$

If the organ is inhomogeneously irradiated the probability of damage for each FSU p_i is not constant, but depends on the dose D_i to the FSU. At the outset it was stated that the dose dependence $p(D)$ is unknown. They assumed, however, that it exists and could be found in some functional form. Let there be n_i FSUs with dose D_i. It follows that the probability prob_m (and hence the NTCP) depends only on the risk histogram $n_i(p_i)$.

Within each dose bin the probability that m_i and only m_i FSUs, from the total n_i in that bin, die is given by the binomial distribution

$$B(n_i, m_i, p_i) = b(n_i, m_i)p_i^{m_i}(1 - p_i)^{(n_i - m_i)}. \qquad (5.60)$$

The probability $P(m_1, m_2, \ldots, m_q)$ that a given number die is the product over all q bins of the probabilities of death in each bin, i.e.

$$P(m_1, m_2, \ldots, m_q) = \prod_{i=1}^{q} B(n_i, m_i, p_i). \qquad (5.61)$$

Hence prob_m is given by

$$\text{prob}_m = \sum_{m_1=0}^{n_1} \sum_{m_2=0}^{n_2} \cdots \sum_{m_q=0}^{n_q} P(m_1, m_2, \ldots, m_q)\delta\left(\sum_{i=1}^{q} m_i - m\right) \qquad (5.62)$$

where the term $\delta(\sum_{i=1}^{q} m_i - M)$ constrains the sum of $m_1 + m_2 + \ldots m_q = m$ as required. This equation provides a direct way to compute prob_m and hence NTCP if p_i is known. However, a direct evaluation appears impossible because of the enormous number of terms in equation (5.62). Jackson *et al* (1993a) show that through a central-limit theorem a very good functional approximation is the Gaussian

$$\text{prob}_m = (1/\sqrt{2\pi\sigma^2})\exp[-(\langle m \rangle - m)^2/(2\sigma^2)] \qquad (5.63)$$

and

$$\text{NTCP} = (1/2)(1 + \text{erf}[(\langle m \rangle - L)/(\sqrt{2}\sigma)]), \qquad (5.64)$$

where

$$\langle m \rangle = \sum_{i=1}^{q} n_i p_i \qquad (5.65)$$

and

$$\sigma^2 = \sum_{i=1}^{q} n_i p_i (1 - p_i). \qquad (5.66)$$

It is seen that these equations provide an algorithmic computation of NTCP for parallel architecture and inhomogeneous irradiation once the risk histogram $n_i(p_i)$ and the dose dependence $p_i(D_i)$ are known. The former can be got from the 3D treatment plan but the latter is unknown and must be modelled.

Unfortunately, as was the case for the parallel architecture model with uniform irradiation, when p_i is estimated from the L-Q model of clonogenic cell kill, the predicted NTCP curve is too steep by several orders of magnitude and indeed is almost a step function. Jackson *et al* (1993a) have to invoke population averaging over the distribution of functional reserve L and radiosensitivity α to explain the observed shallower slope of the NTCP curve.

5.6. RANKING 3D TREATMENT PLANS USING PROXY ATTRIBUTES

There has always been interest in finding some simple way to say whether one treatment plan is better than another. The DVH is a form of simplifying 3D dose data. Calculating TCP and NTCP is another level of simplification. Shalev's 'regret' is another.

Jain *et al* (1994) have presented a flexible approach to determine a figure-of-merit (FOM) defined as

$$\text{FOM} = \prod_{i}^{attributes} (1 - (1 - \text{utility}_i) \times \text{weight}_i) \qquad (5.67)$$

where attributes have objectives representing clinical issues and there are i of them per plan. Clinical attributes (hard to define) are replaced by proxy attributes. For example, a proxy attribute could be the percentage of target volume receiving the prescription dose or the percentage of normal-tissue volume receiving the tolerance dose. The utility$_i$ of the ith proxy attribute is a number between 0 and 1 where 0 implies the objective of that attribute is not met and 1 implies it is met perfectly. The weight$_i$ of the ith proxy attribute is also a number between 0 and 1. A weight of 0 implies small importance to that attribute and 1 implies the attribute is important. The product in equation (5.67) is a convenient way to combine the contributions from several proxy attributes.

For example, suppose v_{pd} is the percentage of target volume that receives at least the prescription dose. We can see that if $v_{pd} = 100$, the objective is met so the utility is 1. If conversely, $v_{pd} \leqslant v_0$ the utility is 0 if v_0 is a volume below which an oncologist would reject the treatment plan for inadequate target coverage. A clinician would be asked to define v_{50} the volume receiving at least the prescription dose for which the utility is considered to be 0.5. A curve of utility versus v_{pd} might be sigmoidal rising with increasing v_{pd}.

Similar arguments are presented by Jain *et al* (1994) for building the curve of utility of the proxy attribute for normal tissue, the percentage v_{td} of volume receiving at least the tolerance dose. The utility curve now has to *fall* with increasing v_{td} since large values of v_{td} correspond to *low* utility.

The determination of the weight$_i$ will depend on the importance the clinician puts on any particular morbidity. A nice feature of the method is that individual clinicians can have different weights for the same attribute, preserving their traditional freedom of clinical judgement.

5.7.   ERROR ESTIMATION IN DETERMINING DVH, TCP AND NTCP

3D treatment plans are usually evaluated by computing the integral DVH and the TCP and NTCP using biological models and data. These tasks involve sampling the dose-space. Lu and Chin (1993a) have considered the problem of whether it is more accurate to sample with a regular Cartesian grid of equispaced points or instead a set of points randomly chosen in space. An earlier paper (Niemierko and Goitein 1990) suggested the latter was more accurate. The work of Lu and Chin (1993a) yielded the opposite answer. The result was found to be completely independent of the form of the 3D dose distribution and characteristic DVH. For example, to achieve the same accuracy of estimating an NTCP as with a thousand gridpoints, some 20 times more random points would be needed.

Lu and Chin (1993a) presented analytic expressions for the accuracy of determining a DVH, the TCP and the NTCP and then verified these with numerical examples. To understand the reasoning consider the following example.

Suppose the task is to estimate a volume of interest V. Let this volume be surrounded by a larger volume V_0. In the method of random sampling let N_0 uncorrelated points be generated within V_0 and count the number N falling within V. An estimate of the volume of interest is then

$$V = V_0(N/N_0). \tag{5.68}$$

The random point falling into the volume of interest obeys a binomial distribution with a probability

$$p = V/V_0 \tag{5.69}$$

so the variance of N (standard error of N) will be

$$(\Delta N)^2 = N_0 \frac{V}{V_0} \left(1 - \frac{V}{V_0}\right). \tag{5.70}$$

The relative error is thus

$$\frac{\Delta V}{V} = \frac{\Delta N}{N} = N^{-0.5}\sqrt{1 - \frac{V}{V_0}}.$$ (5.71)

The standard error is for the distribution of results of repeated calculations.

Now suppose instead the volume of interest is evaluated by a sampling method using equally spaced points on a rectangular Cartesian grid. The position *and orientation* of the grid are randomly decided before the calculation. Clearly points at the centre of the volume cannot contribute to the variance which is determined entirely by the number $N^{2/3}$ of points near the surface of the volume, since these may be either inside or outside the volume depending on how the grid is placed. As a result the relative error is

$$\frac{\Delta V}{V} = C\frac{\sqrt{N^{2/3}}}{N} = CN^{-2/3},$$ (5.72)

where C turned out from experiments to be $\simeq 0.65$ (see below).

Comparing equations (5.71) and (5.72) it is clear that the method of sampling with uniformly spaced grid points when $V/V_0 = 0.5$ is more accurate than with random gridpoints. The number N_{random} required to achieve the same accuracy as with N_{grid} grid points is

$$N_{random} \simeq N_{grid}^{4/3}.$$ (5.73)

For example, the random sampling points must be 20 times more numerous to achieve the same results that 10 000 grid points would achieve. There is a caveat that clearly if the volume of interest becomes comparable to the grid size the relative errors will become the same. Lu and Chin (1993a) also consider so-called quasi-random number sampling whereby points 'maximally avoid' each other by 'remembering' where preceding points were and distributing themselves randomly in the remaining space. Quasi-random sampling leads to relative errors which are between the two extremes of random and uniform grid sampling.

Lu and Chin (1993a) went on to show that the errors in estimating a DVH, TCP and NTCP all have the same functional form of requiring an estimate of means. Having thus reduced these problems conceptually to the same as the above simple problem, the same conclusion was invoked.

They then performed some numerical experiments. Firstly, the volume estimation problem was tackled. The results were found to be in good agreement with the above formulae (and this yielded the constant $C = 0.65$). Then they constructed some model DVHs and showed the gridpoint sampling was consistently better whatever the form of the DVH.

They used the NTCP formulae

$$NTCP = 1 - \prod_{i=1}^{N}[1 - NTCP(D_i)]^{1/N} \qquad (5.74)$$

with

$$NTCP(D) = \frac{1}{1 + (D_{50}/D)^k} \qquad (5.75)$$

where i labels the volume element where the dose is D_i, N is the number of elemental volumes, D_{50} is the dose giving 50% NTCP to the whole volume and $k = 10$ is the volume dependence. These equations were used with both sampling schemes to give predictions of the NTCP from sampled data. These were compared with the *known* NTCP for the analytically specified dose distribution to predict the error in estimating the NTCP. It was again found that the method of uniform sampling was some 20 times more efficient than random sampling.

A useful outcome of the study was the prediction that sampling on a regular grid of spacing 5 mm led to an inaccuracy of only 2% in DVH and resulting biological predictions. Niemierko and Goitein (1989) had also considered this.

The conclusions of Lu and Chin (1993a) have come in for some criticism. Jackson *et al* (1993b) pointed out that there can be problems with the use of a regularly spaced grid when the volume being sampled contains flat surfaces which may align with planes of the grid. This causes grid-sampling errors to be sensitive to both the shape of the volume and the grid orientation and hence be unreliable. Jackson *et al* (1993b) pointed out that a desirable property of any point-placement scheme for estimating volume is that the placement should be uncorrelated so samples are statistically independent and any errors add in quadrature. The regular grid does not satisfy this criterion. However, the extent to which this problem will arise in assessing treatment plans is largely unknown. Lu and Chin (1993b) replying, do not feel that the problem will arise much in practice because volumes generally have curved surfaces. Jackson *et al* (1993b), however, recommended use of the quasi-random sampling scheme which is uncorrelated and incorporates self-avoidance.

Niemierko and Goitein (1993) also criticised the work of Lu and Chin (1993a), pointing out that because of the correlated nature of uniform-grid sampling, the error in using grid sampling depends on the shape of the volume sampled and its relation to the grid. In fact in a numerical experiment estimating the volume of a cylinder inclined to the coordinate axes, they found a $N^{-0.52}$ law for one particular volume estimation instead of the $N^{-0.67}$ predicted by Lu and Chin for a sphere. Lu and Chin (1993b) regard this result as not inconsistent with their findings since an individual case will differ from the behaviour determined from the full spectrum of possible cases.

Niemierko and Goitein (1993) also stated that the accuracy of estimating a DVH depends only on the number of samples and not on the grid size, and errors in estimating TCP and NTCP depend also on the degree of dose inhomogeneity in the dose distribution and not simply on the grid size. They thus concluded that it is 'dangerous' to recommend that a particular grid size will always lead to a particular accuracy in predicting biological response. In fact they recommended the use of 400 quasi-random points per volume. Lu and Chin (1993b) consider that with respect to DVH, Niemierko and Goitein (1993) have misunderstood their work and thus come to the wrong conclusion. In summary Lu and Chin (1993b) stuck to the conclusions presented in Lu and Chin (1993a) and so there is clearly an on-going debate (van't Veld and Bruinvis 1994). It may be fortuitous that as computers become ever faster, there may soon be no need to sample at all and it will be possible to access all computed voxels in forming the DVH and derived biological parameters (Niemierko and Goitein 1995)

5.8.   CONTROVERSIES CONCERNING REPORTING VOLUMES

There has always been controversy concerning how to report the predictions of treatment planning. It is largely accepted that DVHs are generated in the PTV and converted to predictions of TCP through some models such as the ones discussed here. The PTV is constructed from the gross-tumour volume (GTV) and clinical target volume (CTV) via the procedures laid down by ICRU 50 (ICRU 1993). In summary the GTV, a medical concept, has had added to it a biological margin to account for microscopic spread, to create the CTV which is thus an anatomical concept. To the CTV is added a margin whose function is twofold. Firstly it accounts for the possibility of tissue movement (see Chapter 1). Secondly it accounts for setup error (see Chapter 4).

Recently Shalev (1995) has challenged this. He argues that it is reasonable to extend the CTV by a physical margin (which he calls the internal planning margin) to account for tissue movement. The addition of such a margin to the CTV, he calls the IPTV, the internal planning target volume. He argues that it is for the CTV alone that DVHs, and dose statistics should be reported and converted into biological predictions of TCP. He proposes that setup errors which can be studied, and hopefully reduced, by the use of EPIDs, should not be included in the definition of volumes. Instead these are beam concepts and margins should be added to beams using the BEV and possibly with margins changed as the course of treatment progresses. There is also the open question of whether it is reasonable to assume the density of clonogenic cells is constant throughout the CTV. It will be interesting to observe whether there is continued debate on these issues. Clearly individual clinics cannot be left to make up their own reporting schemes which would lead to nonuniform patterns of reporting and possibly wrong conclusions from clinical trials.

Finally, Shalev (1995) points out that tissue movement and indeed setup error can also affect the irradiation of OAR and thus the predictions of NTCP, as well as those of TCP, should include some measure of the uncertainties associated with these effects, at present rarely performed.

5.9. SUMMARY

A great deal of mathematical work has been done to try to generate understandable formulae to predict TCP and NTCP. TCP formulae derive naturally from the statistical basis of cell kill. Attempts have been made to create NTCP models on a similar physical basis. The former assume a tumour is perfectly controlled when *all* clonogenic cells die; the latter invoke the concept that some fraction of normal cells needs to die before a complication is recorded.

Both models, in their simplest form, get into difficulties predicting oversteep probability curves and, in both cases, averaging over a population of varying radiosensitivity has been invoked to provide a way of forcing the models to fit the observed clinical data. Because of this difficulty for NTCP, two empirical equations have been offered instead, being simply functional forms which fit the data.

When the dose distribution is inhomogeneous, the question arises of the best way to bin the data into a DVH which provides the starting point for computing TCP and NTCP when dose is non-uniform. There is considerable controversy in this area.

It should always be remembered that computing biological response is a probabilistic task and whilst many of the equations attain considerable complexity, the user should always bear in mind the limiting assumptions and not expect too much when the basis of the formalism is overstretched. In particular it must be remembered that the data for the dose response of NTCP were largely the result of pooled clinical experience. Since most therapies are by definition designed to generate small, of the order 5% or less, NTCP it is hardly surprising that there is doubt expressed over the credibility of those parts of the sigmoid response curve corresponding to much larger NTCP values. Equally, the dose-response of TCP is not well known for all tumours, as discussed in the introduction. Predicting biological response is an inexact science.

One might recall the maxim 'garbage in–garbage out' and become too pessimistic. Despite these difficulties it is becoming more and more common to report predictions of TCP and NTCP. The reasons are that these can assist the choice of plans when competing plans have overlapping DVHs; they can help to answer the question 'is optimisation worthwhile?'; they can assist with the setting of modal dose (the dose corresponding to 100% on a dimensionless 3D dose map) and they 'give a feel for' the ballpark area in which NTCP lies. The jury is probably still out on the fundamental question of the reliability of predicting probabilities.

REFERENCES

Brenner D J 1993 Dose, volume, and tumour control predictions in radiotherapy *Int. J. Radiat. Oncol. Biol. Phys.* **26** 171–9

Deasy J O 1995a Tumour control probability models and heterogeneities in clonogen radiosensitivities, initial clonogen numbers and tumour doubling times (private communication)

—— 1995b Clinical implications of alternative TCP models for nonuniform dose distributions *Abstracts of the 22nd PTCOG Meeting (San Francisco, 1995)* (Boston: PTCOG) p 14

—— 1995c Clinical implications of alternative TCP models for nonuniform dose distributions (Proc. ESTRO Conf. (Gardone Riviera, 1995)) *Radiother. Oncol.* **37** Suppl. 1 S6

Emami B, Lyman J, Brown A, Coia L, Goitein M, Munzenrider J E, Shank B, Solin L J and Wesson M 1991 Tolerance of normal tissue to therapeutic irradiation *Int. J. Radiat. Oncol. Biol. Phys.* **21** 109–22

Hamilton C S, Chan L Y, McElwain D L S and Denham J W 1992 A practical evaluation of five dose-volume histogram reduction algorithms *Radiother. Oncol.* **24** 251–60

ICRU 1993 Prescribing, recording and reporting photon beam therapy *ICRU report 50* (Bethesda: International Commision on Radiation Units and Measurement)

Jackson A, Kutcher G J and Yorke E D 1993a Probability of radiation induced complications for normal tissues with parallel architecture subject to non-uniform irradiation *Med. Phys.* **20** 613–25

Jackson A, Mohan R and Baldwin B 1993b Comments on 'Sampling techniques for the evaluation of treatment plans' [*Med. Phys.* **20** 151–61 (1993)] *Med. Phys.* **20** 1375–6

Jain N L, Kahn M G, Graham M V and Purdy J A 1994 3-D conformal radiation therapy: V. Decision-theoretic evaluation of radiation treatment plans *The Use of Computers in Radiation Therapy: Proc. 11th Conf.* ed A R Hounsell *et al* (Manchester: ICCR) pp 8–9

Johnson C R, Thames H D, Huang D T and Schmidt-Ullrich R K 1995 The tumour volume and clonogen number relationship: tumour control predictions based upon tumour volume estimates derived from computed tomography *Int. J. Radiat. Oncol. Biol. Phys.* **33** 281–7

Kutcher G J and Burman C 1989 Calculation of complication probability factors for non-uniform normal tissue irradiation: the effective volume method *Int. J. Radiat. Oncol. Biol. Phys.* **16** 1623–30

Kutcher G J, Burman C, Brewster L, Goitein M and Mohan R 1991 Histogram reduction method for calculating complication probabilities for three-dimensional treatment planning evaluations *Int. J. Radiat. Oncol. Biol. Phys.* **21** 137–46

Lu X-Q and Chin L M 1993a Sampling techniques for the evaluation of treatment plans *Med. Phys.* **20** 151–62

—— 1993b Further discussion on sampling methods—a response to Letters to the Editor [*Med. Phys.* **20** 1375–6 (1993)] and [*Med. Phys.* **20** 1377–80 (1993)] *Med. Phys.* **20** 1381–5

Lyman J T 1985 Complication probability as assessed from dose volume histograms *Rad. Res.* **104** S-13–S-19

Lyman J T and Wolbarst A B 1987 Optimisation of radiation therapy 3. A method of assessing complication probabilities from dose-volume histograms *Int. J. Radiat. Oncol. Biol. Phys.* **13** 103–9

—— 1989 Optimisation of radiation therapy 4. A dose-volume histogram reduction algorithm *Int. J. Radiat. Oncol. Biol. Phys.* **17** 433–6

Nahum A E and Tait 1992 Maximising tumour control by customized dose prescription for pelvic tumours (Proc. ART91 (Munich, 1991) (abstract book p 84)) *Advanced Radiation Therapy: Tumour Response Monitoring and Treatment Planning* ed A Breit (Berlin: Springer) pp 425–31

Niemierko A and Goitein M 1989 The influence of the size of the grid used for dose calculation on the accuracy of the dose estimation *Med. Phys.* **16** 239–47

—— 1990 Random sampling for evaluating treatment plans *Med. Phys.* **17** 753–62

—— 1991 Calculation of normal tissue complication probability and dose-volume histogram reduction schemes for tissues with a critical element architecture *Radiother. Oncol.* **20** 166–76

—— 1993 Comments on 'Sampling techniques for the evaluation of treatment plans' [*Med. Phys.* **20** 151–61 (1993)] *Med. Phys.* **20** 1377–80

—— 1995 Dose sampling for dose-volume displays and for TCP/NTCP calculations (Proc. ESTRO Conf. (Gardone Riviera, 1995)) *Radiother. Oncol.* **37** Suppl. 1 S6

Peters L J, Brock W A and Geara F 1995 Radiosensitivity testing *Proc. Röntgen Centenary Congress 1995 (Birmingham, 1995)* (London: British Institute of Radiology) p 73

Schultheiss T E, Orton C G and Peck R A 1983 Models in radiotherapy: volume effects *Med. Phys.* **10** 410–5

Shalev S 1995 On the definition of beam margins in radiation therapy *Proc.*

19th H Gray Conf.—Imaging in Oncology (Newcastle, 1995) (Newcastle: L H Gray Trust)

Steel G G 1993 The radiobiology of tumours *Basic Clinical Radiobiology for Radiation Oncologists* ed G G Steel (London: Arnold) pp 108–19

Steel G G, Deacon J M, Duchesne G M, Horwich A, Kelland L R and Peacock J H 1987 The dose-rate effect in human tumour cells *Radiother. Oncol.* **9** 299–310

Steel G G and Stephens T C 1983 Stem cells in tumours *Stem Cells: Their Identification and Characterisation* ed C S Potten (Edinburgh: Churchill Livingstone)

Suit H, Skates S, Taghian A, Okunieff P and Efird J T 1992 Clinical implications of heterogeneity of tumour response to radiation therapy *Radiother. Oncol.* **25** 251–60

Trott K R 1989 Relation between cell survival and gross endpoints of tumour response and tissue failure *The Biological Basis of Radiotherapy* ed G G Steel, G E Adams and A Horwich (Amsterdam: Elsevier) pp 65–76

—— 1996 Dose-volume effects *Proc. Symp. Principles and Practice of 3-D Radiation Treatment Planning (Munich, 1996)* (Munich: Klinikum rechts der Isar, Technische Universität)

van't Veld A A and Bruinvis I A D 1994 Accuracy in grid-based volume calculations *The Use of Computers in Radiation Therapy: Proc. 11th Conf.* ed A R Hounsell *et al* (Manchester: ICCR) pp 76–7

Webb S 1993 The effect on tumour control probability of varying the setting of a multileaf collimator with respect to the planning target volume *Phys. Med. Biol.* **38** 1923–36

—— 1994 Optimum parameters in a model for tumour control probability including inter-patient heterogeneity *Phys. Med. Biol.* **39** 2229–46

—— 1995 Optimum parameters in a model for tumour control probability including inter-patient heterogeneity (Proc. ESTRO Conf. (Gardone Riviera, 1995)) *Radiother. Oncol.* **37** Suppl. 1 S6

Webb S and Nahum A E 1993 The biological effect of inhomogeneous tumour irradiation with inhomogeneous clonogenic cell density *Phys. Med. Biol.* **38** 653–66

West C M L and Hendry J H 1995 Radiosensitivity testing for human tumours *Proc. Röntgen Centenary Congress (Birmingham, 1995)* (London: British Institute of Radiology) p 73

Wheldon T 1988 *Mathematical Models in Cancer Research* (Bristol: Institute of Physics)

Yorke E D, Kutcher G J, Jackson A and Ling C C 1993 Probability of radiation induced complications in normal tissues with parallel architecture under conditions of uniform whole or partial organ irradiation *Radiother. Oncol.* **26** 226–37

Zagars G K, Schultheiss T E and Peters L J 1987 Inter-tumour heterogeneity and radiation dose-control curves *Radiother. Oncol.* **8** 353–62

CHAPTER 6

PROGRESS IN PROTON RADIOTHERAPY

6.1. NEW FACILITIES

Proton therapy began over 40 years ago when Robert Wilson, disillusioned with the Manhattan Project, envisioned the use of high-energy-physics cyclotrons producing protons for medical radiotherapy. His background at Los Alamos ideally suited him for this change of direction. Raju (1994) has provided a fascinating history, largely from personal perspective, of the development of centres for proton therapy from that time to the present day.

The number of centres able to deliver radiotherapy with protons has been gradually increasing. There are now 18 centres in 10 countries (USA 5, former USSR 3, France 2, Japan 2, Belgium 1, Sweden 1, South Africa 1, Switzerland 1, UK 1, Canada 1) and there are 16 planned new or upgraded facilities (Sisterson 1994, 1996). The gestation time of new centres is lengthy and there is at least a five-year lead time on the new or upgraded centres. The present worldwide clinical experience is summarised in tables 6.1 and 6.2 (which update tables 4.2 and 4.3 of the companion Volume). The main source of these data are the excellent twice-yearly reports issued by Sisterson, at the Harvard Cyclotron Centre/Massachusetts General Hospital, called the *'Particles'* newsletter. The data in tables 6.1 and 6.2 are up to date as of July 1996 (Sisterson 1996). The Proton Therapy Cooperative Group (PTCOG) is the forum for the exchange of experience between centres with proton accelerators for medical purposes. The first meeting of PTCOG took place in January 1983 at Batavia, Illinois, since when, to date, there have been 24 meetings at different centres (Goitein 1995a, Sisterson 1996).

Why are there so few proton centres? There are many theories. Firstly, most of the present centres are medical facilities taking advantage of existing high-energy accelerator physics equipment, on which there has been limited time for treating patients and associated experimentation, and which have been largely static beam lines, without the facility to rotate the beam relative to the patient. Secondly, the birth of proton therapy coincided almost exactly with the development of megavoltage photon therapy which provided strong competition. The historical development and modern practice of proton radiotherapy has been recently reviewed by Raju (1994, 1995). Today there are a few purpose-built clinical proton therapy facilities. However, there is still the perception, real or artificial, that 'protons cost an order of magnitude more than photons' and given that photon radiotherapy itself is already expensive, the financial cost may be prohibitive.

Chu (1994), however, has argued that the real cost of proton therapy is actually *less* than (or certainly no more than) photon therapy when the economics are properly analysed. The basis of his argument is the following:

(i) the capital cost of a proton synchrotron is ten times that of an electron linac; but

(ii) a proton accelerator can serve several, maybe four therapy rooms;

(iii) the useful life of a synchrotron is two to three times that of an electron linac;

(iv) conformal proton therapy can be set up in half the time of conformal photon therapy;

Chu (1994) further argues that when local failures of photon therapy are taken into account, the cost of proton therapy for the same cancer-cure efficacy will be lower than that for photons. Clearly these calculations are hard to do with certainty but there is scope for a wider debate than simply inspecting initial capital costs. Not everyone will accept these premises and the problem of making reliable calculations of cost when one is considering prototype equipment bedevils the subject.

It has also been argued that the clinical need for protons has not been conclusively established and, of course, until there are such centres with proton facilities, there is a poor prospect of rapidly gaining the necessary clinical data to justify them. There is a 'Catch 22' situation here. It is also not sufficient to compare highly selected patients treated with protons with literature averages of photon-treated patients. In these circumstances it is possible to resort to comparisons of planning data although not everyone would accept this as sufficient justification for investment. The argument is that planning studies should, if successfully favouring proton therapy, spawn phase 1 and phase 2 clinical trials. Also the reliability of planning comparisons is questionable, unless the best conformal proton therapy is compared with the best conformal photon therapy. To date very few studies have done this (Lee 1995, Lee *et al* 1995).

The clinical indications for proton therapy have traditionally been where tumours cannot be otherwise treated, or which have low cure rates or high treatment complications with photon therapy. Most of the over 17 500 proton-treated patients worldwide have had tumours of the brain, brainstem in the region of the myelon, in the choroida of the eye or in other regions close to radiation-intolerant structures (Munkel *et al* 1993, Smith *et al*

Table 6.1. *Proton particle facilities for medical radiation therapy.*
(Table extended from the companion Volume, table 4.2 with data from Sisterson (1996).)

Installation	Location and country	Accelerator type	First date of therapy use	Energy
Lawrence Berkeley Laboratory, University of California	Berkeley, CA, USA	Synchrocyclotron	1955	910 MeV
Gustav Werner Institute	Uppsala, Sweden	Synchrocyclotron	1957/1988	185 MeV 100 MeV
Harvard Cyclotron Laboratory	Cambridge, MA, USA	Synchrocyclotron	1961	160 MeV
Joint Institute for Nuclear Research	Dubna, (former) USSR	Synchrocyclotron	1967/1987	70–200 MeV
Institute of Theoretical and Experimental Physics	Moscow, (former) USSR	Synchrotron	1969	70–200 MeV
Konstantinov Institute of Nuclear Physics	Gatchina, St Petersburg, (former) USSR	Synchrocyclotron	1975	1 GeV
National Institute of Radiological Sciences	Chiba, Japan	Sector-focused cyclotron	1979	70 MeV
Particle Radiation Medical Centre	Tsukuba, Japan	Synchrotron	1983	250 MeV
Swiss Institute for Nuclear Physics (Paul Scherrer Institute)	Villigen, Switzerland	Synchrotron and sector-focused cyclotron	1984	70 MeV 200 MeV
Clatterbridge Hospital	Clatterbridge, UK	Cyclotron	1989	62 MeV
Loma Linda University, California	Loma Linda, USA	Synchrotron	1991	70–250 MeV
Université Catholique de Louvain	Louvain la Neuve, Belgium	Isochronous cyclotron	1991	90 MeV
Centre Antoine Lacassagne	Nice, France	Cyclotron	1991	65 MeV
Centre de Protonthérapie	Orsay, France	Synchrocyclotron	1991	200 MeV
National Accelerator Centre	Faure, South Africa	Cyclotron	1993	200 MeV
Indiana Cyclotron Facility	Indianapolis, USA	Cyclotron	1993	200 MeV
UC Davis	Davis, CA, USA	Cyclotron	1994	67.5 MeV
TRIUMF	Canada	Cyclotron	1995	70–120 MeV

Table 6.2. *Clinical experience to date with proton therapy facilities.*
(Table extended from the companion Volume, table 4.3 with data from Sisterson (1996).)

Installation	Location and country	Patients treated	Year started
Lawrence Berkeley, Laboratory, University of California	Berkeley, CA, USA	30 by 1957	1955
Gustav Werner Institute	Uppsala, Sweden	73 by 1976	1957
(2nd installation)		81 by March 1996	1988
Harvard Cyclotron Laboratory	Cambridge, MA, USA	6785 by June 1996	1961
Joint Institute for Nuclear Research	Dubna, (former) USSR	84 by 1980	1967
(2nd installation)		40 by June 1996	1987
Institute of Theoretical and Experimental Physics	Moscow, (former) USSR	2838 by May 1996	1969
Konstantinov Institute of Nuclear Physics	Gatchina, St Petersburg Former (USSR)	969 by Dec 1995	1975
National Institute of Radiological Sciences	Chiba, Japan	86 by June 1993	1979
Particle Radiation Medical Centre	Tsukuba, Japan	462 by July 1995	1983
Swiss Institute for Nuclear Physics (Paul Scherrer Institute)	Villigen, Switzerland	2054 by Dec 1995	1984
Clatterbridge Hospital	Clatterbridge, UK	698 by June 1996	1989
Loma Linda University, California	Loma Linda, USA	2000 by July 1996	1991
Université Catholique de Louvain	Louvain la Neuve, Belgium	21 by Nov 1993	1991
Centre Antoine Lacassagne	Nice, France	636 by Nov 1995	1991
Centre de Protonthérapy	Orsay, France	673 by Nov 1995	1991
National Accelerator Centre	Faure, South Africa	130 by March 1996	1993
Indiana Cyclotron Facility	Indianapolis USA	1 by Dec 1994	1993
UC Davis	Davis, CA, USA	71 by May 1996	1994
TRIUMF	Canada	5 by Dec 1995	1995

Total proton beams worldwide: 17737 by July 1996

1994a). A French study recently concluded that proton therapy was two to three times more costly than conformal photon therapy and, whilst acceptable as routine treatment for uveal melanoma and skull-based tumours, should be an open question for other cancers (Fleurette *et al* 1995).

The proton centres group into two, those with low-energy protons (around 70 MeV) and those with high-energy protons (around 200–250 MeV). The former can be used only for tumours at small depths and are predominantly directed at the treatment of uveal melanoma. The latter, with greater penetration, can be used in comparable roles to high-energy photons. In the next nine sections only the newer purpose-built high-energy centres are reviewed as well as the UK Facility.

Some of the proposed centres have published elaborate proposals, for example the Italian TERA (Fondazione per Adroterapia Oncologica) project (Amaldi 1994, 1995, Amaldi and Silari 1994). The TERA project is centred around a proposed Italian network for hadrontherapy called RITA (Rete Italiana Trattamenti Adroterapici). It is intended that this will comprise a 'hub' hadron therapy centre (figure 6.1) and a number of smaller proton therapy centres which will be connected together through a telematics data link for sharing both skills and data (Amaldi 1994, 1995) (figure 6.2).

Plans are well advanced for the new Northeast Proton Therapy Centre in Boston, USA, serving 38 million people living within 250 miles of Boston (Smith *et al* 1994a,b). This will be built on the site formerly occupied by the Charles Street Jail (NPTC 1996). This centre will have an Ion Beam Applications (IBA) high-field cyclotron proton accelerator and is due to begin treatments at the end of 1998 (figure 6.3). The accelerator will have a fixed energy, 235 MeV, capable of delivering 300 nA continuously. An energy selection system, consisting of a degrader and energy analyser will permit rapid (2 s) energy changes. IBA are teamed with General Atomics (USA) and the accelerator is being built in collaboration with Sumitomo Heavy Industries of Japan. The new Northeast Proton Centre will have three treatment rooms, two with a gantry and the other with a fixed beamline for eye therapy (Goitein *et al* 1996). The gantry is a 'large-throw, in-plane isocentric gantry' with a 45° upward bend, then a number of quadrupole magnets followed by a 135° bending magnet (Smith *et al* 1996b). The gantry will provide an isocentric facility with the virtual source being 2.3 m from the isocentre (Flanz *et al* 1995). The facility will also be able to deliver intensity-modulated proton therapy using scanned pencil beams (Smith 1996a) in addition to the 'usual' passive scattering system (figure 6.4). The second isocentric gantry was procured in January 1996; the cyclotron is scheduled to arrive in February 1997 and the beam transport system in the Spring of 1997. With time allocated for commissioning, the patient treatments are expected to begin in the Autumn of 1998 (Smith 1996b).

A feature of the purpose-built proton-therapy centres is the availability of one or more rotating gantries. This increases the possibilities for directing the

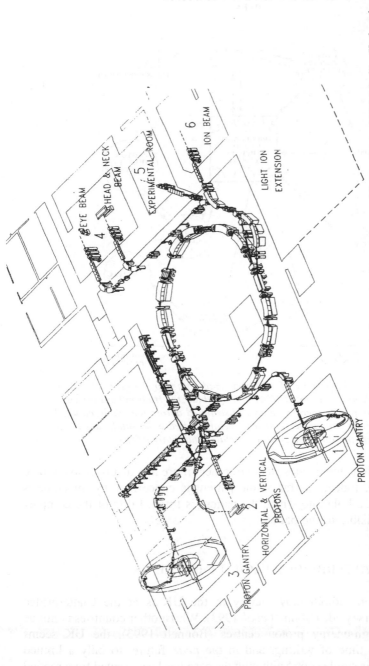

Figure 6.1. *The layout of the main underground floor of the bunker which contains the synchrotron of the proposed TERA Italian Hadrontherapy Centre. There are six treatment rooms shown, together with the injectors for H⁻ and for ions and the beam lines. The figure shows: (i) two proton rooms (1 and 3) which have an 11 m diameter rotating gantry for protons of energy not larger than 250 MeV; (ii) one room (2) which has one vertical and one horizontal beam for protons up to 250 MeV; (iii) one room (4) which is served by two horizontal proton beams, one for eye treatment (energy lower than 70 MeV) and the other for tumours of the head and neck (energy lower than 250 MeV). Also shown is a light-ion room and an experimental room. (From Amaldi and Silari 1994.)*

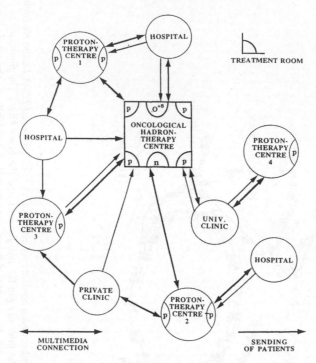

Figure 6.2.   *The purpose of the Italian Hadrontherapy Project is to build, by the year 2000, a network (RITA) which will link the Hadrontherapy Centre with various proton therapy centres (distributed across the whole national Italian territory) and many associated centres (sited in hospitals, university hospitals and private clinics). (From Amaldi and Silari 1994.)*

beam at the patient and creates an environment more like that of conventional photon therapy. Indeed, as far as the patient is concerned, the machine is little more than a hole into which the patient is fitted. Little of the complex equipment is visible to the patient.

6.2.   THE CLATTERBRIDGE FACILITY IN THE UK

The only proton radiotherapy facility in the UK is at the Clatterbridge Hospital on Merseyside (Bonnett *et al* 1993). Whilst other countries continue to plan for high-energy proton centres (Bonnett 1993), the UK seems destined at the time of writing, and in the *near* future, to only a limited expansion of this single centre although the case has been argued for a second facility (Dearnaley *et al* 1993, Nahum *et al* 1994) and there is interest among

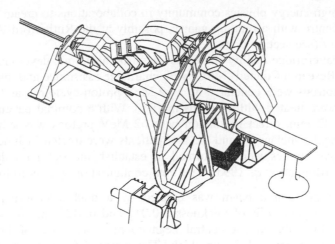

Figure 6.3. *A conceptual illustration of an isocentric gantry for proton irradiation from IBA. (Courtesy of IBA.)*

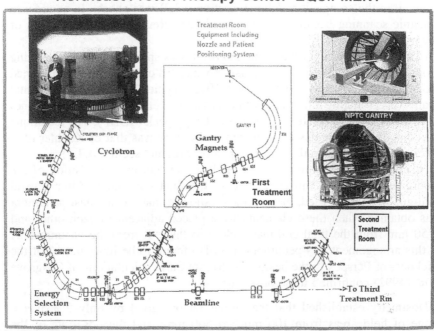

Figure 6.4. *A schematic diagram of the new Northeast Proton Centre at Boston. (From NPTC 1996 courtesy of Dr A Smith.)*

the UK high-energy physics community in collaborations to create a proton therapy centre with a rotating gantry, possibly at the Rutherford–Appleton Laboratory (see section 6.10).

The Clatterbridge facility started as a neutron beamline, developed by the Medical Research Council (MRC) with charitable assistance and installed in 1983. Neutrons were generated by a 62 MeV proton cyclotron and the first patients were treated with neutrons in 1986. With a nominal penetration in water of 32 mm, direct irradiation with 62 MeV protons was suitable for treating ocular melanoma and the first patients were treated in June 1989.

Extensive studies were performed to establish the optimum design of beamline (Bonnett *et al* 1993). The final configuration was as follows.

(i) The broad-area beam was created by a double-scattering system with two tungsten foils of thicknesses 0.017 and 0.0127 mm respectively, separated by 300 mm. A central beamstopper was made of brass rod, 6 mm in diameter and 7 mm thick. The distance from the first foil to the final collimator was 2200 mm and the first 700 mm of the beamline was under vacuum to minimise energy loss of the protons. A beam iris of 5 mm diameter immediately upstream from the first foil gave an acceptable penumbra (about 1.8 mm). Passive scattering (Koehler *et al* 1977; described in detail in the companion Volume, section 4.3.2) was chosen in contrast to dynamic scanning because it was considered safer despite a small loss of beam energy.

(ii) The Bragg peak of the elemental beam was spread out using a rotating propeller which comprised four Perspex vanes (in contrast to the two-vaned propellers described by Koehler *et al* 1975). Typically some ten elemental beams (necessitating ten different thicknesses of Perspex segments per vane) were superposed to create the spread out Bragg peak (SOBP). The technique for computing the angles of each step in the vane was that of Koehler *et al* 1975 (described in detail in the companion Volume, section 4.3.1).

(iii) The beam penumbra was strongly dependent on the position of the modulator and range shifter relative to the final collimator. After several configurations were tried, measurements showed that the smallest penumbra was obtained when these elements were placed adjacent to each other and 1250 mm from the final collimator close to the exit from the vacuum line. In this arrangement this penumbra was also found to be independent of the thickness of Perspex in the beam. The Monte Carlo calculations of Cosgrove *et al* (1992) confirmed these observations.

Dosimetry established that the output factor varied by only 2% over a range of field areas from 100 mm^2 to 500 mm^2 and the dose-rate of the beam was 25 Gy min^{-1}. Isodose distributions were measured in water for both wedged and unwedged fields.

At present the proton facility is used solely for ocular irradiation. Planning software, called EYEPLAN, developed at Clatterbridge, is used in other ocular irradiation centres in the world. To improve the dosimetry there are plans to incorporate CT and MRI images into the treatment-planning process at both Clatterbridge and Nice (Chauvel *et al* 1995). The proton beamline at Clatterbridge is at present being considered for upgrading to a higher energy (Bonnett 1993). The proposal is that a linac booster with a pulsed beam structure will be matched to the continuous beam from the cyclotron, accepting this beam and accelerating protons to an even higher energy.

6.3. THE CLINICAL PROTON FACILITY AT LOMA LINDA, USA

Although the Harvard Cyclotron Centre/Massachusetts General Hospital) (HCC/MGH) has been the pioneer of clinical proton therapy and accrued the largest patient numbers (currently 40.1% of the world total) (Munzenrider 1994), the world's first *purpose-built* clinical proton facility is at Loma Linda University Medical Centre (see the companion Volume, section 4.2). The facility comprises a beamline which can be directed to several treatment rooms (figures 6.5 and 6.6). The first patient (an ocular tumour) was treated on 23 October 1990. The treatment of brain tumours began in March 1991 and treatments using one of the rotating gantries on 26 June 1991.

By 1993 the following disease states were being regularly treated: melanomas, pituitary adenomas, acoustic neuromas, meningiomas, craniopharyngiomas, astrocytomas and other brain tumours, chordomas and chondrosarcomas, cancers of the head and neck, prostatic and other pelvic neoplasms, paraspinal tumours and soft-tissue sarcomas. Arteriovenous malformations are also treated. Over 50% of the treatments were for adenocarcinoma of the prostate. The second and third gantries were commissioned in spring 1994 and by summer 1994 all three rotating gantries were able to treat patients and a research room was also commissioned for radiobiological, physics and engineering research. The 'uptime' of the synchrotron and these delivery systems has been in excess of 98% (Slater 1995). The ultimate aim has been to have the beam rapidly switched from one room to another, to continuously vary the energy if required and to develop a scanning beam (Slater 1993, Preston 1994, Slater *et al* 1994).

Loma Linda can achieve 90–100 Gy m^{-1} on a 20 cm diameter field and most treatment times are less than 2 min per fraction, considerably less than the setup time (Coutrakon *et al* 1994a,b).

In December 1994 the Loma Linda University Medical Proton Centre signed a memorandum of agreement with the National Aeronautics and Space Administration (NASA) to use the facilities to study the effects on the body of solar protons emitted by the sun during solar flares. This novel collaboration with simulated space-capsule experiments does not involve human subjects nor does it interrupt the treatment of patients (Slater 1995).

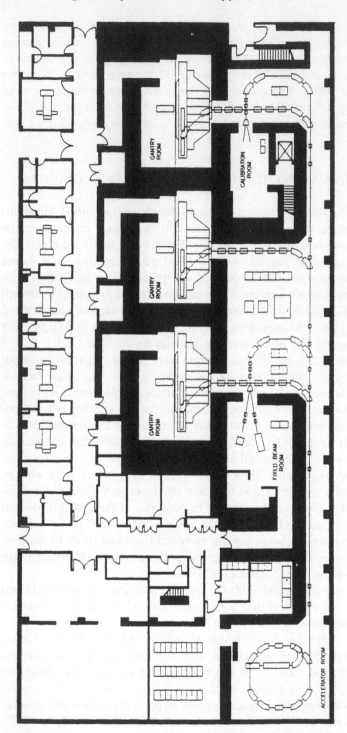

Figure 6.5. *A plan view of the Loma Linda University Medical Centre. The centre has rotating isocentric corkscrew gantries. (From Amaldi and Silari 1994.)*

Figure 6.6. *The proton medical accelerator at Loma Linda. The accelerator is a zero-gradient synchrotron utilizing a 2 MeV radiofrequency (RF) injector followed by a debuncher to reduce the beam's momentum spread at injection. This design was chosen for its compactness. The beam may be extracted at any energy between 70 and 250 MeV. (Reprinted from Coutrakon et al 1994b, with kind permission from Elsevier Science-NL, Sara Burgerhartstraat 25, 1055 KV Amsterdam, The Netherlands.)*

6.4. THE CLINICAL PROTON FACILITY AT ORSAY, PARIS, FRANCE

The proton facility at Orsay, Paris (figure 6.7) is based on the use of a synchrocyclotron, originally constructed in 1958 for physics research and reconstructed in 1977. For medical use it produces 200 MeV protons at up to 3 μA. The beam is split into two lines, the lower of which at 73 MeV, for opthalmic applications, is produced by a graphite degrader inserted into the beam transport system before the beam is split to the two treatment rooms. With the scattering system and for full modulation, the beam current at the isocentre is about 0.4 nA, giving a dose-rate of about 15 Gy m^{-1} for a beam diameter of 4 cm at the isocentre.

A 2 m optical bench has been installed at the end of the opthalmological line. Five brass collimators give a beam diameter of 4 cm at the isocentre and a last collimator is adapted for each patient using the EYEPLAN software from Clatterbridge. A rotating lucite wheel is used as modulator to produce a SOBP. One particularly interesting development is the attempt to sychronise the pulse generation from the accelerator with the rotation of the modulator to 'skip' some of the sectors of the modulator and thus to

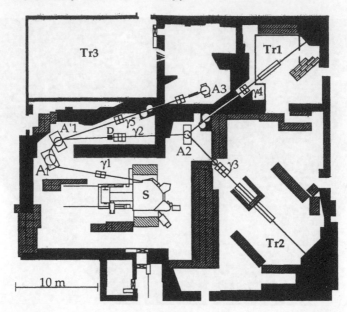

Figure 6.7. *The centre for proton therapy at Orsay, Paris. S is the synchrocyclotron, A are bending magnets, γ are quadrupole lenses, D is a degrader. The beam is split into two beamlines for treatment rooms Tr1 and Tr2. (Courtesy of Dr A Mazal.)*

produce several different modulations for each wheel (Baratoux *et al* 1995). A single-scattering system has been developed to produce a beam which is flat to 1%. This comprises an elliptical 0.1 mm-thick lead ring which only reduces the intensity by 21% and the energy by 0.5 MeV (Nauraye 1995, Nauraye *et al* 1995). For eye irradiations a bite-block is used for head immobilisation in conjunction with a robotic patient-positioning system with six degrees of freedom. The line can also accommodate a stereotactic frame and couch for irradiation of brain tumours (Mazal *et al* 1991, Mazal and Habrand 1993). The second treatment room will be ready for use in June 1997 (Mazal 1996).

Proton dosimetry has been performed mainly with ionisation chambers and Faraday cups. The centre has also performed many radiobiological studies (Loncol *et al* 1994), the main conclusion of which was that the relative biological efficiency (RBE) was around 1.15.

6.5.   THE CLINICAL PROTON FACILITY AT PSI, SWITZERLAND

Choroidal and uveal melanomas have been treated for many years using low-energy synchrotron-generated protons at the Swiss Institute for Nuclear Physics (Paul Scherrer Institute (PSI)) at Villigen. In 1993 it was reported

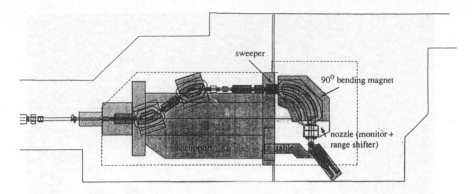

Figure 6.8. *A schematic cross section of the PSI gantry. (Reprinted from Pedroni 1994b, with kind permission from Elsevier Science-NL, Sara Burgerhartstraat 25, 1055 KV Amsterdam, The Netherlands.)*

that over 1500 patients had been treated, the second highest number worldwide at that time (the HCC/MGH, Boston group leading) (Blattmann and Munkel 1993). Further to this, a purpose-built medical proton radiation facility has been constructed based on a rotating gantry (figures 6.8, 6.9 and 6.10). Apart from at Loma Linda, where there are three gantries, this is the only purpose-built rotating-gantry proton facility in the world at present (Blattmann *et al* 1994, Pedroni 1994b, Pedroni *et al* 1995).

Protons with an energy of 200 MeV (but variable) are accelerated by a sector cyclotron and delivered via a 4 m-diameter rotating gantry. The method of spot scanning has been implemented (see section 6.11). The displacement of the spot (elementary Bragg peak) is performed preferentially with a sweeper magnet, secondly with a range shifter and finally by translations of the patient couch as the slowest and least frequent motion (Pedroni *et al* 1993a). Typically 10 000 spots can be delivered in a few minutes. The total weight of the rotating gantry is 120 tons ($\sim 1.22 \times 10^5$ kg) and the isocentre is engineered to lie within a sphere of 1 mm radius.

The facility has a single beam port and both horizontal beam line and isocentric treatments can be performed because the patient couch can also be moved with two rotations and three translations (Pedroni *et al* 1993b). The couch provides translations of 50 cm vertically, 50 cm laterally and 80 cm longitudinally with a precision of 0.1 mm. The beamline is nominally 1.5 m above the floor so the gantry rotates into a pit below floor level. The gantry also carries the detectors, proximal and distal to the patient for proton radiography (see section 6.13) as well as an x-ray imager at 90° to the beam. A dedicated CT unit has been modified to simulate the same patient positioning as at the time of therapy (Pedroni *et al* 1993d). The patient is set up outside the treatment room at the CT facility and then transferred supine into the treatment room lying in the custom mould. In order for precise registration between the CT room and the treatment room, the patient couch

Figure 6.9.   *The PSI gantry during assembly in December 1993. (Reprinted from Pedroni 1994b, with kind permission from Elsevier Science-NL, Sara Burgerhartstraat 25, 1055 KV Amsterdam, The Netherlands.)*

is supported by a removable transporter and is locked into position in each of the two facilities (Pedroni 1993). The first beam through the system was transmitted on 25 April 1994 (Pedroni 1994a).

Munkel *et al* (1994) have emphasised the need for precise immobilisation when spot scanning is in use. The method of immobilisation must be: (i) rigid against breakage; (ii) resistant to everyday use; (iii) comfortable and painless for patients; (iv) not claustrophobic; (v) inexpensive. The choice of immobilisation could depend on the mobility, age and state of health of the patient. It has been suggested that a small thoracic or abdominal compression can be well tolerated and lead to smaller movements of internal organs with breathing. Treatment planning for spot scanning must include an assessment of likely movements.

6.6.   THE CLINICAL PROTON FACILITY AT FAURE, SOUTH AFRICA

The first high-energy proton facility in the African Continent has been established at the South African National Accelerator Centre (NAC) at Faure (figure 6.11). The NAC is about 35 km from downtown Cape Town. It is the only centre in the world where both high-energy neutrons and high-energy protons are used for patient treatment. The main accelerator is a

Figure 6.10. *The PSI gantry during assembly with the 90° magnet and patient treatment console. (Reprinted from Blattmann et al 1994, with kind permission from Elsevier Science-NL, Sara Burgerhartstraat 25, 1055 KV Amsterdam, The Netherlands.)*

variable-energy separated-sector cyclotron, capable of accelerating protons to a 200 MeV horizontal beamline. Patients are referred to the NAC through one of the local university teaching hospitals, either the Groote Schuur Hospital (University of Cape Town) or the Tygerberg Hospital (University of Stellenbosch).

Initial treatments used crossfire techniques but an SOBP has been produced by a double-scatterer plus occluding ring system. The beam-delivery system was designed for a maximum field diameter of 10 cm. A rotating variable-thickness absorber, a propeller made of acrylic, spreads out the Bragg peak (Schreuder *et al* 1996a). For SOBP treatments additional acrylic degraders are inserted in the beam to achieve the desired range. The distance from the first set of steering magnets to the patient isocentre is in excess of 9 m (figure 6.12). The beam is controlled by two computerised feedback systems acting on data sent from two ionisation chambers to two sets of x–y magnets.

The patient is positioned using stereophotogrammetry, accurate to 0.5 mm (Jones *et al* 1995b). This works as follows. At the time of scanning the patient, small radio-opaque targets, 1 mm in diameter, are positioned on a custom-made mask which precisely fits the patient's head. From the scans, the 3D location of these markers is found with respect to some

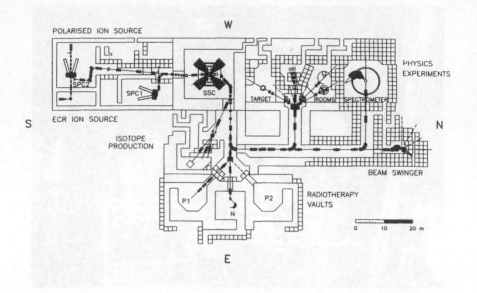

Figure 6.11. *A plan view of the NAC in South Africa. (From Amaldi and Silari 1994.)*

fixed coordinate point, usually the isocentre of the treatment. During the patient positioning stage, retroreflective markers are attached to these same locations, viewed by three charge-coupled TV cameras which capture video images of the markers. These images are then analysed to determine the locations of the retroreflective markers with respect to the isocentre. Where there are differences between the treatment position and the position at scanning, signals are sent to a computer-controlled chair to reposition the patient until the treatment position matches the scan position and the proton beam is collinear with the planned beam vector. Additionally, during treatment, the same information is used to close down the beam if the treatment position deviates more than is acceptable from the planned treatment. These techniques were spawned by developments in land surveying (Jones *et al* 1995b).

The first patient was treated on 10 September 1993 and the early treatments were confined to the brain. Both brain metastases and arteriovenous malformations were treated, each with four to five fields in one to three fractions (Jones 1992, 1994, 1995a,b, Jones *et al* 1993, 1994, 1995a, Stannard 1993).

There are also plans to develop an isocentric gantry (Jones *et al* 1993, 1995a). Treatment planning is done using the VOXELPLAN 3D treatment-planning system from DKFZ and a modified form of the RMNHST proton algorithm (Lee *et al* 1993). The combined system is known as

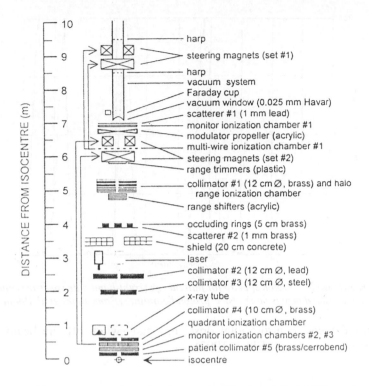

Figure 6.12. *A schematic diagram of the 200 MeV horizontal beamline for proton therapy at the NAC at Faure, South Africa. (From Jones et al 1995a.)*

PROXELPLAN (Jones 1995a, Schreuder *et al* 1996b) and runs on a DEC ALPHA machine.

Radiobiological measurements have established that the proton RBE is 1.0 in the plateau of both monoenergetic and modulated beams whilst values of 1.07 and 1.16 have been obtained in the middle and distal edge respectively of a 10 cm SOBP (Jones *et al* 1995a). Proton spectral measurements have shown, for a clinical beam, a very small low-energy component (figure 6.13).

6.7. THE CLINICAL PROTON FACILITY AT INDIANA UNIVERSITY, USA

The third high-energy clinical proton facility in the USA has been established at Indiana University at Indianapolis (Morphis and Bloch 1994). A 200 MeV beam is generated by a cyclotron and the first patient was treated on 28 September 1993. The patient is aligned using lasers and the position verified by an x-ray port verification system. The effort is directed towards automated dose delivery. A real-time computer system observes all the beam

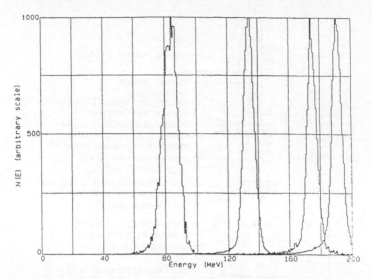

Figure 6.13. *Proton spectra for full-energy (right-hand-side spectrum) and for degraded beams on the NAC proton beamline. (From Jones et al 1995a.)*

monitor devices, controls total dose and can switch off the beam if any mispositioning occurs (Bloch 1992).

6.8. THE CLINICAL PROTON FACILITY AT UCSF DAVIS, USA

The clinical proton facility at UCSF Davis is based on the 67.5 MeV isochronous proton cyclotron at the Crocker Nuclear Laboratory and is used for ocular treatments. The clinical programme was established to act as a substitute for the decommissioned Lawrence Berkeley National Laboratory Bevalac. A beam with peak current of 30 μA is arranged horizontally and can produce a 30 mm field diameter with a uniformity of 2.5%. The range of the beam after passing through various measurement systems is 30 mm in water with a peak-to-entrance ratio of 3.8:1 and a full-width at half-maximum (FWHM) of 5 mm. Variable-thickness propellers are used to spread out the Bragg peak across the target volume. The residual range is varied using a movable water column. The facility came into clinical operation in May 1994 (Daftari *et al* 1995, 1996). 50 patients were treated in the first 18 months.

6.9. THE CLINICAL PROTON FACILITY AT TRIUMF, CANADA

The clinical proton therapy facility at TRIUMF (The Tri-University Meson Facility), Canada, came online with the treatment of the first patient on 21 August 1995. The therapy is a cooperative project between the British

Columbia Cancer Agency (Vancouver Cancer centre), the Eye Care centre and TRIUMF. The accelerator is the 520 MeV cyclotron at TRIUMF which normally produces a beam of intensity 10 μA. The beam is reduced in intensity to the necessary 3–10 nA by means of a pepper-pot limiting device. A fixed horizontal beam is generated and spread out with a passive-scattering system. Treatment of ocular melanoma with a 70 MeV beamline is planned with software (EYEPLAN) from the Clatterbridge centre in England. The only site currently being treated is choroidal melanoma. The patient alignment method for choroidal melanoma does not significantly differ from that used in other centres. The patient treatment chair was made by a local engineering company called Benchmark Engineering.

A number of radiobiological measurements have been made on this beamline which showed the RBE varying from 1.2 in the proximal part of the SOBP to 1.3 in the distal part at high dose and from 1.5 to 1.6 at lower dose. This spatial dependence of the dose-dependent RBE has dosimetric implications.

Since the beamline can also deliver 120 MeV protons it is planned to investigate the treatment of shallow-depth arteriovenous malformations via an upgrading of the eye-treatment facility (Ma and Lam 1996). Even this update is a little short for the treatment of deep-seated tumours but consideration is being given to expanding the facility for the treatment of head-and-neck malignancies (Pickles 1996).

6.10. THE PROPOSED UK HIGH-ENERGY PROTON FACILITY

There is a proposal for a national collaboration to develop 70–300 MeV proton radiotherapy in the UK; the proposal is known as PROTOX since it emanates from the Oxford Radcliffe Hospital (Cole *et al* 1996). The proposal is to utilise the existing 800 MeV proton synchrotron (ISIS) at the Rutherford–Appleton Laboratory at Chilton as the basis for developing a National Referral centre. The intention is to build three treatment rooms, at least two of which would house a rotating isocentric gantry, no more than 5 m in diameter. The remaining room would house a fixed horizontal beamline. The gantry may be of the same design as that at PSI and again, as at PSI, spot scanning would be developed. The beam would be used with particular emphasis on the treatment of prostate cancer. It would not be intended to treat ocular cancers since the Clatterbridge centre already provides for this in the UK. 3D imaging facilities for proton treatment planning would be established for use with the HELAX planning system. The present 800 MeV beamline is generated by injecting the proton synchrotron from a 70 MeV H$^-$ linac. The high-energy beam would be degraded but still maintain sufficient energy to pass through the body so proton radiography can be established. A detailed project proposal has recently been prepared (Spittle 1996).

6.11.   FIXED- VERSUS VARIABLE-MODULATION TREATMENT MODES

6.11.1.   Physical description

There are two basic ways in which a charged-particle beam can create a high-dose treatment volume. In the first method, known as 'fixed modulation', a large-area beam is created by a double-foil scattering technique (see the companion Volume, section 4.3.2). Fixed-range modulation is then created by making a SOBP using a propeller (see the companion Volume, section 4.3.1). The width of the SOBP is made equal to the maximum thickness of the target volume in the direction of the beam. Then a compensator is constructed to tailor the distal surface of the treatment volume to the distal surface of the target volume (see the companion Volume, section 4.4.2). The disadvantage of this technique is the direct consequence that the *proximal* surface of the treatment volume will not follow the proximal surface of the target volume. Proximal normal tissue will be undesirably irradiated. On the other hand this is a tried and trusted method (figure 6.14). It is relatively simple and reliable. A disadvantage is that the scattering method requires a long distance between scatterer and patient, hence the radius of the gantry is large. A large amount of individualised hardware such as compensators and collimators is required (Pedroni 1994b).

In the second technique, known as variable modulation, both the area of the beam and the width of the Bragg peak remain small. An elementary pencil beam is created which has the elementary narrow Bragg peak or a very small SOBP. The depth of the Bragg peak is changed by inserting a range shifter of an appropriate thickness to pull the elementary Bragg peak more proximal. The pencil is electromagnetically scanned in the other two directions and in this way the target is covered. There are variations on the technique. In raster scanning a slice of the target is scanned with fixed modulator. In principle the scanning movement alone can control the area of the slice but sometimes a backup MLC is also used. Then the modulator is changed to identify a different slice and the scanning repeated. By starting at the most distal slice and progressing anteriorly the whole target volume is covered. This technique is more complicated, relies on electropneumatic components and electromagnetic scanning but will deliver a lower dose to proximal tissues (figure 6.14). It also requires fewer patient-specific pieces of hardware (Pedroni *et al* 1995) such as compensators and collimators. It would be possible to deliver a non-uniform dose to the target volume. Renner *et al* (1994) describe how the 'occupation function', giving the weighting for each pencil beam, can be deduced by deconvolving the beam profile from the desired dose distribution. The gantry size can be smaller. Potential disadvantages are sensitivity to organ movement and increased complexity of technique.

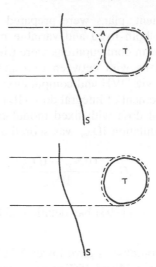

Figure 6.14. *The difference between fixed and variable modulation. In the upper part of the figure, the large-area proton beam has a fixed width SOBP, the distal surface of which is tailored to the distal surface of the target volume T. However, an unwanted consequence is that the proximal surface of the target volume will not follow the proximal surface of the treatment volume and some normal structures will be undesirably irradiated (region A). In the lower part of the figure is illustrated variable modulation which leads to a lower proximal dose. The dotted curve represents the high-dose region. Line S represents the patient surface.*

Castro *et al* (1992) have described the construction of a facility with variable modulation at the University of California, Lawrence Berkeley Laboratory (UC LBL) for both (initially) heavy ions and protons. The elementary beam has a FWHM of 1 cm and a 1 cm wide SOBP. The beam scans fast horizontally (1–2 cm ms^{-1}) and slow vertically (each slice is completed in under 1 s which is the length of the Bevatron beam pulse). The scan speed and the spill rate of the accelerator are variable. A MLC with a resolution of 9 mm was developed for shaping each layer in the target.

6.11.2. Clinical comparison

Because variable modulation can give lower doses to proximal normal structures it is inevitable that, at fixed TCP, lower NTCPs must result. The only question is how much lower and is this worth the extra complexity of the technique? This is a very important question and had been addressed some time ago using models by Goitein and Chen (1983). Daftari *et al* (1993) have completed a detailed study for ten patients, three with biliary tract tumours (normal structures liver and gut nearby), five pancreatic tumours (normal structure gut) and two oesophagus tumours (normal structure lung nearby). For each patient, plans were prepared for irradiation with neon ions and protons with both fixed and variable modulation. Moreover, for each patient, several beam configurations were tried, some comprising more beams than others. Plans with between two and five beams were made. The plans were scored via DVH and computation of NTCP with the Lyman model and also measurement of integral dose (ID). A 'gain factor' G defined in terms of the integral dose with fixed modulation ID_{fm} and the integral dose with variable modulation ID_{vm} was scored also:

$$G = \frac{(ID_{fm} - ID_{vm})}{ID_{fm}}. \tag{6.1}$$

The paper of Daftari *et al* (1993) has detailed tables of G and NTCP. They concluded that:

(i) variable modulation always gave lower NTCPs than fixed modulation;
(ii) gain factors between 14 and 25% were recorded for proton irradiations;
(iii) the NTCP for a two-beam irradiation with variable modulation was generally comparable with the NTCP for the same structure for a five-beam irradiation with fixed modulation;
(iv) the gains recorded were greater than those predicted by Goitein and Chen (1983), this being put down to the fact that the tumours had a very irregular shape whereas those in the model study were elliptical. They concluded that variable modulation is likely to have its main role when the targets have very irregular shapes.

Unfortunately this study did not make baseline comparisons with what can be achieved with photon therapy but work along these lines has been done by Lomax *et al* (1993a,b), Lee *et al* (1995) and others (see section 6.12).

The PSI facility can also plan and deliver variable modulation. Munzenrider *et al* (1995) have made a comparison of fixed- versus variable-modulation proton planning for cases of nasopharyngeal cancer. They created plans which met strict dose constraints in the PTV. On inspection it was found that the proton plans with variable modulation led to greater sparing of dose to OAR. Specifically, dose reductions to the spinal cord and brainstem were achieved of some 5–10 CGE (cobalt-Gray-equivalent) for approximately 60% of each structure. Optic nerve doses were also reduced by 1 CGE for 50–80% of the volume. Parotid glands were spared better with variable than with fixed modulation.

6.11.3. *Treatment planning for spot scanning*

The spot scanning technique is predicated on the superposition of elementary pencil beams, the position of the elementary Bragg peak being scanned through the PTV. The planning problem is to compute the beamweight for each elementary pencil. Since each pencil contributes dose to a large number of voxels, mainly those along the path to the Bragg peak but also voxels away from the line-of-sight path due to scatter, each voxel receives a contribution from each pencil, albeit a very small contribution for some. At PSI a computer algorithm has been developed to compute the beamweights recursively minimising a χ^2 parameter representing the difference between delivered dose and prescription dose. Simultaneous optimisation of multiple ports is suited for spot scanning and the optimisation can be biologically based (see Chapter 1) (Pedroni *et al* 1993c).

6.11.4. *Proton tomotherapy and distal-edge tracking*

In Chapter 2 we reviewed the concept of tomotherapy in which a beam of photons is collimated to a narrow slit but the intensity of radiation across the slit is modified by the time-variable presence of stubby collimators moving at right angles to the slit. As the radiation is delivered the gantry rotates about the patient. Deasy *et al* (1995a,b) have recently proposed a similar arrangement for proton radiotherapy and this represents a third class of treatment delivery to add to the above two. At present the idea is only a concept.

The notion is that a proton beam, also collimated to a fan, has its intensity similarly modulated as the gantry rotates about the patient in a helical pattern. The proton range is modulated so that at any given instant the proton pencil-beam Bragg peak matches the distal edge of the PTV. Although from any one beam the dose to the rest of the PTV is low, the PTV is 'filled in' by the set of beams as the rotation progresses. The treatment has been simulated and shown to give good results (90–50% dose fall-off all round the tumour in approximately 5 mm).

6.12.   PHOTONS OR PROTONS?

As well as consideration of the costs of a proton facility, the key clinical question is 'are protons better than photons?'. An inherent difficulty in responding to this question is that, because there are so few clinical proton facilities, there is little comparative clinical data. When the PTV is extremely close to OAR, e.g. choroidal melanomas and sarcomas abutting the central nervous system, it has been argued that proton therapy is justified without the need for randomised trials comparing it with photon therapy because of the intrinsic sparing of normal structures which can be arranged to be beyond the range of protons. The question of whether protons can give superior results for *common tumours in the abdomen and pelvis* has not been widely addressed clinically.

Hence what data there are come from comparative planning studies. Urie *et al* (1993) argue that, for the comparison to be fair, the proton and photon beams should be oriented the same when comparing plans, and they conducted a study with this feature. An optimised photon plan was constructed for a large horseshoe-shaped abdominal mass surrounding the spinal cord and in close proximity to the right kidney and gut. Then the photon beams at the selected orientations were replaced by proton beams. The finding was that the DVH in the PTV was similar for photons and protons but that in the proton plan the normal structures received less dose. It was concluded that, for the same NTCP, the dose to the PTV could be increased by some 10 Gy.

Lee *et al* (1995) took a different approach and compared the predicted TCP and NTCP results of parallel-opposed proton plans with a 'conventional' three-field and six-field arrangement of photon beams for prostate cancer. The proton plans were for a passive-scattering system. Twenty patients were studied but it was predicted that, due to part of the rectum being in the PTV the predicted TCP was only 2–3% greater for protons than with photons, for the same NTCP. This is somewhat disappointing. Slightly more optimistic comparisons were made for tumours of the lung. Collier *et al* (1995) also found superior results for proton irradiation of lung tumours.

Lomax *et al* (1993a,b) compared a four-field photon box, rotated relative to the AP direction with a three-field proton plan (three of the four directions used for photons) for a large pelvic tumour in close proximity to the rectum and large bowel. The proton plan was for the PSI spot-scanning system. Better dose conformation was demonstrated using comparative DVHs for the target and dose–*surface* histograms for the rectum and large bowel.

Lin *et al* (1995) compared photon and proton plans for nasopharyngeal cancer with advantages for the latter. In practice in Taiwan a combined photon and proton treatment protocol has been proposed.

It is debatable whether any of these approaches to comparative treatment planning is strictly correct. Surely what needs to be known is whether the *optimum* arrangement and weighting of a proton treatment is better (and by how much) than the *optimum* photon treatment. Because teams working on photon optimisation generally do not have access to proton planning facilities, these studies are not done even at the planning stage. There is scope for progress in this area.

The subject of proton treatment planning is nothing like so well developed as the corresponding photon treatment planning. Major manufacturers market photon planning systems (admittedly all with somewhat different capabilities) and there are several well-developed 'in house' University-Hospital systems. These have developed to meet the need of planning many hundreds of thousands of photon irradiations. Because of the dearth of proton irradiation facilities, algorithms for proton planning have tended to be 'in-house' developments. Generally these have been guarded jealously because they have not been subject to the same rigorous quality control as applied to photon therapy-planning systems. When algorithms have been shared, they have been passed on with recommendations for the secondary user to take full responsibility for their clinical use.

Against this background there is a pressing need to conduct photon–proton planning comparisons with the best candidates on each side of the scales. For photon treatment planning this must include the possibility to plan patient cases with IMBs. At present most 3D treatment-planning systems do not have this facility available and are restricted to planning uniform-shaped fields. Verhey (1995) has taken a clinical case of a nasopharynx patient from the Loma Linda clinic and planned the patient using the NOMOS PEACOCKPLAN system (see Chapter 2) with intensity-modulated photon fields. In this specific case the photon treatment plan met the planning constraints. In view of the lack of both types of treatment-planning facilities within the same computer system, comparative treatment planning with the full range of options is unlikely to be possible in a single radiation physics centre and the best hope of achieving such comparisons lies in the sharing of clinical cases and the sharing of skills and tools, with interchange facilitated by some form of telematics communication. Plans are well advanced in Europe for such cooperative effort. Lomax *et al* (1995) have gone further and taken the same clinical cases through spot-scanned proton therapy planning and photon planning with intensity modulation. In order to do this the group at PSI have teamed up with the group at the DKFZ, Heidelberg to gain access to the facilities for each type of planning (Smith *et al* 1996a).

Goitein (1995b) has argued that it is possible to show by a thought experiment that when equal doses are delivered to a PTV, a proton plan will always deliver less integral dose to the rest of the body than a photon plan, this conclusion being independent of the number of beams or whether, and the manner in which, they are intensity modulated. He followed on to say that the important question is whether this matters and by how much. Remembering that 'every cell in the body is there for a reason' it was argued that scoring dose in OAR and converting these to measures of complication probability may not be all that matters. It may be important to assess the effects of integral dose on other parts of the body, not considered to be OAR.

6.13. PROTON RADIOGRAPHY

It is important in proton radiotherapy to have high precision of patient positioning with respect to the beam. Proton radiography was originally studied in the 1970s as a low-dose alternative to x-ray imaging but perhaps was discontinued because of the success of x-ray CT (Cookson 1974). The team at PSI (Villigen) make use of the proton beam itself as a diagnostic tool to assist patient positioning (Schneider *et al* 1993a,b, Schneider and Pedroni 1994). The energy of the beam is increased, so there is a transmitted beam, and the intensity is simultaneously reduced. An energy-sensitive detector images the proton beam which has penetrated the patient. Because the range–energy relationship is well known for water, the measured drop in energy of the proton beam traversing the patient can be translated into a measure of effective pathlength in water. This measurement automatically accounts for tissue inhomogeneities and uncertainties in the proton path due to multiple scatter. The spatial resolution is a problem in proton radiography but 1 mm can be achieved. With this resolution there is a 10–20 times dose advantage, however, compared with conventional x-radiography (Schneider *et al* 1993a,b).

The proton radiograph has many uses. Qualitatively it can be used in the same way as an x-ray portal image, i.e. a DRR is made from the x-ray CT data of the patient and can be compared with the proton radiograph. The proton radiograph is then translated relative to the DRR until the spatial correlation is highest, giving the required adjustment to make in the patient positioning at the time of treatment.

More quantitatively, the experimental proton radiograph has a second use, the prediction of a method of converting from CT data to proton stopping power (Schneider *et al* 1993a,b). A theoretical proton radiograph was computed by Monte Carlo techniques using: (i) the 3D CT data; (ii) a calibration curve converting Hounsfield units into relative proton stopping power. A first estimate of a calibration curve was derived from measurements of tissue-equivalent CT samples and calculation. When the experimental and theoretical proton radiographs of a sheep's head were compared, range errors, some larger than 10 mm, were observed. The calibration curve was then adjusted and a new theoretical proton radiograph computed. The adjustment was terminated when the new theoretical radiograph and the experimental proton radiograph had much lower discrepancies. In this way a more accurate calibration was obtained. Schneider *et al* (1995) call this 'stoichiometric calibration'.

A third use of proton radiographs is to check on range uncertainties. By inspecting the proton radiograph, those regions where there are sudden changes of range can be identified. These will, for example, arise when the beam passes close to interfaces between bone and soft tissue. Because the patient may move slightly it could then be advantageous to add an extra security margin and replan.

6.14. SUMMARY

The pace of growth of proton radiotherapy is increasing. This is reflected both by the number of new centres being constructed or planned and also in the number of treatments in existing centres. Proton radiotherapy has been carried out for over 40 years, yet 44.6% of the present world total of

17 737 treatments have taken place in the last five years. There is a move towards a treatment configuration with a rotating gantry similar to that of photon irradiation. There is no consensus on treatment techniques, different centres adopting different ways of producing a spread-out beam of finite area and depositing this beam in the patient. The number of treatments is still too small for the advocates of proton therapy to have solid evidence for its value. Hence there is still a 'Catch 22 situation' that there never will be enough evidence until there are more centres and vice versa. There is a geographical imbalance in the location of centres, perhaps rather surprisingly few in the USA. However, proton therapy is a good example where the old barriers with the former USSR seemed to present few difficulties with collaborative studies. Treatment planning is highly developed but with little standardisation across centres. There is a strong expectation that proton therapy will continue to thrive but conversely it is not without its critics who favour the distribution of resources to improving conformal therapy with photons.

REFERENCES

Amaldi U 1994 The Italian hadrontherapy project *Hadrontherapy in Oncology* ed U Amaldi and B Larsson (Amsterdam: Elsevier) pp 45–58
—— 1995 The Italian hadrontherapy project (Proc. ESTRO Conf. (Gardone Riviera, 1995)) *Radiother. Oncol.* **37** Suppl. 1 S42
Amaldi U and Silari M 1994 *The TERA Project and the Centre for Oncological Hadrontherapy* (Frascati Roma: INFN-LNF Divisione Ricerca, SIS Ufficio Pubblicazioni PO Box 13)
Baratoux D, Mazal A, Ferrand R, Delacroix S, Nauraye C and Habrand J L 1995 Variable modulation by pulse gating *Orsay Internal Report*
Blattmann H and Munkel G 1993 Radiation medicine *Paul Scherrer Institute Life Sciences Newsletter 1993. Annexe 2—annual report 1993* p 3
Blattmann H *et al* 1994 The Swiss protontherapy program *Hadrontherapy in Oncology* ed U Amaldi and B Larsson (Amsterdam: Elsevier) pp 122–9
Bloch C 1992 News from Indiana University Cyclotron Facility, Bloomington, USA *Particles Newsletter* ed J M Sisterson No 10 p 5
Bonnett D E 1993 Current developments in proton therapy: a review *Phys. Med. Biol.* **38** 1371–92
Bonnett D E, Kacperek A, Sheen M A, Goodhall R and Saxton T E 1993 The 62 MeV proton beam for the treatment of ocular melanoma at Clatterbridge *Brit. J. Radiol.* **66** 907–14
Castro J R, Petti P L, Daftari I K, Collier J M, Renner T, Ludewigt B, Chu W, Pitluck S, Fleming T, Alonso J and Blakely E 1992 Clinical gain from improved beam delivery systems *Radiat. Environ. Biophys.* **31** 233–40

Chauvel P, Sheen M and Kacperek A 1995 Problems encountered during eye treatment planning and modifications needed in EYEPLAN *Particles Newsletter* ed J M Sisterson No 15 p 7

Chu W T 1994 Cost comparison between proton and conventional photon facilities *Particles Newsletter* ed J M Sisterson No 13 pp 8–9

Cole D J, Weatherburn H, Walker G and Williams P 1996 PROTOX—a proposal to develop high energy proton beam radiotherapy in Oxford, UK *Abstracts of the 24th PTCOG Meeting (Detroit, MI, 1996)* (Boston: PTCOG)

Collier J M, Choi N and Niemierko A 1995 A treatment planning comparison of protons versus photons in non-small cell lung cancer *Abstracts of the 22nd PTCOG Meeting (San Francisco, 1995)* (Boston: PTCOG) pp 6–7

Cookson J A 1974 Radiography with protons *Naturwissenschaften* **61** 184–94

Cosgrove V P, Aro A C A, Green S, Scott M C, Taylor S C, Bonnett D E and Kacperek A 1992 Studies relating to 62 MeV proton cancer therapy of the eye (Proc. 7th Symp. Neutron Dosimetry (Berlin, 1991)) *Rad. Protection Dosimetry* **44** 405–9

Coutrakon G, Hubbard J, Johanning J, Maudsley G, Slaton T and Morton P 1994b A performance study of the Loma Linda proton medical accelerator *Hadrontherapy in Oncology* ed U Amaldi and B Larsson (Amsterdam: Elsevier) pp 282–306

Coutrakon G, Johanning J, Maudsley G, Slaton T and Morton P 1994a Intensity measurements and upgrade paths for the Loma Linda accelerator *Abstracts of the 19th PTCOG Meeting (Cambridge, MA, 1993)* (Boston: PTCOG) p 15

Daftari I, Petti P L, Collier J M, Castro J R and Pitluck S 1993 Evaluation of fixed- versus variable-modulation treatment modes for charged-particle irradiation of the gastrointestinal tract *Med. Phys.* **20** 1387–98

Daftari I K, Renner T R, Verhey L J, Singh R P, Nyman M, Petti P L and Castro J R 1996 New UCSF proton ocular beam facility at the Crocker Nuclear Laboratory Cyclotron (UC Davis) *Nucl. Instrum. Methods* **380** 597–612

Daftari I K, Verhey L J, Renner T R, Nyman M, Singh R P, Petti P L and Castro J R 1995 Establishment of a new UCSF proton ocular beam facility at Crocker Nuclear Laboratory cyclotron (UC Davis) *Med. Phys.* **22** 1543

Dearnaley D, Nahum A E and Steel G G 1993 The case for proton beam therapy at the Royal Marsden Hospital and the Institute of Cancer Research *Internal document*

Deasy J O, Mackie T R and DeLuca P M 1995a Conformal proton tomotherapy using distal-edge tracking *Abstracts of the 22nd PTCOG Meeting (San Francisco, 1995)* (Boston: PTCOG) pp 17–8

—— 1995b Conformal proton tomotherapy using distal-edge tracking (Proc. ESTRO Conf. (Gardone Riviera, 1995)) *Radiother. Oncol.* **37** Suppl. 1 S43

Flanz J, Durlacher S, Goitein M, Smith A, Woods S, IBA Staff and Bechtel Staff 1995 The Northeast Proton Therapy Center at Massachusetts General Hospital *Proc. 5th Workshop on Heavy Charged Particles in Biology and Medicine (Darmstadt, 1995)* (Darmstadt: GSI)

Fleurette F, Charvet-Protat S, Metral P and Roche B 1995 Proton radiation in cancer treatment: clinical and economic outcomes *Particles Newsletter* ed J M Sisterson No 16 13–4

Goitein M 1995a PTCOG meetings; an historical minute *Particles Newsletter* ed J M Sisterson No 15 p 5

—— 1995b Comparison of proton and photon dose distributions (Proc. ESTRO Conf. (Gardone Riviera, 1995)) *Radiother. Oncol.* **37** Suppl. 1 S43

Goitein M and Chen G T Y 1983 Beam scanning for charged particle therapy *Med. Phys.* **10** 831–40

Goitein M, Smith A, Flanz J, Durlacher S, Woods S and Tarpey C 1996 Status report: the Northeast Proton Therapy Centre at Massachusetts General Hospital, Boston, MA *Particles Newsletter* ed J M Sisterson No 17 pp 10–2

Jones D T L 1992 Status of particle therapy at the NAC, South Africa *Particles Newsletter* ed J M Sisterson No 10 p 6

—— 1994 Proton therapy commences at NAC, South Africa *Particles Newsletter* ed J M Sisterson No 13 p 10

—— 1995a Operation of the NAC particle therapy facilities, Faure, South Africa *Particles Newsletter* ed J M Sisterson No 16 p 12

—— 1995b NAC—The only proton therapy facility in the Southern Hemisphere *Ion Beams in Tumor Therapy* ed U Linz (London: Chapman and Hall) pp 350–9

Jones D T L, Schreuder A N and Symons J E 1993 Status of the particle therapy facilities at the National Accelerator Centre *Abstracts of the 19th PTCOG Meeting (Cambridge, MA, 1993)* (Boston: PTCOG) p 10

—— 1995a Particle therapy at NAC: physical aspects *Proc. 14th Int. Conf. on Cyclotrons and their Applications (1995)*

Jones D T L, Schreuder A N, Symons J E, Van der Vlugt G, Bennett K F and Yates A D B 1995b Use of stereophotogrammetry in proton radiation therapy *Proc. Int. FIG Symp. (Commission 6) University of Cape Town (Cape Town, 1995)* pp 138–53

Jones D T L, Schreuder A N, Symons J E and Yudelev M 1994 The NAC particle therapy facilities *Hadrontherapy in Oncology* ed U Amaldi and B Larsson (Amsterdam: Elsevier) pp 307–28

Koehler A M, Schneider R J and Sisterson J M 1975 Range modulators for protons and heavy ions *Nucl. Instrum. Methods Phys. Res.* **131** 437–40

—— 1977 Flattening of proton dose distributions for large field radiotherapy *Med. Phys.* **4** 297–301

Lee M 1995 Comparison of proton and megavoltage x-ray conformal therapy planning for cancer treatment *PhD Thesis* University of London

Lee M, Nahum A and Webb S 1993 An empirical method to build up a model of proton dose distribution for a radiotherapy treatment planning package *Phys. Med. Biol.* **38** 989–98

Lee M, Wynne C, Webb S, Nahum A E and Dearnaley D 1995 A comparison of proton and megavoltage x-ray treatment planning for prostate cancer *Radiother. Oncol.* **33** 239–53

Lin F J, Lin P J, Moyers M F, Miller D W, Slater J D and Slater J M 1995 Comparative study of proton therapy and photon therapy for nasopharyngeal cancer *Abstracts of the 22nd PTCOG Meeting (San Francisco, 1995)* (Boston: PTCOG) p 3

Lomax A, Munkel G, Bortfeld T, Dykstra C, Blattmann H and Debus J 1995 A treatment planning intercomparison of radiation therapy using spot scanned protons and intensity modulated photons *Abstracts of the 22nd PTCOG Meeting (San Francisco, 1995)* (Boston: PTCOG) p 7

Lomax A, Scheib S, Munkel G and Blattmann H 1993a The comparison of spot scanning proton radiotherapy with conventional photon therapies *Paul Scherrer Institute Life Sciences Newsletter 1993. Annexe 2—annual report 1993* pp 8–9

—— 1993b The comparison of spot scanning proton radiotherapy with conventional photon therapies *The use of computers in radiation therapy: Proc. 11th Conf.* ed A R Hounsell *et al* (Manchester: ICCR) pp 366–7

Loncol T, Cosgrove V, Denis J M, Gueulette J, Mazal A, Menzel H G, Pihet P and Sabattier R 1994 Radiobiological effectiveness of radiation beams with broad LET spectra: microdosimetric analysis using biological weighting functions *Radiat. Protect. Dosimetry* **52** 347–52

Ma R and Lam G 1996 News from the TRIUMF proton therapy facility, Vancouver, Canada *Particles Newsletter* ed J M Sisterson No 17 p 12

Mazal A 1996 Private communication (letter and papers to S Webb; 21 April 1996)

Mazal A and Habrand J L 1993 La protonthérapie: le centre D'Orsay *Patholog. Biol.* **41** 122–5

Mazal A *et al* 1991 Orsay: status report *Proc. 4th Workshop on Heavy Charged Particles in Biology and Medicine (Darmstadt, 1991)*

Morphis J III and Bloch C 1994 First patient treated at the Indiana University Proton Therapy Centre, Indiana, USA *Particles Newsletter* ed J M Sisterson No 13 p 5

Munkel G, Lomax A, Scheib S, Pedroni E and Blattmann H 1993 Indications for proton therapy *Paul Scherrer Institute Life Sciences Newsletter 1993. Annexe 2—annual report 1993* pp 4–5

Munkel G and the Group Radiation Medicine 1994 Patient positioning and tumour outlining *Hadrontherapy in Oncology* ed U Amaldi and B Larsson (Amsterdam: Elsevier) pp 425–7

Munzenrider J E 1994 Proton therapy with the Harvard cyclotron *Hadrontherapy in Oncology* ed U Amaldi and B Larsson (Amsterdam: Elsevier) pp 83–101

Munzenrider J E, Adams J, Munkel G, Liebsch N and Smith A 1995 Fixed versus variable modulation in proton beam therapy of advanced nasopharyngeal cancer: a comparative treatment planning study *Abstracts of the 22nd PTCOG Meeting (San Francisco, 1995)* (Boston: PTCOG) pp 4–5

Nahum A E, Dearnaley D P and Steele G G 1994 Prospects for proton-beam radiotherapy *Europ. J. Cancer* **30** 1577–83

Nauraye C 1995 Optimisation des caracteristiques physiques et dosimétriques d'un faisceau de protons de 200 MeV et mise en oeuvre d'une ligne de faisceau adaptée a la radiotherapie *PhD Thesis* Toulouse University Paul Sabatier

Nauraye C, Mazal A, Delacroix S, Bridier A, Chavaudra J and Rosenwald J-C 1995 An experimental approach to the design of a scattering system for a proton therapy beam line dedicated to opthalmological applications *Int. J. Radiat. Oncol. Biol. Phys.* **32** 1177–83

NPTC 1996 *A brief description of the Northeast Proton Therapy Centre* NPTC-53 April (courtesy of Dr A Smith)

Pedroni E 1993 Status of the development of the 200 MeV proton facility *Paul Scherrer Institute Life Sciences Newsletter 1993. Annexe 2—annual report 1993* pp 13-6

—— 1994a Beginning of the commissioning of the 200 MeV proton therapy facility at PSI *Particles Newsletter* ed J M Sisterson No 14 p 6

—— 1994b Beam delivery *Hadrontherapy in Oncology* ed U Amaldi and B Larsson (Amsterdam: Elsevier) pp 434–52

Pedroni E, Bacher R, Blattmann H, Böhringer T, Coray A, Lomax A, Lin S, Munkel G, Scheib S, Schneider U and Tourovsky A 1995 The 200-MeV proton therapy project at the Paul Scherrer Institute: conceptual design and practical realisation *Med. Phys.* **22** 37–53

Pedroni E, Blattmann H, Böhringer T, Coray A, Lin S, Lomax A, Munkel G, Scheib S and Schneider U 1993a News from PSI, Villigen *Particles Newsletter* ed J M Sisterson No 11 p 9

—— 1993b Installation of the PSI compact gantry for spot scanning *Abstracts of the 18th PTCOG Meeting (Orsay and Nice, 1993)* (Boston: PTCOG) p 11

—— 1993c Treatment planning for spot scanning: present and future developments at PSI *Abstracts of the 18th PTCOG Meeting (Orsay and Nice, 1993)* (Boston: PTCOG) p 7

Pedroni E, Blattmann H, Böhringer T, Coray A, Lin S, Lomax A, Munkel G, Scheib S, Schneider U and Tourovsky A 1993d The PSI proton unit for medical applications: a status report *Abstracts of the 19th PTCOG Meeting (Cambridge, MA, 1993)* (Boston: PTCOG) p 9

Pickles T A 1996 private communication (letter to S Webb; 15 April 1996)

Preston W 1994 News from Loma Linda University Medical Centre, USA *Particles Newsletter* ed J M Sisterson No 13 p 6–8

Raju M R 1994 Hadrontherapy in a historical and international perspective *Hadrontherapy in Oncology* ed U Amaldi and B Larsson (Amsterdam: Elsevier) pp 67–79

—— 1995 Proton radiobiology, radiosurgery and radiotherapy *Int. J. Radiat. Biol.* **57** 237–59

Renner T R, Chu W T and Ludewigt B A 1994 Advantages of beam scanning and requirements of hadrontherapy *Hadrontherapy in Oncology* ed U Amaldi and B Larsson (Amsterdam: Elsevier) pp 453–61

Schneider U, Lomax A, Pedroni E, Pemler P and Schaffner B 1995 The calibration of CT units to proton stopping power for proton therapy treatment planning *Abstracts of the 22nd PTCOG Meeting (San Francisco, 1995)* (Boston: PTCOG) pp 14–5

Schneider U and Pedroni E 1994 Proton radiography as a tool for quality control in proton therapy *Hadrontherapy in Oncology* ed U Amaldi and B Larsson (Amsterdam: Elsevier) pp 67–79

Schneider U, Pedroni E and Tourovsky A 1993a Proton radiography *Paul Scherrer Institute Life Sciences Newsletter 1993. Annexe 2—annual report 1993* pp 17–20

Schneider U, Tourovsky A and Pedroni E 1993b Proton radiography: a tool for quality control in proton therapy *Abstracts of the 19th PTCOG Meeting (Cambridge, MA, 1993)* (Boston: PTCOG) p 8

Schreuder A N, Jones D T L, Symons J E, Fulcher T and Kiefer A 1996a The NAC proton therapy beam delivery system *Abstracts of the 23rd PTCOG Meeting (Cape Town, 1995)* (Boston: PTCOG) p 27

Schreuder A N, Symons J E, de Kock E A, Jones D T L, Hough J K, Wilson J, Vernimmen F J, Schlegel W, Hoess A and Lee M 1996b The PROXELPLAN system used at NAC for proton treatment planning *Abstracts of the 23rd PTCOG Meeting (Cape Town, 1995)* (Boston: PTCOG) p 33

Sisterson J M 1994 World wide proton therapy experience: where are we now in 1994? *Abstracts of the 20th PTCOG meeting (Chester, 1994)* (Boston: PTCOG) p 19

—— (ed) 1996 *Particles Newsletter* No 18, July

Slater J 1993 News from Loma Linda University Medical Centre, USA *Particles Newsletter* ed J M Sisterson No 11 pp 6–7

—— 1995 News from Loma Linda University Medical Centre, USA *Particles Newsletter* ed J M Sisterson No 16 pp 9–10

Slater J M, Archambeau J O, Dicello J F and Slate J D 1994 Proton beam irradiation: towards routine clinical utilization *Hadrontherapy in Oncology* ed U Amaldi and B Larsson (Amsterdam: Elsevier) pp 130–7

Smith A 1996a private communication (letter to S Webb; 6 June 1996)

—— 1996b Status report: the Northeast Proton Therapy Center at Massachusetts General Hospital, Boston, MA, USA *Particles Newsletter* ed J M Sisterson No 18 p 8

Smith A, Adams J, Munzenrider J and Liebsch N 1996a Comparative treatment planning for nasopharynx tumours *Abstracts of the 23rd PTCOG Meeting (Cape Town, 1995)* (Boston: PTCOG) p 2

Smith A, Goitcin M, Durlacher S, Flanz J, Levine A, Reardon P and Woods S 1994a The Massachusetts General Hospital Northeast Proton Therapy Centre *Hadrontherapy in Oncology* ed U Amaldi and B Larsson (Amsterdam: Elsevier) pp 138–44

Smith A, Goitein M, Suit H, Munzenrider J, Durlacher S, Flanz J, Gall K, Levine A, Rosenthal S and Woods S 1996b The Massachusetts General Hospital Northeast Proton Therapy Center, Boston, Massachusetts *Proc. NIRS Int. Seminar on the Applications of Heavy Ion Accelerators to Radiation Therapy of Cancer in Connection with the 21st PTCOG Meeting (1994)* (Boston: PTCOG) pp 50–8

Smith A, Urie M and Niemierko A 1994b Protons—therapy for the future (Proc. World Congress on Medical Physics and Biomedical Engineering (Rio de Janeiro, 1994)) *Phys. Med. Biol.* **39A** Part 1 491

Spittle M 1996 PROTOX—a national collaboration to develop 70–300 MeV proton radiotherapy in the UK *Project proposal from Oxford Radcliffe Hospital* February

Stannard C 1993 Status report on proton therapy at NAC *Abstracts of the 18th PTCOG Meeting (Orsay and Nice, 1993)* (Boston: PTCOG) p 8

Urie M, Niemierko A and Smith A R 1993 Optimised photons versus protons *Abstracts of the 19th PTCOG Meeting (Cambridge, MA, 1993)* (Boston: PTCOG) p 7

Verhey L J 1995 Optimized, intensity-modulated treatment plan for standardized nasopharynx case using the Peacock planning system *Abstracts of the 22nd PTCOG Meeting (San Francisco, 1995)* (Boston: PTCOG) p 2

CHAPTER 7

IMAGE FORMATION, UTILISATION AND DISPLAY, THE ROLE OF 3D AND THE CT-SIMULATOR

7.1. TOMOGRAPHY 'DRIVES' CONFORMAL RADIOTHERAPY

Chapter 1 has already dealt with the very important topic of creating 'volumes of interest' from imaging data. The determination of the PTV and the OARs is the vital first step in the 3D planning process. Before 1972 radiation therapy relied wholly on the use of transmission radiographs and Röntgen's discovery of the x-ray was as important for planning and delivering radiotherapy as it was for diagnosis (figure 7.1) (Mould 1993, 1995a,b). It has been rightly argued that one of the most significant factors on which recent improvements in conformal radiotherapy have depended is the ability to determine more precisely the spread of disease and the geometrical relationships between the PTV and the nearby OAR (Hendee 1995).

The first major development was the use of CT data in the early 1970s (figures 7.2 and 7.3) since which time tomographic MRI, SPECT and PET have all been harnessed to assist planning (figure 7.4) (Webb 1990a, 1995). Cormack (1994), at a conference to celebrate 75 years of the Radon Transform, has described how his desire to improve the physics of radiotherapy planning in South Africa led to the first laboratory developments of CT scanning, work for which he was subsequently honoured, sharing the 1979 Nobel Prize for physiology and medicine with Hounsfield. However, it is probably fair to say that it is still x-ray CT which is the workhorse of volumetric imaging for 3D treatment planning. The information from the other three modalities is generally regarded as complementary rather than substitutive. In Chapter 4 we considered the special role of electronic portal imaging. In this Chapter we highlight some of the ways in which medical images are used in the planning process to interactively assist *choices* made by the planner at that stage, to interactively

Figure 7.1.   *Wilhelm Conrad Röntgen on 10 February 1923 when aged 77. He died the same year. The discovery of the x-ray was as important for planning treatment as it was for tumour diagnosis. (From Schedel and Keil 1995.)*

guide the user to understand the *results* of planning and to *visualise* the treatment geometry. Firstly, we begin with a short review of progress in one particular form of x-ray tomography which uses radiotherapy equipment, rather than a diagnostic CT scanner.

7.2.   CROSS SECTION IMAGING ON A RADIOTHERAPY SIMULATOR

There has been continued progress towards developing CT on a radiotherapy simulator (Webb 1990b). The first suggestions for the use of a simulator for CT and one of the first practical devices came from Newcastle in 1978 where effort in this direction still continues today (Kotre 1995). Developments up to 1992 were reviewed in the companion Volume. In this section we bring the review up to date. Like the early developments single-section imagers are still being built by some groups. However, a more exciting and practical recent initiative is the construction of cone-beam CT devices which can provide volumetric data for 3D treatment planning. All cross section imaging on a simulator works on the same principle, that of recording projections

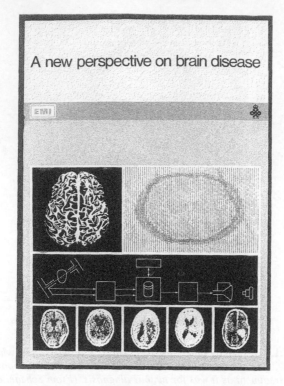

Figure 7.2.   *The front cover of the 1973 commercial brochure for the EMI Scanner. Note there is no number attached to the scanner since at this time this was the one-and-only commercial CT scanner. (EMI Company literature.)*

(usually 1D) as the gantry rotates, followed by image reconstruction. The major differences between machines relate to the detecting mechanism and whether the imaging is single-slice or multi-slice.

Mallik and Hunt (1993a,b) and Mallik *et al* (1993, 1995) have reported extensive clinical use of the Varian CT-simulator (figure 7.5). The term CT-simulator is used in two contexts which should be clearly distinguished. It can sometimes mean 'simulation performed with CT data' (see section 7.5) or, alternatively (as here), 'CT performed on a radiotherapy simulator'. The latter, but not the former, is also sometimes called simulator-CT. With the prototype they evaluated, the reconstruction circle was 50 cm in diameter and the aperture 87 cm. The detector was an image intensifier viewed by a TV camera together with a 32 channel solid-state photodiode outrigger. The data were recorded with 19 bits giving a dynamic range of 500 000. The spatial and contrast resolution in the 512^2 images could be varied by changing the distance from the source to the detector. When the scan circle was 48 cm in diameter, the spatial and density resolutions were 1.5 mm and

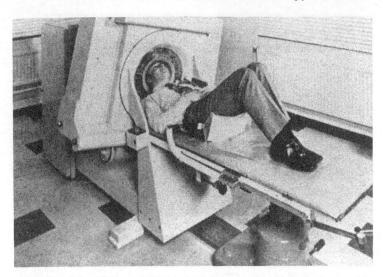

Figure 7.3. *The prototype EMI CT scanner, showing the patient in position. This machine can now be seen in the Science Museum, London. (From Hounsfield 1973.)*

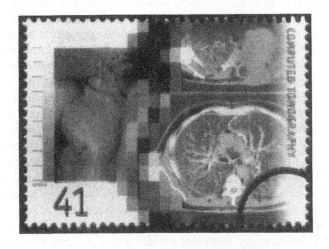

Figure 7.4. *The commemorative UK stamp issued 27 September 1994 as part of a set of stamps to mark the significant impact of imaging in medicine. This one shows a CT scan of the thorax. X-rays and radiology actually provide a popular thematic feature on postage stamps. The mammoth historical work by Grigg (1965) is full of them.*

1% respectively; when the scan circle was 25 cm in diameter, the spatial and density resolutions were 1.0 mm and 0.5% respectively. In the first three years of operation over 500 patients were imaged.

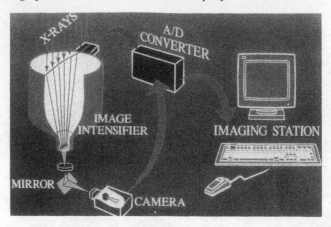

Figure 7.5. *A schematic diagram of the arrangement for detecting a fan-beam of x-rays in the Varian CT-simulator. An outrigger of photodiodes is attached to an image intensifier and the responses are matched. This increases the field-of-view. CT images are reconstructed from multiple fan-beam projections. (From Mallik et al 1993.)*

In the late 1970s the Oldelft company manufactured an analogue CT device attached to their radiotherapy simulator. The concept was somewhat overtaken by the development of digital simulator-CT (Webb 1990b). Recently this same company, now merged with Nucletron, manufacture a digital CT-simulator called CT-Extension. This makes use of an offset image intensifier with no extra outrigger. The irradiation pattern on the intensifier is re-imaged from its anode screen on to a linear photodiode array by suitable mirrors and glass optics. 3000 transmission profiles are measured in a single rotation taking 1 minute. The spatial resolution was determined to be better than 2 mm with an American Association of Physicists in Medicine (AAPM) phantom and the density resolution was better than 2%. A full reconstruction took 3 min on an HP-Pentium (Van der Giessen *et al* 1995). Images obtained with the CT-Extension were compared very favourably with diagnostic CT scans for the same cross section of the same patient in the same treatment position (figure 7.6). The diagnostic unit for comparison was a General Electric PACE CT scanner. The performance of simulator CT was considered adequate for radiotherapy treatment planning and the data were passed to a Theraplan 500 planning system.

Whilst the first commercial product using digital reconstruction from Varian Associates has consolidated, and the Nucletron digital machine and a machine from Philips have also become available, there have been further 'one-off' developments. Diallo *et al* (1994) have constructed a linear solid-state detector which has been added to a Mecaserto–Siemens simulator to collect projection images. The prototype detector comprises a scintillation layer of gadolinium oxysulphide, viewed by two banks of 768 silicon photodiodes (each 0.8 mm long × 1.6 mm wide) such that the detection

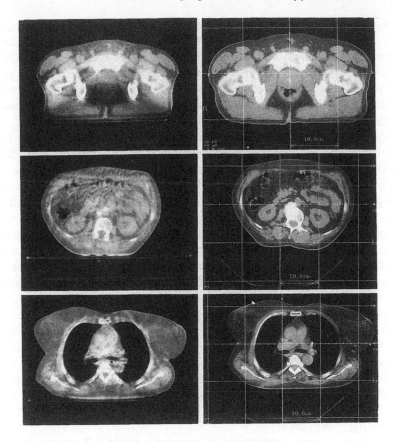

Figure 7.6. *A comparison of images obtained with the CT-Extension CT-simulator of Nucletron (Oldelft) with diagnostic CT scans taken with a General Electric PACE CT scanner for the same cross section of the same patient. For each pair of images the CT-simulator image is on the left and the PACE image is on the right. (From Van der Giessen et al 1995.)*

sampling interval along the projection is 1.6 mm and the slice thickness is 3.2 mm. Each approximately 60 cm-long projection thus comprises 384 pixels. The 12-bit signal acquisition is performed through a board developed by the Thomson Company and inserted into a PC computer. 900 projections are taken over a 360° rotation in 1 min. 110 kV, 3 mA x-rays are used with an additional filtering of 0.5 mm of copper and a wedge compensating filter. The prototype 256^2 pixel reconstruction is obtained by a fan-filtered back-projection algorithm. The efficiency of x-ray detection was estimated to be 70%, roughly equal to that of the image intensifier used in a previous development, and which the array replaced. The new array does not suffer of course from distortion or the effects of the earth's magnetic field unlike the image intensifier. The dynamic range of the detector was 5000, some five times that of the image intensifier, and is linear. The patient aperture has a diameter of 67 cm and the reconstruction circle has a diameter of 45 cm. A commercial version can create a 512^2 image and has a 100 cm diameter aperture. Impressive images have been obtained. There is still a need to construct sets of tomographic images in one rotation of a simulator and this would require a series of such detectors, at present a somewhat expensive requirement (Chavaudra 1994).

Engelbrecht *et al* (1995) have reported another 'in-house' development. They offset the caesium iodide image intensifier of a Philips' MAXIMUS CM80 simulator to create half-projections through 360°. The x-rays were collimated to a fan with a slice width of 4 mm and ten TV lines were averaged and binned to create effectively 144 full projections each with 190 pixels. The CT sections were then reconstructed using a convolution and backprojection (CBP) algorithm with a Shepp and Logan filter embodied in the well-known Donner Laboratory reconstruction software package. It was determined that the spatial resolution was 4.05 mm and the outlines of structures could be extracted with a mean accuracy of 3 mm with a maximum difference in the region of sharp corners of 10 mm.

A unique feature of this development was the following. Since, even with an offset detector the reconstructed field-of-view diameter was only 30 cm, they further offset the detector to create an 'outside' set of half-profiles, i.e. the detected radiation filled an annulus with a hole in the centre as the simulator rotated through 360°. The 'hole' was filled by acquiring an 'inside' set of projections with the detector offset reduced. The acquisition of the two sets of half-projections required two complete rotations through 360°, each taking 1 min 20 s.

An issue of some importance if simulator-based CT scans are to be used for radiotherapy treatment planning, including the effects of tissue inhomogeneities, is the accuracy of the CT numbers generated. Hartson *et al* (1995) have studied this problem for the Oldelft CT-Extension comparing the CT numbers generated for a number of standard phantoms in different scanning geometries with the corresponding results for a Picker 1200 SX diagnostic CT scanner. Detailed tables and figures were shown indicating varying accuracy depending on circumstances. However, there was good agreement between the results for small phantoms. For larger phantoms the agreement varied with the materials inserted into the phantoms.

A second issue of importance when considering the use of CT in treatment planning is that of whether the patient is, for the planning scan, truly in the position for therapy. In Chapter 1 we have already commented on issues of patient movement and organ motion etc. However, a potential source of gross error is that the patient may require a treatment position in which imaging with a commercial CT scanner cannot be done. Figure 7.7 shows an example of the gross distortions which may then appear. This was certainly the rationale behind early developments of CT-simulators for

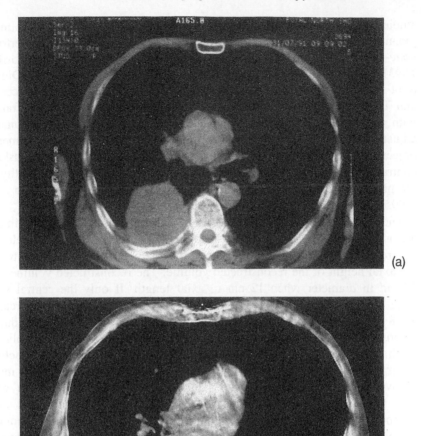

Figure 7.7. *(a) A diagnostic CT scan of a mesothelioma of the chest wall taken with the arms down. (b) A scan taken on the Varian CT-simulator using the same field centre with arms up in the intended treatment position. The tumour has moved by some 3 cm. (From Varian product literature.)*

breast imaging such as that constructed at the RMNHST, reviewed in the companion Volume, Chapter 8.

Another limitation of most of the early and present commercial simulator-based CT systems is that they produce only one slice per rotation. Since rotations are also quite slow (of the order of a minute), multiple-rotation studies are impractical. Yet volumetric imaging is required for 3D treatment planning. A number of authors have suggested the solution is to perform *cone-beam* CT using an area detector (Silver *et al* 1992, Cho and Griffin 1993). Cho *et al* (1995) describe the use of a Philips Medical Systems Digital Spot Imager (DSI) which consists of an image-intensifier digital processor and image display and transfer facilities. The DSI can acquire frames with a maximum rate of eight frames per second. The DSI has automatic adjustment of x-ray flux with gantry position for non-circularly-symmetric objects, variable dynamic range compression and automatic unsharp-masking contrast variation and these non-linear transformations limited the utility of the projection data for quantitative CT reconstruction. However, Cho *et al* (1995) showed very good qualitative reconstructions.

The area detector was offset to create 'half-projections' since the image intensifier was not wide enough to span the whole patient. Since the detector is circular there is a trade-off between the diameter of reconstruction and the axial height of the reconstructed volume. The reconstructed volume was 32 cm in diameter when 12 cm in axial length. If only the central slice were reconstructed (axial length = 0) the reconstructed diameter increased to 48 cm. Conversely an axial length of 35 cm was available with a reduced reconstruction diameter. Cho *et al* (1995) described a method of splicing the half-projections in which artefacts due to the sudden edges of each of the half-projections are avoided. Basically the half-projections were made to overlap slightly and each was padded out prior to convolution with data from nearest opposing rays. After convolution the data were only retained for reconstruction at filtered-projection pixel sites corresponding to measurement positions. Reconstruction made use of the well-known Feldkamp algorithm (Feldkamp *et al* 1984). Strictly the Tuy-completeness condition (the condition that all reconstructed planes must intersect the source trajectory at least once) was violated but this was shown to be unimportant for data with a long source-to-isocentre distance. Other workers using the Feldkamp algorithm for other applications have also noted that the condition can be violated and yet still give good reconstructions (Machin and Webb 1994).

The data taking was calibrated to take account of the periodic displacement of the image intensifier in the axial direction as a function of the gantry angle (sinewave with amplitude 1.5 mm), and for distortions introduced by the intensifier and for rotations and translations introduced by the camera. Experimental projection data were recorded at two frames per second with 100 or 120 kV x-rays, images captured at 512×512 pixels resolution. Projections were recorded approximately at every 1° interval in 360°.

From a series of measurements with phantoms it was established that (using 'width-truncated data') the geometric fidelity of the reconstruction was 1.08 mm at worst. A 1% contrast enhancement was visible in a 1.25

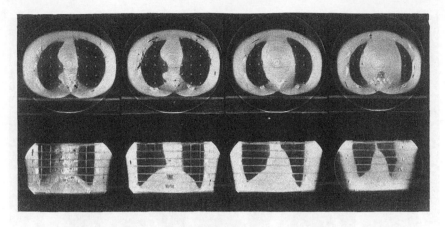

Figure 7.8. *Cone-beam reconstruction of a Rando chest phantom using 360° width-truncated projections. Top row: transaxial planes. From left to right the planes are at distances $z = -3.5, -1.5, 1.5$ and 3.5 cm axially from the $z = 0$ plane in which the source rotates. Bottom row: coronal planes. From left to right $x = -3.4, -1.7, 1.7$ and 3.4 cm with the plane $x = 0$ containing the rotation axis of the scanner. (From Cho et al 1995.)*

cm diameter tube and the spatial resolution was established, using crow's-foot and line-set phantoms, as a visible 5 cycles cm^{-1}. Cho *et al* (1995) successfully imaged both head and chest anthropomorphic phantoms and whilst some ring artefacts were present in the images (figures 7.8 and 7.9) the quality was clearly acceptable for a treatment-planning system which uses segmented regions-of-interest with assigned density values. No attempt to generate truly quantitative data was made because of the non-linearities described above in the image projection-data-taking chain.

Whilst not CT with ionising radiation, optical techniques have also been proposed to create the external contour of a patient lying on the couch of a simulator. Using three lasers and two cameras mounted on walls and ceiling, multiple contours may be obtained and indeed used to create surface images (Wilks 1993). The system is marketed as OSIRIS.

7.3. MULTIMODALITY IMAGING AND IMAGE COMPRESSION

In Chapter 1 the complementary role of different imaging modalities was discussed. Routinely x-ray CT is still by far the most common tomographic imaging method feeding into conformal radiotherapy planning. But MRI and functional SPECT and PET are playing increasing roles (Anselmi and Andreucci 1995, Chen and Pelizzari 1995, Feldmann *et al* 1996, Kamprad *et al* 1996, Mohan *et al* 1995). In order to make proper use of these other modalities, which do not of course give electron density on a voxel-by-voxel

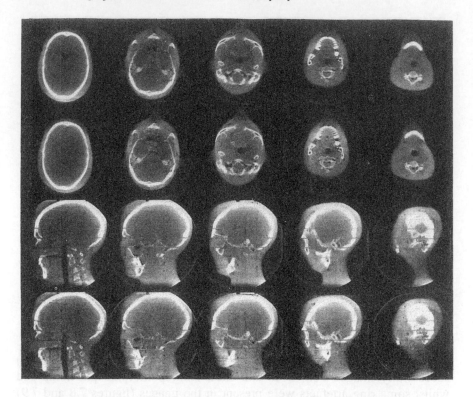

Figure 7.9. *Cone-beam reconstruction of a Rando head phantom from 337 views. Cylindrical holes used for TLD placement are visible in some of the planes. Top row: transaxial planes of reconstruction using full-width data (455×455 pixel^2). From left to right $z = -5.5$, 0, 1.6, 4.7 and 8.6 cm. Second row: transaxial planes of reconstruction using width-truncated data (248×455 pixel^2). Third row: sagittal planes of reconstruction using full-width data. From left to right, $y = 0$, 2.3, 4.7, 7, 11.7 cm. Bottom row: sagittal planes of reconstruction using width-truncated data. (From Cho et al 1995.)*

basis, these data must be registered spatially with x-ray CT data. A variety of methods have been proposed including two which are now well established; (i) landmark-based registration; (ii) surface registration. These were reviewed in detail in the companion Volume, Chapter 1. Landmark-based registration, as the name suggests, requires identification of corresponding anatomical or physical landmarks (points) inside the patient and/or the placement of point markers on the patient's skin, which can be imaged and identified in each modality. Then, the corresponding rotation, translation and scaling operators are determined to transcribe one dataset onto another. Because the point identifications or placements are never perfect, due to human error and/or the limited spatial resolution of the imaging equipment, these registrations are never perfect but can be accurately quantitated (Kessler 1989, Kessler *et al* 1991). Surface-based registration, or, as it has become known, 'hat-head' fitting (since it has been almost exclusively applied to image data of the head), succeeds by minimising some function, for example the mean-squared distance, of the distance between two surfaces identified in each data-set (Pelizzari *et al* 1991, 1992). SPECT and x-ray CT images have also been correlated using chamfer matching to provide a method to monitor the effects of radiation on lung tissue (Kwa *et al* 1995) and it was shown that this technique is more accurate than the use of external markers. Chamfer matching is a developing technique acquiring a wide range of applications. For example Van Herk *et al* (1995) have shown that the position of the apex of the prostate can be established more accurately if transaxial CT scans are registered by chamfer matching to coronal MR scans. Recently, multimodality image registration has been used to study and quantitate intra-lesional radiotherapy (Sgouros *et al* 1990, 1993, Flux *et al* 1994, Flux 1995, Lazzari *et al* 1995).

3D medical images occupy a great deal of storage space in computers. This is especially so when there are a large number of CT slices, each with a large number (e.g. 512^2) of pixels and especially when there are multiple CT studies per patient and/or multimodality image data-sets. For this reason image compression is often sought. Lossless compression is achieved by searching for regions of uniform density; it can achieve compression ratios of up to 3:1 and is identically reversible. Non-recoverable image compression in which certain information is discarded as irrelevant or assumed to be noise can achieve 15:1 compression ratio and studies have shown that recovered images are totally appropriate for radiation-therapy applications (Mohan *et al* 1995).

7.4.   3D DISPLAY OF ANATOMY AND RADIATION QUANTITIES

A great deal has been written over the years about the power of 3D display of image data (Robb 1995, Schlegel and Bendl 1996). Until relatively recently (about the mid 1980s) software to generate 3D medical images was not commercially available and individual workers created their own display packages (e.g. Christiansen and Stephenson 1985 (the MOVIE.BYU package), Hohne *et al* 1987 and Ward 1995 (VOXELMAN), Udupa 1982 (DISPLAY-82 from MIPG), Udupa and Herman 1991, Webb *et al* 1986). Towards the present time the number of such individually produced 3D display programmes has grown beyond the point where it is possible to include a full bibliography. The book by Udupa and Herman (1991) contains tutorial review material.

Today, 3D displays are relatively fast to compute and users expect to find software to form, manipulate and display 3D images in treatment-planning systems. Excellent colourplate examples may be found, for example, illustrating several treatment-planning systems, in the special issue of *Seminars in Radiation Oncology* **2** (4) (1992) pp 291–300. The most common form of input is a set of 2D transaxial x-ray CT or MRI slices, closely spaced, so the 3D images do not contain surface furrows and so fine detail can be captured. Spiral CT data are particularly suitable. Shaded-surface images of the patient are produced by first extracting the external contour from the CT data, followed by tiling the bands between adjacent contours and finally illuminating the surface as if by appropriately placed computer 'lights', a technique in which the surface reflectivity is adjusted in terms of the angles of the incident 'light' and the viewing direction. Similar methods have been used to display functional SPECT and PET data in 3D although these are far less common and are somewhat coarser scale images because of the inherently poorer spatial resolution of nuclear medicine images (Webb *et al* 1986).

There is no doubt that the view of a shaded-surface 3D image is intuitively pleasing since the image 'looks like' the patient, whereas clearly tomographic slice images more resemble anatomical slices. However, shaded-surface images alone are of very little use in 3D treatment planning. The development which has made 3D display much more useful is the ability to show shaded-surfaces of external contour and internal anatomy *in the same image*. To do this one of the following choices must be arranged:

(i) the external contour could be rendered translucent so the viewer sees through it towards the 3D shaded-surface images of one or more internal structures (figure 7.10);

(ii) alternatively the external contour can be rendered as a 'wire frame' (being simply the contours from which the tiling would have progressed) through which the internal structures are visible (figure 7.10).

It is becoming common to have the ability in 3D treatment-planning systems to achieve the following.

(i) The composite 3D image can be rotated about any axis in space to better facilitate the viewing of the relative disposition of structures. This requires a substantial computer memory and fast hardware. In the VIRTUOS treatment-planning system from DKFZ, for example, (Bendl *et al* 1994) all structures are switched into wire-frame mode during this rotation and revert to shaded-surface mode as soon as the 3D image is once again static, a feature known as 'fast-draw'. The very act of movement itself adds to the perception of the 3D nature of structures perhaps reflecting the innate habit of humans to add to their binocular vision of a hand-held object by rotating it when making an inspection. In PEACOCKPLAN (the 3D planning system from the NOMOS Corporation) a wire-frame schematic patient rotates in real-time and is only rendered when at the required orientation. The Hamburg system VOXELMAN first developed for displaying diagnostic images in

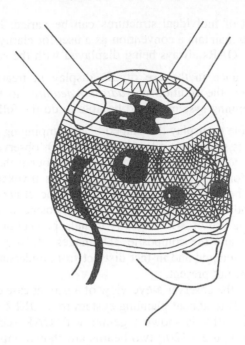

Figure 7.10. *A schematic attempt to illustrate some concepts of 3D display in radiation treatment planning. The external surface of the patient is represented as a wire-frame, tiled between the contours which have been extracted from CT or MRI tomographic data-sets. For clarity, only some segments have been shown tiled. The internal structures are represented as 3D shaded surfaces. Those shown are the tumour, the ventricles, the brainstem, the orbits and the optic nerves. In a practical system these would be shaded different colours for easy identification and could be toggled on and off (as could the wire frame for the skin surface). A physician's-eye-view of two (of possibly more) beams is also shown with the beam entrances at the skin surface. Again these would in practice be shaded surfaces and could be toggled on and off at will. This arrangement, when combined with the ability to see the whole moving dynamically and statically from various directions, leads to a practical tool to investigate the 3D relation between beams, tumour and sensitive structures. 3D treatment-planning systems also have the ability to display the 3D dose distribution in relation to these features, for example as isodoses or as surfaces shaded with surface dose or as ribbon contours. Because of the inevitable complexity, it is necessary to be able to toggle these features also and for clarity they are not shown here.*

3D has recently been adapted to accept structures and dose maps from VOXELPLAN (Schmidt and Frenzel 1995).

(ii) Individual structures can be switched on and off by computer toggles. This can aid the inspection of a structure which may find itself behind another in some particular viewing orientation.

(iii) The colours of individual structures can be varied. However, it is sometimes useful to maintain a convention as a user for clarity (e.g. tumours shown red, specific OARs always being displayed with the same colour).

The greatest advance in the use of 3D display in treatment-planning systems is, however, the addition of *radiation features* to the display of anatomy. A 3D planning system should be able to do the following.

(i) It should be able to display the shaped beams impinging on the surface of the patient. This is sometimes referred to as the 3D 'observer's-eye-view or 'physician's-eye-view' (PEV) (figure 7.11). Of course these beams are 'locked to the anatomy' so also rotate when this is invoked. The beams should be able to be toggled on and off just as the structures and their colours should be adjustable. These are more than cosmetic considerations. The rationale for the use of 3D images is to help perception of the whole treatment geometry and the composite 3D images are very 'busy'. They can easily become cluttered and in fact distract from understanding if these toggling features are not present.

Figure 7.12 shows the observer's-eye-view of a patient case displayed with the VOXELPLAN 3D treatment-planning system from DKFZ at Heidelberg. In figure 7.12(*a*) the PTV is shown together with OAR such as the eyes and optic nerves. In figure 7.12(*b*) two beams are shown impinging on this target. The display clearly shows to the observer the direction of orientation of the beams. However, at the same time it obscures some of the anatomy. In figure 7.12(*c*) we see a useful feature which has been programmed to overcome this difficulty. A clipping plane can be set on one or more of the beams. The figure shows the beam from anterior to posterior clipped to display the right eye and optic nerve hidden in figure 7.12(*b*). On the 3D planning system the features are of course all in colour and the effect is more striking.

(ii) The PEV terminology distinguishes from the beam's-eye-view (BEV) which can also be a 3D shaded-surface or wire-frame image (figure 7.13). The BEV is quite simply the view down the beam from the source, generally arranged so the centre of the image is the central ray. If the beam is geometrically shaped, the field edge will be irregular. The region outside the field is shielded and through the port is seen the structures just as in the PEV. In fact the BEV *is* a PEV along one particular beam (the physician being as if in the treatment head!).

Figure 7.14 shows the BEV facility of VOXELPLAN. In figure 7.14(*a*) the irregular field is shown around the PTV as well as the location of a rectangular field created by backup jaws on the accelerator complementing the MLC. In figure 7.14(*b*) the setting of the MLC leaves is shown with respect to such an irregular contour.

(iii) A very useful feature is the ability to simultaneously display the 3D dose distribution together with the 3D anatomy, possibly even with the 3D

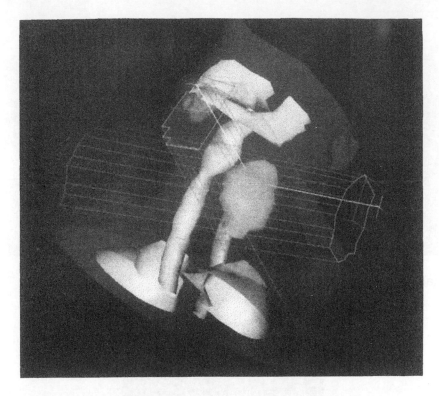

Figure 7.11.   *The PEV or observer's-eye-view. The structures of interest are shown as shaded-surface structures on which two beams may be seen to impinge. The location of the structures with respect to the external contour can be appreciated and rays may be seen schematically. (From Sauer and Linton 1995.)*

beam display. There are several possibilities. One is to shade the surface of a particular 3D volume-of-interest (VOI) with a colour representing dose. Then it becomes clear whether all of this VOI is in some particular dose band (high for PTV; low for OAR) or whether there is a dose inhomogeneity. A second option is to display isodose 'bands' around the PTV. Then if all the PTV is raised to a dose greater than some particular value the corresponding isodose band will be complete. On the other hand, if part of the PTV falls below some isodose value then that part of the PTV 'pokes out' through the band which appears to go inside the PTV. A third option is to display 3D isodoses but this can lead to a very cluttered picture unless individual isodoses can be toggled on and off. In some systems (e.g. PEACOCKPLAN) the user can toggle between shading 3D structures by their structure colour or by the dose at their surfaces.

Figure 7.15 shows a typical example of the dose display in VOXELPLAN. Coloured isodoses can be toggled on and off in three orthogonal views

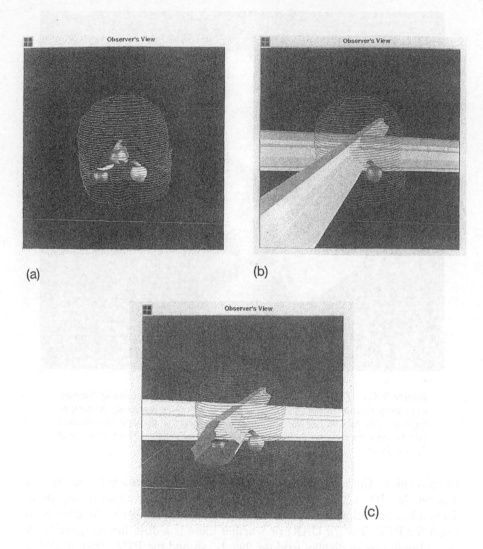

(a)   (b)

(c)

Figure 7.12. *(a) 3D features displayed on the 3D treatment-planning system VOXELPLAN. The beams are toggled off; (b) as (a) but with the beams toggled on and some of the vision of anatomical structures is lost; (c) the effect of applying a clipping plane is to allow simultaneous observation of the directions of the beams with respect to the planning volume whilst still allowing vision of some of the anatomical structures.*

simultaneously. A cursor can be dragged in any one view to change the slice number of the other two views. The beams themselves can be toggled on and off. The figure shows isodoses superposed on CT data but the facility to display colour wash, isodoses alone or a dose image alone is in this system

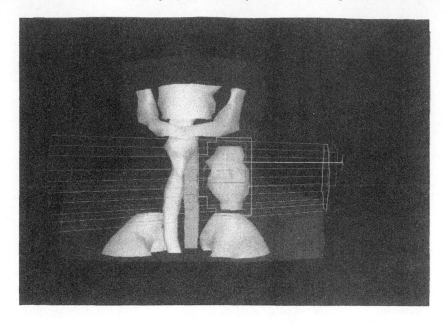

Figure 7.13.   *The BEV of one of the beams in figure 7.11. The shape of the beam can be seen collimated by a MLC. It can be perceived that this shaped beam does not directly irradiate OAR. The second beam is seen coming from the patient's left. This and figure 7.11 are from the PLATO 3D treatment-planning system. (From Sauer and Linton 1995.)*

as is the ability to display dose on the surface of 3D volumes in the PEV.

Wolff *et al* (1996) have defined a simple but effective index of dose conformation which does not directly involve images. This is the ratio of the treated volume covered by the 80% isodose surface and the PTV. In a study of 68 plans of brain-tumour treatments this so-called conformation index had a mean value of 2.4 for coplanar treatments and 1.7 for non-coplanar treatments.

A further necessary feature is that there must be some way of linking the information in a 3D view to the set of 2D slice images that are simultaneously displayed. One linking feature is that whatever attribute is being toggled should toggle in the slice images too.

Altogether, the use of 3D display is becoming more common and more 'taken for granted' in 3D treatment-planning systems. In publications it is not so easy to fully demonstrate the power of 3D imaging since this results from a synergistic addition of (i) movement, (ii) computer toggling of features and (iii) interaction with 2D images, to the basic 3D image which one might display in a review such as this (figure 7.10).

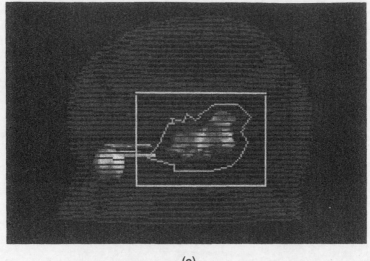

(a)

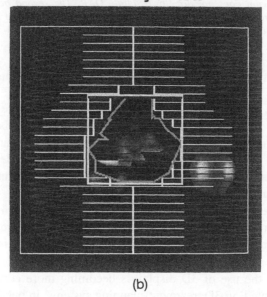

(b)

Figure 7.14. *The BEV facility within VOXELPLAN (a) showing the location of an irregular contour and also the backup collimators (b) showing also how the location of the MLC leaves can be displayed with respect to an irregular contour.*

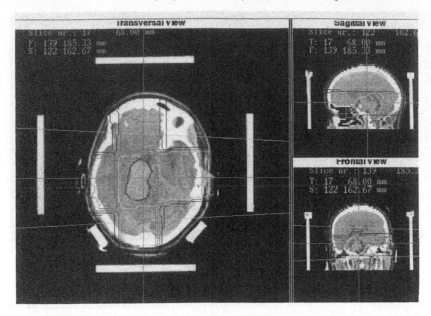

Figure 7.15. *Display of isodoses with respect to CT data in three orthogonal views in the VOXELPLAN 3D treatment-planning system (see text for details).*

7.4.1. 'Tumour's-eye-view'

Many treatment planners would select the beam orientations manually using the display tools available in 3D treatment-planning systems. A common goal is to select those orientations which avoid OAR. Two groups have developed a so-called 'spherical view' or 'tumour's-eye-view' tool to assist. This is like a picture of the globe in which gantry angles look like meridians and couch angles look like parallels of latitude on a globe (figure 7.16). The structures (PTV and OAR) are then mapped onto this surface as well as any areas which correspond to collisions of gantry and couch. The goal is then to select practically achievable areas of the spherical surface which 'see' the PTV but which do not see overlapping OAR and direct the radiation from these orientations. The DKFZ team have incorporated this concept into VIRTUOS (Bendl *et al* 1994) and the University of Michigan team have it in UMPLAN (McShan 1990). If a swathe is drawn through such acceptable areas then the latter team have christened this 'the river of desire' (because it is shaded blue on that system).

Figure 7.17 shows another nice feature of VIRTUOS; a schematic view of the linac in relation to the couch which can be screened at the same time as the spherical view. Then as a cursor drags a circle containing a cross through the spherical view this schematic is simultaneously updated providing a very real 'feel' for what is actually going to happen in the treatment room. In the

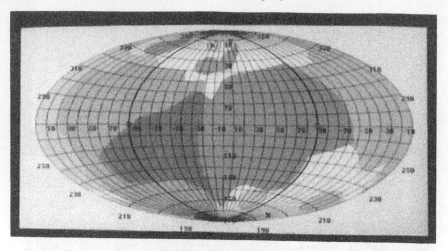

Figure 7.16. *The 'Spherical View' as incorporated into the VOXELPLAN 3D treatment-planning system (see text for details). (Courtesy of R Bendl.)*

example shown this cursor is deliberately put into a forbidden zone and the schematic shows the corresponding collision.

VIRTUOS actually contains a number of different types of this feature. In figure 7.18 is shown the 'world view' in a different method of display of meridians and latitude parallels. Now the cursor has been put into an acceptable location and the corresponding schematic of the linac and couch shows this too.

7.5. THE CT-SIMULATOR

Traditionally, patient treatment setups are simulated in a machine, a simulator, which has all the features of the treatment device but which uses a diagnostic x-ray tube to create projection radiographs or image-intensifier images. It is these images which can be correlated with megavoltage portal images to assess patient positioning at the time of treatment. Of course the patient has to be in the treatment position on the simulator. The simulator is used to (i) localise the target and OAR; (ii) mark the field locations on the patient; (iii) check block positions and ensure the feasibility of the treatment, specifically collision avoidance; (iv) assess mobile volumes using the image intensifier.

Recently the concept of CT-simulation has been developed in which instead of, or complementary to, a visit to the simulator, some or all of the above tasks can be performed in a special CT scanner, designed for their achievement (Ragan *et al* 1993, Galvin 1995). Such CT-simulators are manufactured by Siemens and Varian (Picker) and other manufacturers including General Electric and Philips have systems under development.

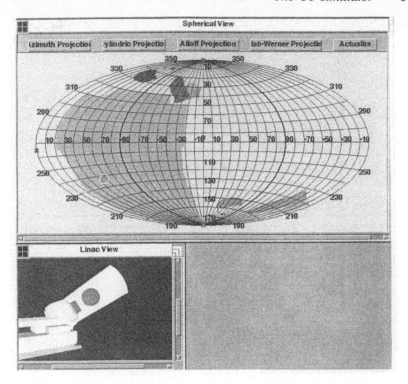

Figure 7.17. *The spherical view can be linked to the linac view as shown here in VOXELPLAN. The cross within a circle cursor on the spherical view deliberately shows a beam orientation which would cause collision (the cursor is in the forbidden shaded region) and the linac view shows this simultaneously.*

The term CT-simulator is used in two contexts which should be clearly distinguished. It can sometimes mean 'CT performed on a radiotherapy simulator' (see section 7.2) or, alternatively (as here), 'simulation performed with CT data'.

The CT-simulator can give better localisation of target and OAR, since these may be defined from tomographic rather than projection data. If, however, x-ray CT data are obtained with the patient in a position other than the treatment position, then the data may still be used but must be combined with a visit to the conventional simulator.

CT scanners also produce a 'scout view'. This is the projection radiograph of the patient translated through the aperture of the scanner without the source moving. The field diverges transaxially but not longitudinally. The scout view shows the location of the individual CT slices. The scout view of the CT-simulator cannot be used as a portal radiograph because it has no field divergence in the longitudinal direction and a much greater field divergence transaxially. Instead the DRR is constructed (see Chapter 4) and

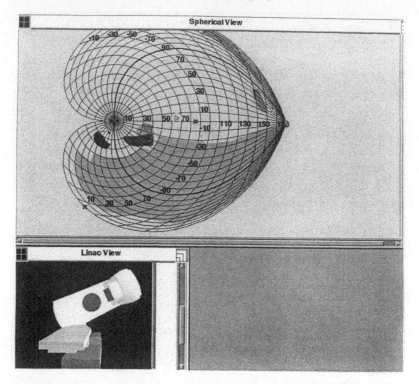

Figure 7.18. *The VOXELPLAN spherical view in a different mode, this time showing a favourable beam orientation which is reflected in the linac view.*

field marks, cross hairs etc are superposed. The spatial resolution of the DRR depends on the slice thickness which should be very small, putting large demands on the tube cooling requirements. If a large volume is to be imaged the time taken will be large and the patient may move during this time. Although in some ways the DRR has more desirable properties than a conventional radiograph (contrast can, for example, be varied artificially and one can compute an artificial DRR in which only certain structures appear, e.g. the bones) in other ways it may be inferior. It also takes a long time to compute DRRs, particularly if trilinear interpolation is used (Galvin *et al* 1993).

There are many ways of using a CT-simulator, reviewed by Galvin (1995). The simplest to describe is that in which following CT scanning, the patient is contoured, planned, organs and targets are identified and special lasers on the CT-simulator are used to mark the patient, all without the patient leaving the equipment (figures 7.19 and 7.20). However, whilst this may be the most consistent approach it takes a long time and other methods in which the patient makes several visits to the CT-simulator have been devised.

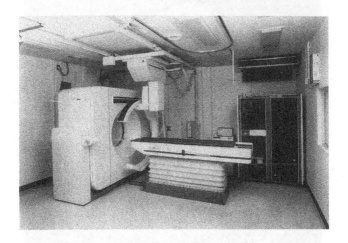

Figure 7.19. *The scanning room of the CT-simulator in Kyoto. As well as the CT scanner the room is equipped with a laser-beam field projection system. (From Nagata 1993.)*

Whereas the conventional simulator can determine the envelope of motion of mobile volumes by use of the image intensifier this is less easy with the CT-simulator. If there is no breath control the slices will fail to register properly because of organ motion during breathing. The appearance of the edge of structures may give a clue as to organ motion.

7.6. INFORMATION COORDINATION

An issue of some importance is the persisting relative difficulty of communicating information between different parts of the therapy process. For example, image formats differ from manufacturer to manufacturer and image data may be hard to install in 3D planning systems. CT images coming from different modalities may have different numbers of pixels, different pixel sizes and different labelling conventions. The slice widths may be different and the technique for labelling the superior–inferior slice localisation may vary. It is also useful to be able to transfer image sets, plan files (data of beams), contours and dose maps from one 3D planning system to another, specifically when a plan may be optimised on a 3D treatment-planning system in a research setting but the plan requires validation on a commercial 3D treatment-planning system. This apparently simple requirement can be a far from trivial computational task, particularly since it requires careful verification. All these potential difficulties are in addition to the requirement to, and problems of, image registration when multimodality imaging is used to aid treatment planning. The output from such 3D planning

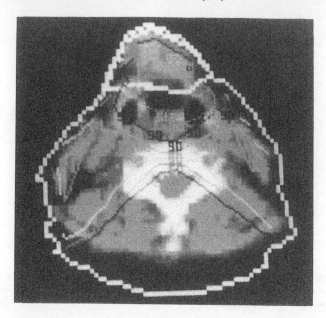

Figure 7.20. *An example of Virtual Simulation. CT scans were taken through the volume of interest, in this case a massive papillary adenocarcinoma of the thyroid involving the neck, thoracic inlet and upper mediastinum. A slice at the neck level is shown. The target volume was designated along with the critical normal structures. The dose distribution was then computed for a particular field arrangement and the treatment portals were transferred on to the patient's skin using a laser system. (From Hussey 1993.)*

systems needs to communicate with the treatment devices, specifically the MLC. It is perhaps a consequence of the somewhat sophisticated nature of all of these pieces of equipment that data transfer is more of an obstacle than it should be. Hospitals would also like to integrate scientific information about the patient's diagnosis and treatment with administrative details. Some efforts towards solving these problems of communication, data transfer and the associated safety issues are underway (Neumann and Richter 1995, Gorlatchev *et al* 1996). The field is moving towards adoption of the DICOM-3 (Digital Image Communications-3) standard for image formats for picture archiving and communications (ACR 1993).

7.7. COMPUTATIONAL ADVANCES THAT ASSIST RADIATION THERAPY

A recurrent theme throughout this text has been the intimate relationship between advances in radiation therapy and advances in computing ability.

Indeed one might argue that most of what we do today was dreamt of by our predecessors who only lacked the technology to achieve their dreams. Today, to use an overworked phrase, 'we have the technology'.

Optimisation is possible because of the development of parallel computing architecture and fast processors. Real-time interactive treatment planning is possible because of the development of advanced graphics cards and the ability to interact with screens via mouse-driven commands in a Windows environment or even touch-screen. Cheap memory allows storage and interaction with huge quantities of image data. High-speed datalinks, both as local-area networks and as wide-area networks, allow data from one hospital department to be accessed by another. Advances in control software and hardware permit real-time movement of treatment devices, for example of MLCs and the NOMOS MIMiC. Touch-screen hardware is also emerging in these fields. It will be fascinating to speculate on the role which may be played by virtual reality in the future. Those close to the developments may cynically curse that computer developments throw up as many new problems as they solve (see section 7.6 again) but on the whole the benefits outweigh the labours of development and movement is in the positive direction.

7.8. SUMMARY

The starting point for achieving 3D CFRT is the accurate spatial localisation of the patient's disease. Until the early 1970s the only way to do this was from interpreting x-radiographs. Now, some 25 years on, we have the ability to obtain exquisitely detailed tomographic information. X-ray CT is still the predominant method of identifying the target although the use of MRI is catching up. Although functional imaging has a lot to offer and can actually show that the spatial extent of disease may be quite different from its anatomical extent, the use of SPECT and PET for planning purposes is almost non-existent.

There is a need to be able to easily scale, de-distort, register and transfer all these 3D tomographic images into the 3D treatment-planning computer. This taxes issues of format, data-transfer speed, storage and cooperation between different imaging departments and it is probably fair to say that the delay on integrating 3D information for planning is more man-made than natural. There are certainly difficult issues concerning extraction of detail but they tend to be dominated by these impediments.

However, those centres who have successfully merged data and made use of the new information now have available a wealth of tools to make use of it. Computational hardware has now improved to the point whereby routine use of 3D shaded-surface images, taken together with slice data, is possible for choosing beam orientations, designing fields and studying the interaction of radiation with 3D structures through the display of dose distributions in relation to anatomy.

Some centres are opting for CT simulation in which the role of the simulator is being reduced or even eliminated. Other centres adopt the opposite tactic of further exploiting the simulator itself through its CT capability. It is inevitable that as more developments are made in medical imaging for diagnostic purposes, those same developments will impact on the planning and evaluation of 3D radiation therapy.

REFERENCES

ACR 1993 American College of Radiology—National Electrical Manufacturers Association *Digital Imaging and Communications, Version 3.0* National Electrical Manufacturers Association, 2101 L Street NW, Washington, DC 20037

Anselmi R and Andreucci L 1995 Tumour and normal structure volume localisation and quantitation in 3D radiotherapy treatment planning (Proc. ESTRO Conf. (Gardone Riviera, 1995)) *Radiother. Oncol.* **37** Suppl. 1 S33

Bendl R, Pross J, Hoess A, Keller M A, Preiser K and Schlegel W 1994 VIRTUOS—A program for virtual radiotherapy simulation and verification *The Use of Computers in Radiation Therapy: Proc. 11th Conf.* ed A R Hounsell *et al* (Manchester: ICCR) pp 226–7

Chavaudra J 1994 private communication (Rio; 25 August 1994)

Chen G T Y and Pelizzari C A 1995 The role of imaging in tumour localisation and portal design *Radiation Therapy Physics* ed A Smith (Berlin: Springer) pp 1–17

Cho P S and Griffin T W 1993 Single-scan volume CT with a radiotherapy simulator *Med. Phys.* **20** 1292

Cho P S, Johnson R H and Griffin T W 1995 Cone-beam CT for radiotherapy applications *Phys. Med. Biol.* **40** 1863–83

Christiansen H M and Stephenson M 1985 *MOVIE-BYU* (Provo, UT: Community)

Cormack A M 1994 My connection with the Radon Transform *Proc. Conf.: 75 Years of the Radon Transform* pp 32–5

Diallo I, Gavoille A, Aubert B, Chebbi K, Prieur-Drevon P and Chavaudra J 1994 Evaluation of a linear solid state detector for radiotherapy simulator-based CT scanner (Proc. World Congress on Medical Physics and Biomedical Engineering (Rio de Janeiro, 1994)) *Phys. Med. Biol.* **39A** Part 1 510

Engelbrecht J S, Duvenage J, Willemse C A, Lötter M G and Goedhals L 1995 Computed tomography imaging with a radiotherapy simulator *Br. J. Radiol.* **68** 649–52

Feldkamp L A, Davis L C and Kress J W 1984 Practical cone-beam algorithm *J. Opt. Soc. Am.* **A1** 612–9

Feldmann H J, Gross M W, Weber W A, Bartenstein P, Schwaiger M and Molls M 1996 *Proc. Symp. Principles and Practice of 3-D Radiation Treatment Planning (Munich, 1996)* (Munich: Klinikum rechts der Isar, Technische Universität)

Flux G D 1995 Multimodality image registration and its application to the dosimetry of intralesional radionuclide therapy *PhD Thesis* University of London

Flux G D, Webb S, Ott R J, Thomas R, Chittenden S, Brazil L and Cronin B 1994 Multimodality imaging to monitor intralesional treatment of recurrent high-grade glioma *Eur. J. Nucl. Med.* **21** S37 abstract 139

Galvin J M 1995 The CT-simulator and the simulator CT: advantages, disadvantages and future directions *Radiation Therapy Physics* ed A Smith (Berlin: Springer) pp 19–32

Galvin J M, Sims C, Dominiak G S and Cooper J S 1993 The use of digitally reconstructed radiographs for 3D treatment planning and CT simulation *Int. J. Radiat. Oncol. Biol. Phys.* **27** Suppl. 1 141

Gorlatchev G, Ksenofontov A, Zhurov Y, Zaitsev R and Kumykov M 1996 Networking oriented software design for radiation therapy planning *Proc. Symp. Principles and Practice of 3-D Radiation Treatment Planning (Munich, 1996)* (Munich: Klinikum rechts der Isar, Technische Universität)

Grigg E R N 1965 *The Trail of the Invisible Light; From X-Strahlen to Radio(bio)logy* (Springfield, IL: Thomas)

Hartson M, Champney D L, Currier J, Krise J, Marvel J, Schrijvershof M and Sensing J 1995 Comparison of CT numbers determined by a simulator CT and a diagnostic scanner *Activity: International Nucletron-Oldelft Radiotherapy Journal, Special Report No 6 Treatment planning, external beam, stereotactic radiosurgery, brachytherapy and hypothermia* pp 37–45

Hendee W R 1995 X rays in medicine *Phys. Today* November 51–6

Hohne K H, Riemer M and Tiede U 1987 Volume rendering of 3D tomographic imagery *Proc. 10th Int. Conf. on Image Processing in Medical Imaging (IPMI) (Utrecht, 1987)* ed M Viergever and C N de Graaf (New York: Plenum)

Hounsfield G N 1973 Computerised transverse axial scanning (tomography): Part 1 Description of system *Br. J. Radiol.* **46** 1016–22

Hussey D H 1993 Clinical assessment of CT Simulation *CT Simulation for Radiotherapy* ed S K Jani (Madison, WI: Medical Physics Publishing) pp 57–72

Kamprad F, Wolf U, Seese A, Wilke W and Otto L 1996 Image correlation and its impact on target volume *Proc. Symp. Principles and Practice of 3-D Radiation Treatment Planning (Munich, 1996)* (Munich: Klinikum rechts der Isar, Technische Universität)

Kessler M L 1989 Integration of multimodality imaging data for radiotherapy treatment planning *PhD Thesis* University of California at Berkeley

Kessler M L, Pitluck S, Petti P and Castro J R 1991 Integration of multimodality imaging data for radiotherapy treatment planning *Int. J. Radiat. Oncol. Biol. Phys.* **21** 1653–67

Kotre C J 1995 Simulator CT: Is there still a gap in the market *Proc. 19th L H Gray Conf. (Newcastle, 1995)* (Newcastle: L H Gray Trust)

Kwa S L S, Theuws J C M, Van Herk M, Muller S H and Lebesque J V 1995 Application of chamfer matching in three-dimensional correlation of CT-SPECT and CT-CT of the lungs (Proc. ESTRO Conf. (Gardone Riviera, 1995)) *Radiother. Oncol.* **37** Suppl. 1 S33

Lazzari S, Giorgetti G and Turci B 1995 Image correlation: meaning in clinical radioimmunotherapy dosimetry for treatment planning (Proc. ESTRO Conf. (Gardone Riviera, 1995)) *Radiother. Oncol.* **37** Suppl. 1 S34

Machin K and Webb S 1994 Cone-beam X-ray microtomography of small specimens *Phys. Med. Biol.* **39** 1639–57

Mallik R and Hunt P 1993a The Royal North Shore Hospital experience with the Varian CT option *Proc. 14th Varian Users Meeting (Waikoloa, HI, 1992)* pp 53–60

—— 1993b The Royal North Shore Hospital clinical experience with the Varian CT option *Proc. 2nd Asian Varian Users Meeting (Hong Kong, 1992)* pp 61–8

Mallik R, Hunt P and Fowler A 1995 The development of a simulator based CT scanner *Proc. 19th L H Gray Conf. (Newcastle, 1995)* (Newcastle: L H Gray Trust)

Mallik R, Hunt P, Seppi E, Shapiro E, Pakovich J and Henderson S 1993 The Royal North Shore Hospital clinical experience with the Varian CT option *CT Simulation in Radiotherapy* ed S K Jani (Madison, WI: Medical Physics Publishing) pp 87–107

McShan D 1990 (private communication)

Mohan R, Rothenberg L, Reinstein L and Ling C C 1995 Imaging in three-dimensional conformal radiation therapy *Int. J. Imaging Syst. Technol.* **6** 14–32 (special issue on Optimisation of the Three-Dimensional Dose Delivery and Tomotherapy)

Mould R F 1993 *A Century of X-rays and Radioactivity in Medicine* (Bristol: IOP Publishing)

—— 1995a Invited review: Röntgen and the discovery of x-rays *Br. J. Radiol.* **68** 1145–76

—— 1995b The early history of x-ray diagnosis with emphasis on the contributions of physics 1895–1915 *Phys. Med. Biol.* **40** 1741–87

Nagata Y 1993 The clinical application of CT Simulation at Kyoto University *CT Simulation for Radiotherapy* ed S K Jani (Madison, WI: Medical Physics Publishing) pp 109–18

Neumann M and Richter J 1995 Current status of the development of an interface language for electronic data exchange in radiotherapy *Medizinische Physik 95 Röntgen Gedächtnis-Kongress* ed J Richter (Würzburg: Kongress) pp 166–7

Pelizzari C A, Chen G T Y and Du J Z 1992 Registration of multiple MRI scans by matching bony surfaces *Proc. IEEE Eng. Med. Biol.* (Piscataway, NJ: IEEE) pp 1972–3

Pelizzari C A, Tan K K, Levin D N, Chen G T Y and Balter J 1991 Interactive patient registration *Information Processing in Medical Imaging* ed A C F Colchester and D J Hawkes (Berlin: Springer) pp 132–41

Ragan D, He T, Messina C F and Ratanatharathorn V 1993 CT-based simulation with laser patient marking *Med. Phys.* **20** 379–80

Robb R A 1995 *Three Dimensional Biomedical Imaging: Principles and Practice* (Weinheim: VCH)

Sauer O A and Linton N 1995 The transition from 2D to 3D treatment planning *Activity: International Nucletron-Oldelft Radiotherapy Journal, Special Report No 6 Treatment Planning, External Beam, Stereotactic Radiosurgery, Brachytherapy and Hyperthermia* pp 3–11

Schedel A and Keil G 1995 *Der Blick in den Menschen Wilhelm Conrad Röntgen und seine Zeit* (Munich: Urban and Schwarzenberg)

Schlegel W and Bendl R 1996 Display techniques in 3D treatment planning *Proc. Symp. Principles and Practice of 3-D Radiation Treatment Planning (Munich, 1996)* (Munich: Klinikum rechts der Isar, Technische Universität)

Schmidt R and Frenzel T 1995 Visualisation of three-dimensional biological dose distributions *Medizinische Physik 95 Röntgen Gedächtnis-Kongress* ed J Richter (Würzburg: Kongress) pp 266–7

Sgouros G, Barest G, Thekkumthala J, Chui C, Mohan R, Bigler R E and Zanzonico P B 1990 Treatment planning for internal radionuclide therapy: three-dimensional dosimetry for nonuniformly distributed radionuclides *J. Nucl. Med.* **31** 1884–91

Sgouros G, Chui S, Pentlow K S, Brewster L J, Kalaigian H, Baldwin B, Daghighian F, Graham M C, Larson S M and Mohan R 1993 Three-dimensional dosimetry for radioimmunotherapy treatment planning *J. Nucl. Med.* **23** 63–74

Silver M D, Yahata M, Saito Y, Sivers E A, Huang S R, Drawert B M and Judd T C 1992 Volume CT of anthropomorphic phantoms using a radiation therapy simulator *Medical Imaging-6: Instrumentation (SPIE 1651)* (Bellingham, WA: SPIE) pp 197–211

Udupa J K 1982 Display-82—a system of programmes for the display of three-dimensional information in CT data *Medical Image Processing Group Technical Report MIPG 67* University of Pennsylvania

Udupa J K and Herman G T 1991 *3D Imaging in Medicine* (Boca Raton, FL: CRC Press)

Van der Giessen P H, Geertse-van Buul H J E M and Vlaun V 1995 Commissioning and applications of a tomographic extension on a conventional simulator *Activity Special Report* (Veenendaal: Nucletron-Oldelft) No 7 pp 66–72

Van Herk M, De Munck J, Lebesque J V and Muller S 1995 Determination of the position of the apex of the prostate by matching axial CT scans with coronal MR scans (Proc. ESTRO Conf. (Gardone Riviera, 1995)) *Radiother. Oncol.* **37** Suppl. 1 S34

Ward P 1995 3D medical images imitate human anatomy *Diagnostic Imaging Eur.* Feb 23–24, 57

Webb S 1990a *From the Watching of Shadows: the Origins of Radiological Tomography* (Bristol: IOP Publishing)

—— 1990b Non-standard CT scanners: their role in radiotherapy *Int. J. Radiat. Oncol. Biol. Phys.* **19** 1589–607

—— 1995 The invention of classical tomography and computed tomography *The Röntgen Centenary: The Invisible Light: 100 years of medical radiology* ed A M K Thomas, I Isherwood and P N T Wells (Oxford: Blackwell Science) pp 61–3

Webb S, McCready V R, Flower M A, Ott R J and Long A P 1986 Three dimensional display of functional images generated by single photon emission computed tomography *Compact News Nucl. Med.* **17** 323–31

Wilks R J 1993 An optical system for measuring surface shapes for radiotherapy planning *Br. J. Radiol.* **66** 351–9

Wolff U, Dieckmann K, Hartl R F E and Pötter R 1996 Conformation index—a tool to determine the quality of 3D treatment plans? *Proc. Symp. Principles and Practice of 3-D Radiation Treatment Planning (Munich, 1996)* (Munich: Klinikum rechts der Isar, Technische Universität)

EPILOGUE

As the millennium approaches the practice of radiotherapy is poised to dramatically change. Computer-controlled linear accelerators together with their computer-controlled collimation can in principle deliver conformal high-dose distributions, enabling dose escalation and greater protection of the normal-tissue functions of the patient. The clinical outcome should be improved local control (cure) of primary disease, less metastatic spread and a greater quality of life for the surviving patient.

The role of physics and engineering has been vital in reaching this point. The improvements have come about because of advances in the medical electrotechnical industries and in both computing hardware and software. These developments have enabled physical scientists to put into practice new techniques which until now have been largely only mathematical or theoretical possibilities. It should become less and less necessary to select radiation orientations, field-shapes and intensities on the basis of experience alone and more and more scientifically based on predictions of the outcome of such choices. An intelligent optimisation algorithm, fed with the physical and biological constraints of the problem should be able to predict best choices.

A major theme for the future is the role of intensity-modulation. Until recently, theoretical studies ran way ahead of practicality but methods to deliver IMBs are becoming available and intensively studied. In turn this focuses attention on other uncertainties such as the stability of the target volume with respect to the beams. The moving patient may well provide the ultimate limitation of applicability of IMB therapy, so monitoring the actual dose delivered via portal and MVCT imaging taken together with transit dosimetric methods will grow in importance.

The radiotherapy community is still making its decisions largely in terms of one single physical variable, dose. Not surprisingly this is because of the long history of understanding this quantity with respect to laboratory experiments and experiential knowledge of the performance of patients irradiated to different dose values. Dose is still central to planning because, whilst there are attempts to understand the relationship between biological outcome and dose delivered, the models and data are regarded suspiciously.

Despite the obvious requirement for such knowledge it is very hard to gather and process reliable data. But efforts to do so must continue.

Proton therapy is expanding. The curve of the growth of proton centres and the number of treatments has a positive second derivative. However, the use of protons is largely confined to certain body sites where there is no question of their superiority with respect to photons. It is still an open question whether proton therapy for more common tumours, especially abdominal ones, is better than photon therapy and of course with few proton centres it is hard to gather the conclusive evidence one way or the other.

It may be still true to say that the greatest impact on modern therapeutic methods is the use of 3D medical imaging data. If this is true the use is still not wide enough. There is growing evidence that functional imaging modalities can provide complementary data to anatomical imaging. Since 'missing the tumour' is the greatest mistake that can be made in the therapeutic chain, it is important to continue development in this area.

Together with the companion Volume, this Volume has attempted to survey the issues, the controversies and the techniques in the developing field of increasing the precision of radiotherapy. If it is a patchy review it is in part because the effort in the constituent areas is itself somewhat patchy. One cannot write a textbook on this subject as one could on say classical mechanics. Whenever workers gather to discuss this subject there are heated debates on the validity and applicability of new suggestions. My aim has been to bring together much of this material into a single place where each problem may be seen in the context of the other elements of the radiotherapy chain. I hope I have succeeded.

DECISION TREES IN CONFORMAL TREATMENT PLANNING

Three-dimensional treatment planning with the aim of achieving conformal radiotherapy is a complex multistage task. There is no single 'universal technique', neither should there be. The protocol which is adopted depends on many factors and this appendix summarises the main decisions that must be taken.

The decisions depend on: (i) the nature of the clinical problem; (ii) the availability of specific equipment and expertise to use it. Many of the techniques reviewed in this book are specific to one particular centre or at too early a stage of development to have been widely adopted.

It is also important to distinguish between what are essentially scientific questions which determine the choice of treatment technique and what are technical limitations which may vary between radiotherapy centres and which may change as technology and techniques become better developed and more widely available.

The principle 'chain of events' is summarised in table A.1, together with issues guiding decisions arising at each link in the chain.

Table A.1. *Sequence of events in treating a patient.*

(i)   Identify the disease, its stage and extent (issues A)
(ii) Collect 3D medical imaging information on the patient's disease (issues B)
(iii) Describe the location of disease to be treated using medical images (issues C)
(iv) Transfer medical images to a 3D treatment-planning system (issues D)
(v)   Decide if the PTV calls for treatment with uniform-intensity beams or
      whether IMBs are required. This is a major branching point
      in the decision-making tree;

Branch 1: Beams are not intensity-modulated	Branch 2: Beams are IMBs
(vi) Determine the beam orientations, field-sizes and beamweights (issues E)	(vi) Compute IMBs (issues F)
(vii) Predict response (issues G)	(vii) Predict response (issues G)
(viii) Treat the patient (issues I)	(viii) Treat the patient (issues H)
(ix) Verify the treatment (issues J).	(ix) Verify the treatment (issues J).

Table A.1.   *(Continued)*

Issues

(A) - Is the disease confined to one anatomical locality or has it metastasised?
 - Is radiotherapy indicated?

(B) - What 3D imaging modalities are available (CT, MRI, SPECT, PET)?
 - Is the tumour volume more clearly visible on one or the other?
 - Do different imaging modalities give complementary or conflicting information?
   Are the anatomical and functional disease volumes identical?
 - Was the patient in exactly the same position for each imaging session?
 - Can the 3D images from different modalities be registered in some way?
 - Are there format incompatabilities? Can image data be transferred
   to a 3D treatment-planning system?
 - Was the patient position at the imaging time the same as it will be
   at treatment time?

(C) - How easy it to draw the contours of the gross-tumour volume (GTV) and
   the clinical-target volume (CTV) on tomographic slices?
 - Are there tools for automatic outlining or must the outlines be drawn by hand?
 - Is there inter- and/or intra-clinician variability?
 - What account is made for tissue movement during the fractionated course
   of therapy?
 - What margins will be added to create the PTV? Are these isotropic
   or anisotropic?
 - Does the definition of PTV include part of an OAR?

(D) - What 3D planning system is available? Is it commercial or developmental?
 - How is the 3D planning system tested for quality control?
 - Can the 3D planning system accept medical images from each and
   any modality in the same/different hospital/department?
 - What mechanism ensures the planning system understands the correct geometry
   of tomographic slices with respect to the treatment source?
 - Can digital-reconstructed radiographs be made from CT data to compare with
   the images of simulating a patient in the treatment position or with portal
   images on the treatment machine?

(E) Direction:
 - What 3D planning system is available?
 - Are there planning tools to assist the choice of beam orientations?
 - Do these take account of the locations of OAR?
 - What are the physical constraints? Can beams be non-coplanar? Are there
   collision possibilities forbidding certain directions?

Geometrical shape:
 - Can beams be shaped to irregular fields?
 - Can these irregular fields be fitted to a pattern of leaf settings for an MLC?
 - What algorithm determines the fitting of leaves?
 - Can the MLC be rotated with respect to the projected contour?
 - Does the planning system deliver the prescription of MLC leaf shapes in
   a form which the computer driving the MLC leaves in the clinic can accept?
 - Does the dose-calculation algorithm properly account for irregular field shapes?
 - Are wedges needed?

Table A.1.   *(Continued)*

Beamweights:
- Has the 3D planning system got a method of automatically optimising
  the beamweights and wedge-angles for geometrically shaped static fields?

(F) - What 3D planning system is available?
- Is there an algorithm installed for creating IMBs?
- If yes, is it an analytic or iterative inverse planning method?
- How fast can the IMB calculation be done?
- What is the spatial and intensity resolution of the IMBs created?
- How many IMBs are required?
- Is the choice of the number and modulation of IMBs dependent
  on the choices of beam orientation?
- What are the expected advantages of using IMBs?

(G) - What model is used for determining the TCP?
- What model is used to determine the NTCP?
- If this involves a histogram reduction which method is used?
- What biological-response data support the calculations and how reliable are they?

(H) - What equipment is available to treat using IMBs?
- Can compensators be designed and constructed?
- If an MLC is available can it be used in multiple-static field mode?
- If an MLC is available can it be used in dynamic-leaf mode?
- Is a tomotherapy device available such as the NOMOS MIMiC and accessories?

(I) - Can multiple fields be treated without operator intervention?

(J) - What portal imaging is available? Is this traditional film or is there an
  electronic portal-imaging device available?
- Are there methods available for automatically extracting the information
  needed to register the portal image with a DRR or a simulator image?
- How convenient and fast are these methods?
- Will intervention to adjust the patient position be attempted?
- Can MVCT data be obtained, either single- or multiple-slice?
- Is transit dosimetry available to quantitate the delivered dose?

APPENDIX B

METHODS OF COMPUTING DOSE TO A POINT AND FEATURES OF EACH METHOD

Table B.1. *Methods of computing dose to a point and features of each method.*

Method	Fundamental 'raw' and derived data	Secondary derived data	2D or 3D	Tissue inhomogeneity
CLASS A: MEASUREMENT-BASED TECHNIQUES				
Matrix methods (primitive form with measured depth-dose data overlaid and summed) e.g. 'hand planning')	Depth-dose data in reference situation for square open and wedged fields on and off central axis (raw data for *all* class A)	Equivalent square data interpolated for rectangular or (crudely) irregular fields	Applied to selected 2D planes (dose is directly given)	No (measurement has no data for these)
Beam generating functions (per cent depth dose + off-axis factors)	Equations fitted to above measured data with interpolation for unmeasured depths and off-axis positions	Tissue–air ratios (TARs) and scatter–air ratios (SARs) derived	2D as applied plane-by-plane (if ratio-TAR method used then these ratios are applied to depth-dose data. If effective-SSD method is used, dose given directly)	Strictly no but water-equivalent depths can be used in the ratio-TAR method or in dose term in effective-SSD method
Beam generating functions + power-law inhomogeneity correction factors	TAR and SAR functions fitted to measured depth-dose data	Interpolation possible for unmeasured depths	2D as above	Uses 'Batho' (power-law) TAR method on CT inhom-ogeneity data (also accounts for surface contour)
Separation of primary and scattered radiation (ETAR)	As above + TARs extrapolated to zero field size fitted to measurements	Derived SARs	'2.5D' since CT planes are collapsed into a single effective plane	Yes by scaling for radiological depth AND scaling for lateral inhomo-genities in ratio-TAR method
CLASS B: MONTE-CARLO-BASED TECHNIQUES Convolution/ superposition integrals				
1. Point kernel or	Monte-Carlo derived mono-energetic kernels in water-cube	Polyenergetic kernels (needs knowledge of spectrum) in water cube	3D	Yes; but only approximately by scaling kernels and scaling fluence
2. Pencil beam	Monte-Carlo derived mono-energetic kernels in water-cube (or computed from convolving above with depth-dependent fluence)	Polyenergetic kernels (needs knowledge of spectrum) in water cube (or computed from convolving above with depth-dependent fluence)	3D	Can only account for 'slab' inhomogeneities (not even at interfaces)
Fully individualised Monte-Carlo techniques	Beam spectrum	None	3D dose map directly calculated	Yes

Table B.1. *(Continued)*

Beam hardening	Scatter	Surface contour	Electron tracks	Presence of blocks	Asymmetric fields	Wedges
CLASS A: MEASUREMENT-BASED TECHNIQUES						
Yes; in fundamental data and accounted for by ray-path calculation	as left	Strictly no (matrix moved *en bloc* to account for non-flat surface)	No	No; simply sets dose below blocks to zero	No	Yes; wedged data used directly
Yes; as above	Attempts to account for scatter by SAR integration (Clarkson method)	Yes; by computing depth in patient and changed SSD (effective-SSD and ratio-TAR methods)	No	Not strictly but can adjust dose under open part for lack of scatter and dose under blocked part for lack of primary	No; but often treated as blocked fields	Yes; wedged data used directly
Yes; as above	As above	Yes; as above	No	As above	As above	Yes; wedged data used directly
Yes; as above	Attempts to account for scatter by lateral scaling	Yes; automatically accounted in ratio-TAR method	No	Partly; zero voxel weights under block accounts for decreased scatter. Will calculate scattered dose under block	Treats as a blocked symmetric field	Wedged TARs used
CLASS B: MONTE-CARLO-BASED TECHNIQUES						
No if polyenergetic kernels used. Yes; if monoenergetic	Yes; but only to the limits of scaling	Yes; accounted for in fluence term	Yes	Yes; accounted for in adjusted fluence	Yes; as left	Yes; accounted for in adjusted fluence
Yes; can be modelled in generating the pencil beam	Not for general 3D inhomogeneous medium	As above	Yes	Yes; as above	Yes; as left	Yes; as above
Yes	Yes	Yes	Yes	Yes	Yes	Yes

Notes: All methods have limitations and inaccuracies.
Commercial treatment-planning systems use one of the above classes.
The 'reference situation' means the geometry in which dose data are measured in a standard geometry of a large cube of water in a tank with the central axis of the beam impinging at right angles to its surface. All planning algorithms must reproduce the measured data in such configurations.

APPENDIX C

QUALITY ASSURANCE OF CONFORMAL THERAPY

Many techniques for conformal therapy are so new that no specific guidelines exist, supported by recognised professional bodies. Quality assurance or quality control is thus generally discussed in relation to the uses of specific familiar components of the 'therapy chain'. Since many of these components form building blocks towards the achievement of conformal therapy a summary of some of the principal issues is given in table C.1. For full details one should refer to the many detailed reports on methods for commissioning new equipment, using new software and monitoring its routine performance.

The following general principles should apply to ensure quality:

(i) Procedures should be carried out according to protocols. It is the duty of professional bodies to advise, convene panels of experts and produce such protocols.

(ii) QA should be documented and there should be a system of control of such documents, uniquely identifying the status of documents, distinguishing them from earlier versions and identifying a particular person who 'signs off' on QA. This is particularly important if the person carrying out QA procedures is new to the procedure or undergoing training. Some form of positive checking should be established where items require ticking off on check lists.

(iii) QA should be referenced to manufacturers' version numbers for equipment and particularly computer software.

(iv) The frequency of QA should be considered in relation to the provision of replacement parts (e.g. on an accelerator), the possibility of gradual but cumulative deteriorations.

(v) As well as consideration of the QA of equipment, software and procedures, the involvement of humans in all the process should be studied with a view to minimising potential sources of error. Ideally tasks should be referenced to designated competences which can be checked.

The table C.1 indicates areas of concern about QA in conformal therapy.

Table C.1. *Areas of concern about QA in conformal therapy.*

Equipment	Potential sources of error; areas where QA is required
Imaging	Remember the patient is represented as a geometric model.
CT scanner	Diagnostic calibration: noise performance, uniformity, calibration of CT numbers in terms of electron density for planning. Effect of artefacts due to movement or presence of dense objects. Slice-width determination. Patient position with respect to treatment position.
MR imager	Geometric fidelity. Possible distortions. Patient position with respect to treatment position. Registration with CT data.
Simulator	Machine tolerances on movements and measuring tools (x-ray field size, isocentric distance, angle of gantry). Fidelity of geometric magnifications. Account for divergent beam. Reproducibility of patient position (effect of immobilisation devices). Determination of patient outline by lead wire, plaster bandages etc. Relation to outline from 3D tomographic data. Ability of simulator to mimic all linac movements and features. Image intensifier performance. Reference marks applied to patient skin. Determination of block positions.
Interpretation of images	
Determination of PTV and OAR	Inter- and intra-observer variability. Movement of organs during course of treatment. Account taken of partial-volume effects due to finite slice width in 3D imaging. Possible differences between anatomical and functional volumes.
Treatment planning	
Beam data	Measurement accuracy; accuracy of transfer to the planning computer; checks of standard fields and arrangements, for example field widths and percentage depth-dose points. Effects of interpolation and extrapolation. Output calibration for treatment machines.
Entry of patient data	From digitiser: geometric accuracy; alignment and linearity of axes. CT/MR image data: format; geometric fidelity; labelling of slices and axes.
Beam algorithms	Effect of assumptions (see Appendix B).
Correction factors built into TP software	Wedges, blocks, compensators, irregular contours, oblique incidence, asymmetric fields, irregular fields.

Table C.1. *(continued)*

Equipment	Potential sources of error; areas where QA is required
Treatment planning (cont.)	
Normalisation	Calculation of absolute dose values.
Hard-copy display	Printer magnification, linearity; documentation of machine data and MUs per field. Manual inspection of printed plans. Accuracy of fieldsizes with respect to the clinical request. Accuracy of recording gantry angles, requests for 'accessories' (e.g. wedges, compensators).
Patient immobilisation	
Stereotactic frame	Transfer of coordinate data between imaging, planning computer and therapy setup. Reproducibility of patient positioning.
Face mask	Movement within mask.
Body shells	Reproducibility of patient positioning.
Treatment	
Setup	Accuracy of repositioning. Reading of printed plans and data entry to treatment machine.
Machine accuracy	Isocentric stability, scales on moving parts (gantry and couch), beam output stability and calibration, accuracy of collimator especially MLC setting.

APPENDIX D

A NOTE ON THE WORK OF GEORGE BIRKHOFF

The theme of the construction of arbitrary dose distributions from a set of appropriately intensity-modulated beams (IMBs) is a strong thread through Chapters 1 and 2. Consider the 2D problem. In the companion Volume, Chapter 2 it was shown (equation (2.1b)) that the transaxial distribution of dose $d(r, \theta)$ is built up from the set of IMBs $f(r_\phi, \phi)$ in the absence of photon attenuation via

$$d(r, \theta) = \left(\frac{1}{2\pi}\right) \int_0^{2\pi} f(r_\phi, \phi) \, d\phi \qquad (D.1)$$

where the dose is specified on polar coordinates (r, θ) and ϕ is the orientation of the IMB in the same polar coordinate system (figure D.1). $r_\phi = r \cos(\theta - \phi)$ labels the position of a beam element or bixel in the IMB with respect to the normal from the projection to the central axis of rotation. The problem in inverse planning is to determine the set of IMBs given some specified dose distribution to be achieved, i.e. given the lhs $d(r, \theta)$ of equation (D.1) it is required to evaluate the function $f(r_\phi, \phi)$ under the integral. In the companion Volume, section 2.1 it was shown that in general this equation can only be analytically inverted if unphysical negative values are allowed in the IMB set and the consequences of that difficulty were examined in depth. To avoid the difficulty, iterative solutions are sought which give the closest approach to the desired dose distribution, whilst maintaining the non-negativity of beam elements.

On the other hand certain very specific distributions of dose can be formed without the need for negative IMBs. For example, in the special case of a circularly symmetric dose distribution $d(r)$ and with $r_\phi = r \cos \phi$, the θ dependence can be dropped and equation (D.1) becomes instead

$$d(r) = \left(\frac{1}{2\pi}\right) \int_0^{2\pi} f(r_\phi) \, d\phi. \qquad (D.2)$$

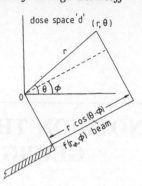

Figure D.1. *The dose space and the relationship between dose at a point (r, θ) and the beam intensity $f(r_\phi, \phi)$ at a particular position r_ϕ in the treatment beam port at angle ϕ.*

In view of the symmetry, all IMBs being the same, this simplifies to

$$d(r) = \left(\frac{2}{\pi}\right) \int_0^{\pi/2} f(r_\phi)\, d\phi. \tag{D.3}$$

Making the variable change $r_\phi = r \cos \phi$ so that

$$dr_\phi = -r \sin \phi\, d\phi = -\sqrt{r^2 - r_\phi^2}\, d\phi \tag{D.4}$$

equation (D.3) becomes

$$d(r) = \left(\frac{2}{\pi}\right) \int_0^r \frac{f(r_\phi)\, dr_\phi}{\sqrt{r^2 - r_\phi^2}}. \tag{D.5}$$

This is the well-known Abel integral discovered in 1828 (Korn and Korn 1968, Barrett and Swindell 1981) and it has the solution

$$f(r_\phi) = \frac{d}{dr_\phi} \left[\int_0^{r_\phi} \frac{d(r) r\, dr}{\sqrt{r_\phi^2 - r^2}} \right]. \tag{D.6}$$

A special case has been examined in depth (Brahme *et al* 1982), that of $d(r) = D$ for $r > r_0$ and $d(r) = 0$ for $r < r_0$ in which case the solution for $r_\phi > r_0$ is

$$f(r_\phi) = \frac{D r_\phi}{\sqrt{r_\phi^2 - r_0^2}}. \tag{D.7}$$

A rather more general case is summarised in the companion Volume where photon attenuation was included but, if we set the linear-attenuation coefficient $\mu = 0$, then equation (D.5) is equation (2.9) of the companion Volume and solution (D.6) is equation (2.21a) and solution (D.7) is equation (2.21)†

It turns out that all these equations can be found in a seminal publication by George Birkhoff (1940). He was interested in examining the problem of whether an arbitrary drawing could be formed by the superposition of straight lines of different greyness (he wrote of 'density of lead or ink deposited') coming from a set of different directions. In this case equation (D.1) represents the superposition. The lhs, $d(r, \theta)$ is the 2D drawing and on the rhs, $f(r_\phi, \phi)$ represents the density of a line drawn at right-angles to the direction specified by angle ϕ and whose normal distance to the origin of the drawing is r_ϕ. The special cases above were presented to introduce what is a much more complex general case. Birkhoff (1940) showed that any drawing could be represented by a formal expansion in 2D polar coordinates and from this he deduced the coefficients of the expansion of the function $f(r_\phi, \phi)$ for any general drawing. He then stated that only if these coefficients were continuous and non-negative could the drawing be made from non-negative pencil lines. He went on to state that, provided negative lines could be used (he called these 'rectilinear erasures'), any arbitrary picture could be created. Of course these negative lines do not erase all in their line of sight. They do not erase lines already drawn; they simply add in 'negative blackness' in the same way that positive lines add in 'positive blackness'. He finally concluded that 'to all intents and purposes one can make any drawing if a single uniform erasure all over the figure be allowed at the end'.

Unfortunately of course just as there are no 'negative pencils', there are no 'negative radiation beams'. So the conclusion is that an arbitrary dose distribution cannot be made from a set of positive IMBs, a result established differently in the companion Volume in the manner of Cormack (1987).

REFERENCES

Barrett H H and Swindell W 1981 *Radiological Imaging: the Theory of Image Formation, Detection and Processing* (New York: Academic) p 406
Birkhoff G D 1940 On drawings composed of uniform straight lines *J. Math. Pures Appl.* **19** 221–36

† To assist the reader comparing with the analysis in the companion Volume and/or inspecting the paper by Birkhoff (1940), the following is added to explain notation. x (the companion Volume) $\equiv r_\phi$ (here) $\equiv s$ (Birkhoff). $m(x)$ (the companion Volume) $\equiv f(r_\phi)$ (here) $\equiv f(s)$ (Birkhoff). $d(r, \theta)$ (here and the companion Volume) $\equiv F(r, \theta)$ (Birkhoff). Birkhoff omits the factor $1/2\pi$ in equation (D.1) and so his versions of equations (D.5), (D.6) and (D.7) differ also by this factor. Also the phase of ϕ (Birkhoff) is advanced by $\pi/2$ with respect to the analysis here and in the companion Volume (which does not affect the argument).

Brahme A, Roos J E and Lax I 1982 Solution of an integral equation encountered in rotation therapy *Phys. Med. Biol.* **27** 1221–9

Cormack A M 1987 A problem in rotation therapy with X-rays *Int. J. Radiat. Oncol. Biol. Phys.* **13** 623–30

Korn G A and Korn T M 1968 *Mathematical Handbook for Scientists and Engineers* (New York: McGraw-Hill) p 15-4-1

INDEX

Numbers in **bold** refer to **figures** and those in *italics* refer to *tables*.

T - #0466 - 101024 - C0 - 229/152/22 - PB - 9780750303972 - Gloss Lamination